PROSTAGLANDINS IN CELLULAR BIOLOGY

ALZA CONFERENCE SERIES

Volume 1 – Prostaglandins in Cellular Biology
Edited by Peter W. Ramwell and Bruce B. Pharriss • 1972

PROSTAGLANDINS IN CELLULAR BIOLOGY

Proceedings of the ALZA Conference on
Prostaglandins in Cellular Biology and the Inflammatory Process
held in Carmel, California, October 24-26, 1971

Edited by
Peter W. Ramwell and Bruce B. Pharriss

ALZA Corporation
Palo Alto, California

PLENUM PRESS • NEW YORK–LONDON • 1972

Library of Congress Catalog Card Number 78-188715

ISBN 978-1-4684-2846-9 ISBN 978-1-4684-2844-5 (eBook)
DOI 10.1007/978-1-4684-2844-5

Softcover reprint of the hardcover 1st edition 1972

A Division of Plenum Publishing Corporation
227 West 17th Street, New York, N.Y. 10011

United Kingdom edition published by Plenum Press, London
A Division of Plenum Publishing Company, Ltd.
Davis House (4th Floor), 8 Scrubs Lane, Harlesden, London, NW10 6SE, England

PREFACE

This book represents the first of the ALZA CONFERENCE SERIES which will deal with a wide variety of topics of biomedical interest.

These Conferences are planned to cover a range of subjects from the molecular level, such as drug receptor and drug membrane interactions, to the organ and organismal levels and the dynamics of drug therapy. This year's topic is "PROSTAGLANDINS."

It is rapidly becoming clear that prostaglandins are of great interest and potential utility in therapeutics. An understanding of the role of prostaglandins in the regulation of cell processes should provide insight into the understanding of a variety of difficult and intransigent fields such as cancer, allergy, and transplant rejection. We hope that this Conference may focus attention on novel approaches to these areas.

We wish to thank our distinguished Chairmen and guests for their attendance. The papers and discussions were prepared by Yvonne Hendrickson, using an ATS/360 Text-Editing System; we acknowledge with gratitude her skill, patience, and hard work.

P.W.R.

B.B.P.

INTRODUCTION

The prostaglandins have been implicated in most physiological processes which have been measured with these agents in mind. If this is so, how could the importance of these lipids have remained so hidden in the previous decade of intensive scientific investigations? How could intercellular and intracellular processes have become so defined and documented without introducing a lacuna that needed to be filled by such extremely potent and ubiquitous materials? In view of the intimate relationship between prostaglandins and cyclic adenosine monophosphate, it seems strange that so much descriptive work could have been done on the latter without implicating a need for the former. Perhaps in their ubiquity they achieved invisibility.

This ubiquity has the interesting role of bringing together many disciplines and uniting them in common uncertainty. The prostaglandins have the ability to stimulate as well as to sow confusion; privately, we are convinced that the role of prostaglandins is to make people go back to school and learn their subjects a little better.

P.W.R.

B.B.P.

CONTENTS

INTRODUCTORY REMARKS TO THE FIRST SESSION

E. W. Salzman
Beth Israel Hospital, Boston, Mass.

As an introduction, it may be useful to review current thinking on the mechanism of platelet function (Fig. 1) (Salzman, E. W., 1971, Fed. Proc. 30:1503). Platelets do not adhere to normal endothelium, but in areas of endothelial denudation, as at the end of a lacerated blood vessel or on an atherosclerotic arterial plaque, they adhere to subendothelial connective tissue, probably to collagen or a collagen-like material, and are then induced to undergo a secretory process, the "release reaction". Not a non-specific disruption of the platelet surface but a selective secretion of the contents of certain intracellular granules, the release reaction leads to expulsion from the platelet of biogenic amines such as catecholamines and serotonin and activation of other substances with biological activity. These include platelet factor 4 (PF 4), an antiheparin, and platelet factor 3 (PF 3), an alteration in the lipid component of the platelet surface that serves to accelerate the intrinsic clotting system. Also released are enzymes and adenine nucleotides, including adenosine diphosphate, a potent stimulant of platelet aggregation. Many of the released substances, such as epinephrine and ADP, are able to induce the release reaction in additional platelets, so the reaction is a self-perpetuating one. Thrombin is simultaneously generated through the intrinsic clotting system as a consequence of contact of plasma proteins with sub-endothelial connective tissue and also induces release.

Thus, injury to the blood vessel lining leads to platelet adhesion and then by way of the release reaction, to platelet aggregation and formation of a hemostatic plug or a platelet thrombus.

In 1966 Kloeze (Kloeze, 1966, J. Proc. Nobel Symp. II, Stockholm, p. 241) reported that PGE(1) was a potent

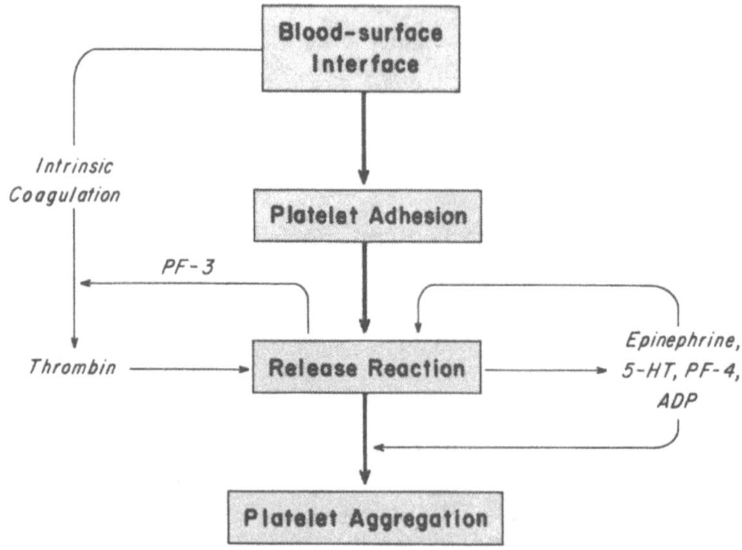

Figure 1. The mechanism of platelet function. Reprinted from the Bulletin of The New York Academy of Medicine (in press).

inhibitor of platelet clumping. This observation has been confirmed in several laboratories. PGE(1) has been shown to stimulate platelet adenylate cyclase and to elevate the platelet content of cyclic AMP. It is likely that this accounts for its affect on platelet aggregation, for there is evidence that many substances known to raise platelet cyclic AMP levels by stimulation of adenylate cyclase or inhibition of phosphodiesterase are associated with inhibition of platelet aggregation. Conversely, reduction of platelet cyclic AMP levels appears in many instances to be accompanied by initiation or augmentation of the release reaction and of platelet clumping (Table I) (Salzman, E. W., and Levine, L., 1971, J. Clin. Invest. 50:131) (Salzman, E. W., Kensler, P. C., and Levine, L., in press, Ann. N. Y. Acad. Sci.).

For example, epinephrine has been shown to inhibit platelet adenylate cyclase and reduce the level of platelet cyclic AMP, an alpha-adrenergic effect inhibited by phentolamine but not by propranolol. The stimulation of platelet aggregation by epinephrine is likewise an alpha-adrenergic effect.

Kloeze also observed that platelet aggregation was stimulated by PGE(2), and we have confirmed this observation. The details of these studies will be presented later in this conference.

Smith and Willis (Smith, J. B., and Willis, A. L., 1971, Nature New Biol. 231:235) have reported that platelets contain PGE(2) and PGF(2-alpha) and that the levels of these materials are increased by thrombin and reduced by aspirin. These observations may have important implications as an explanation for the increase in vascular permeability seen in association with platelet thrombi and for a more general insight into the regulatory role of prostaglandins in platelet function.

An effect of prostaglandins on erythrocytes is also to be discussed at this conference, but here the relationships are less clearly understood. Sutherland in 1962 (Sutherland, E. W., Rall, T. W., and Menon, T., 1962, J. Biol. Chem. 237:1220) reported the activity of adenylate cyclase in nucleated red cells of lower forms but was unable to demonstrate adenylate cyclase activity in the non-nucleated erythrocytes of mammals (dogs). Rosen (Rosen, O. M., and Rosen, S. L., 1968, Biochem. Biophys. Res. Commun. 31:82) subsequently found no adenylate cyclase capable of epinephrine stimulation in human erythrocytes. More recently Sheppard and Burghardt (Sheppard, H., and Burghardt, C., 1969, Biochem. Pharmacol. 18:2576) described adenylate cyclase activity in red cell ghosts prepared from humans, cats, dogs, mice and rats, and found the rodent erythrocytes to be stimulated by norepinephrine. They described stimulation of adenylate cyclase activity in humans, cats and dogs by sodium fluoride. Since they were unable to recognize intact white cells and platelets among their red cell ghosts, they concluded that the enzymatic activity lay in erythrocyte ghosts rather than in contaminating leukocytes or platelets. This conclusion must be regarded as unproven because of the possibility of fragmentation of white cells and platelets during the sample preparation, a possibility not ruled out in their experiments. Perhaps the studies to be reported by Drs. Allen and Rasmussen will shed some light on these interesting questions.

TABLE I

	Aggregation	Cyclic AMP	Adenylate Cyclase	Phosphodiesterase Platelet Lysates	Phosphodiesterase PRP
Epin./Norepin.	+	<	<	0	>
alpha-block	<		>(< Epin.)		
beta-block	>		±<(0 Epin.)		
Serotonin	+		<		
Thrombin	+	<	<	0	>
ADP	+	<	0	0	>
Collagen	+	<	<	0	>
Imidazole	>			>	
Nicotinic acid	>			>	
Kaolin	+	<			>
Centrifugation	+	<			>
PGE(2)	>	<	<		
PGE(1)	<	>	>	0	
Caffeine/theophylline	<	>		<	
Papaverine	<	>		<	
Dipyridamole	<	±		<	

< indicates inhibition of an effect. + indicates initiation of an effect, and >, its augmentation. 0 indicates the absence of an effect. Phosphodiesterase activity in platelet lysates is enhanced by imidazole and nicotinic acid and in PRP by agents that induce the release reaction.

USE OF PROSTAGLANDINS IN THE PREPARATION OF BLOOD COMPONENTS

C. R. Valeri, C. G. Zaroulis, J. C. Rogers, R. I. Handin, and L. D. Marchionni

Naval Blood Research Laboratory
Chelsea, Mass.

INTRODUCTION

With recent increased understanding of the cellular physiology of red cells and platelets, and with development of large scale methods for plasma fractionation, the concept of component therapy in clinical medicine has evolved. This approach involves separation and individual storage of cellular and plasma components. In order to separate whole blood into its components shortly after collection, centrifugation is necessary to prepare red cell concentrates, platelet concentrates, and platelet-poor plasma (PPP); separation can be achieved by either serial or continuous centrifugation (Graw et al, 1971; Jones, 1968; Tullis et al, 1967 and 1968).

Current methods of separation do not permit either optimum separation of red cells from white cells and platelets, or recovery *in vitro* of the total number of platelets in whole blood.

Prostaglandin E(1) has been shown to be a potent inhibitor of platelet aggregation (Emmons et al, 1967; Kloeze, 1967; Shio et al, 1970). Accordingly, we carried out experiments in which PGE(1) was added to whole blood collected in the liquid anticoagulant citrate-phosphate-dextrose (CPD) and to platelet-rich plasma (PRP) prepared from CPD-collected whole blood. We studied effects of this procedure on the recovery *in vitro* of platelets from whole blood, the survival *in vivo* of liquid-stored and previously frozen platelets, and the survival *in vivo* of liquid-stored red cells.

Prostaglandin E(2) has been reported to decrease the filterability and increase the osmotic fragility of human red cells (Allen and Rasmussen, 1971). We studied the physical-structural and metabolic parameters of human red cells stored with PGE(1) and PGE(2).

METHODS

Collection of Whole Blood

From each healthy volunteer 450 ml of whole blood was collected in 63 ml of citrate-phosphate-dextrose (CPD) or in 67.5 ml of acid-citrate-dextrose (ACD, National Institutes of Health, Formula A). CPD-collected blood was used for both *in vivo* and *in vitro* studies, while ACD-collected blood was used only for *in vitro* studies.

In Vitro Recovery of Platelets From CPD-Collected Whole Blood With and Without PGE(1)

Within 4 hr after collection in a CPD JF-25 double pack unit (Fenwal Lab) and after storage at room temperature (22 ± 1°C) an amount of PGE(1) (Alza Corp) was added aseptically to CPD whole blood to give a final concentration of about 8 nanogram/ml whole blood. A stock solution of PGE(1) (4000 nanogram/ml) was prepared aseptically and an amount was added to the unit of whole blood or to the platelet-rich plasma (PRP) using 0.45 micron Millipore filters to give final concentrations that ranged from 4 to 12 nanogram/ml of whole blood or PRP. Units of CPD-collected whole blood *without* PGE(1) were handled in a similar manner. Each unit of whole blood was centrifuged in a refrigerated centrifuge (Sorvall RC3) at 4500 x g for 3 min; the PRP was transferred into the attached plastic bag. The PRP was centrifuged at 4500 x G for 5 min and the platelet-poor plasma (PPP) was removed. The platelets sedimented from the PRP-containing PGE(1) were easily and immediately resuspended in 30 ml of PPP, but the platelet concentrates prepared from PRP *without* PGE(1) had to be stored at room temp for 1 hr before they could be resuspended. Smooth platelet concentrates without clumps were routinely prepared from CPD PRP without PGE(1) and without acidification.

The recovery *in vitro* of the platelets prepared from whole blood with and without PGE(1) was determined. Phase microscopy platelet counts of whole blood were performed by the method of Brecher and Cronkite (1950), and PRP was counted with the Coulter Model B counter equipped with a 70 micron aperture (Bull et al, 1965). The volumes of whole blood and of the platelet concentrate were derived from their weight.

Effects of Addition of PGE(1) to Platelet-Rich Plasma (PRP) with Liquid Storage of Platelet Concentrate at 4°C or 22°C

A unit of CPD whole blood was collected and centrifuged at 4500 x g for 3 min. The PRP was removed, and the red cells returned to the donor. PGE(1) was added to the PRP to achieve the concentrations of 4, 8, or 12 nanogram/ml PRP. Platelet concentrates in 30 ml of plasma were stored at either 4°C or 22°C in TA-2 transfer packs made of the plastic PL-146 (Fenwal Labs), with agitation at 20 cycle/min in an aliquot mixer (Murphy et al, 1970). Other platelet concentrates were prepared without PGE(1) as described above and were stored at either 4°C or 22°C.

Effects of Addition of PGE(1) to Platelet-Rich Plasma Before Freeze-Preservation of Platelets

From healthy volunteers PRP was obtained by plasmaphoresis. Freeze-preservation of platelet concentrates was carried out as outlined in Fig. 1 (Handin and Valeri, 1971a). This method employs: (1) a controlled-rate of adding dimethylsulfoxide (DMSO), (2) a final concentration of 5% DMSO, (3) freezing at 1°C/min and storage in liquid nitrogen at -150°C, and (4) post-thaw dilution and removal of DMSO. PGE(1) was added to a final concentration of 8 nanogram/ml PRP, and the platelets were prepared for freezing as described above. Some platelet concentrates were frozen without PGE(1). After the platelets had been subjected to the freeze-thaw and the washing procedures, recovery *in vitro* was estimated from the change in the total number of platelets; this was estimated from the platelet count measured by the Coulter counter and from the volume of platelet concentrate.

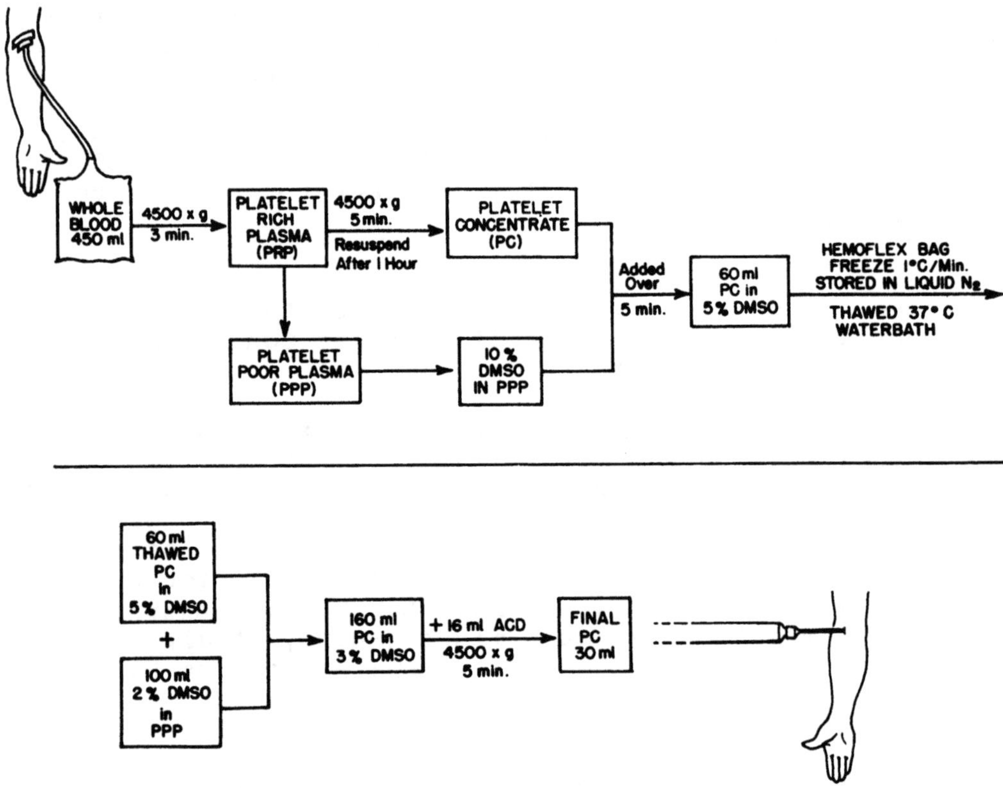

Figure 1. An outline of the procedure for platelet freezing. Studies were performed to evaluate the effect of the addition of PGE(1) to platelet-rich plasma (PRP) (final concentration of 8 nanogram/ml PRP) before freeze-preservation on recovery _in vitro_ and survival _in vivo_.

Platelet Recovery _In Vivo_ and Lifespan

Platelet survival _in vivo_ was determined using ^{51}Cr labelling of autologous platelets as previously described (Murphy and Gardner, 1969; Handin and Valeri, 1971b). Platelet recovery was calculated by comparing platelet radioactivity present 1-2 hr after transfusion with the amount infused. Blood volume was estimated from Evans blue plasma volume and total body hematocrit (peripheral venous hematocrit x 0.90) (Strumia et al, 1958). Lifespan was estimated from the disappearance of radioactive chromium from the peripheral blood.

Effects of PGE(1) on Stored Autologous Red Cells

From each of three healthy volunteers 450 ml of whole blood was collected in CPD anticoagulant. PGE(1) was added to give a final concentration of 8 nanogram/ml whole blood. The PRP was removed, and the concentrated red cells were stored for about one week at 4°C with a maximal hematocrit of approximately 90%, achieved by removing all the visible plasma (approximately 250 ml).

On the day of the autotransfusion, about 150 ml of nonbuffered saline was added to the concentrated red cells. A 50 ml aliquot was then labelled with 50 microcuries of $Na_2{}^{51}CrO_4$ (Squibb Chromoscope) at 37°C for 30 min. Ascorbic acid was not used, and the red cells were not washed in the labelling technique (Valeri, 1968). In healthy volunteers plasma volume was measured with Evans blue. The circulating red cell volume was indirectly estimated from plasma volume and total body hematocrit. Details of plasma volume measurements and calculations of erythroctye survival were similar to those reported previously (Valeri and Runck, 1969).

To determine whether intravascular liberation of hemoglobin occurred in the plasma during removal of the nonviable red cells, benzidine-measured plasma hemoglobin (Crosby and Furth, 1956) and plasma radioactivity levels (Valeri and Runck, 1969) were determined before and 1, 3, 5, 10, 15, and 30 min after the small aliquot (10 ml) autotransfusion.

Effects of the Addition of PGE(1) and PGE(2) On Platelet Concentrates and on Concentrated Red Cells

<u>In Vitro Platelet Measurements</u>. From each healthy volunteer 450 ml of whole blood was collected in ACD. Using a sterile technique, three units of the same ABO and Rh type were pooled in a TA-20 polyvinylchloride plastic bag. The pooled units were then divided equally into three TA-1 transfer pack units. PGE(1) was added to one unit to give a final concentration of 8 nanogram/ml whole blood; PGE(2) to another to give a final concentration of 8 nanogram/ml whole blood, and to the final unit an equal volume of normal saline was added. Each unit was then centrifuged at room temp for 3 min at 4500 x g. About 250 ml of PRP was removed and the concentrated red cells were stored with a maximal hematocrit up to 21 days at 4°C. The PRP was centrifuged at 4500 x g for 5 min, and the PPP was removed. After 60 min at room temp the concentrated

platelets were resuspended in about 30 ml of plasma. Samples were taken for platelet counts, and for measurements of pH and of the response to hypotonic stress. Each of the PGE(1), PGE(2), and saline-treated platelet concentrates was then divided into two equal portions and stored respectively at 4°C and 22°C with agitation at 20 cycle/min. Samples were removed aseptically at 24, 48, and 72 hr for determination of the platelet count, the response to hypotonic stress, and pH (Radiometer). Recovery from hypotonic stress was measured by a modification of the Handin et al (1970) procedure. A dual beam ultraviolet spectrophotometer (Unicam SP-1800) was blanked with PPP and the change in optical density of a 1 ml aliquot of platelet concentrate after 0.5 ml of water had been added was measured at 610 nanometers.

In Vitro Red Cell Concentrate Measurements. Concentrated red cells prepared from units containing PGE(1), PGE(2), or no prostaglandin were studied within 4 hr of collection, and again at intervals after storage at 4°C up to 21 days.

Aliquots of concentrated red cells were assayed for ATP (microM/g Hb) and 2,3 diphosphoglycerate (2,3 DPG, microM/g Hb) as previously described (Valeri and Fortier, 1969). The oxyhemoglobin dissociation curve was measured by the Bellingham and Huehns (1968) procedure using a diluted washed red cell suspension in 0.9% sodium chloride and 0.108 g/100 ml disodium phosphate buffered to pH 7.2.

Osmotic fragility of the red cells was measured in a D-2 Fragiligraph (Kalmedic Inc). Both the cumulative and derivative curves were recorded as previously described (Valeri et al, 1966). For these determinations blood samples were diluted 1:30 with buffered isotonic saline solution, and the temp during the test was maintained at 26°C. Duplicate recordings were analyzed to determine the sodium chloride concentration at which the 25, 50, and 75 % points of the cumulative osmotic fragility occurred. The index of the derivative curve of the osmotic fragility test was analyzed. The index of the unimodal derivative curve was estimated by calculating the ratio of the height of the derivative curve at its peak divided by the width of the curve at the midpoint of the ascending portion (Valeri et al, in press).

Red cell pH of the liquid packed red cells was measured by the freeze-thaw method (Hilpert, 1963), and the intracellular hydrogen ion concentration was recorded as nanoM/l of red cells. The red cell concentrations of potassium ion (mEq/10^{12} red cells) and sodium ion (mEq/10^{12} red cells) were measured as previously described (Valeri et al, 1969).

Filterability of aliquots of stored red cells was studied by a modification of a technique introduced originally by Teitel (1967), and modified by Weed et al (1969), and more recently by Allen and Rasmussen (1971). Aliquots of concentrated red cells were collected in plastic tubes (Falcon). These red cells were washed three times with a pH 7.0 phosphate buffer solution and were then resuspended at a hematocrit value of about 70% with a phosphate buffer containing CaCl2 and glucose. Each sample was incubated at 37°C in a water bath with agitation for 10 min. The washed red cells were filtered by gravity flow through a Schleicher and Schuell #589 filter that had been soaked with the buffer. The volume of blood that passed through the filter at 1, 2, 4, 8, and 16 min was determined by quadruplicate measurements of weight and density.

In Vivo Toxicity of PGE(1)

The toxicity *in vivo* of PGE(1) used in the preparation of the ^{51}Cr-labelled platelet concentrates and of the stored red cell concentrates was determined prior to and for two weeks after the infusion. Hemoglobin concentration, hematocrit level, white cell and platelet counts, blood urea nitrogen level, serum levels of creatinine, uric acid, SGOT, SGPT, LDH, and bilirubin were measured in volunteer subjects.

Bacteriological Studies

Each unit of platelet concentrates and concentrated red cells was cultured on blood agar, thioglycollate and Sabouraud's media prior to storage and again prior to infusion into the volunteer.

RESULTS

All units of platelet concentrates and red cell concentrates were sterile. The addition of PGE(1) in a concentration of 8 nanogram/ml whole blood produced a statistically significant increase in platelet recovery *in vitro* ($T = -3.49$, $p < 0.01$, $df = 33$) (Hays, 1963) (Fig. 2). The mean recovery of platelets (±SD) from 22 units of

CPD whole blood was 64 ± 14%, whereas the recovery of platelets from 17 units of CPD whole blood containing PGE(1) (8 nanogram/ml) was 79.5 ± 11%.

In a group of healthy volunteers autologous platelets prepared from PRP treated with 8 nanogram/ml, and stored as platelet concentrates at 4°C for 24 hr had slightly better ^{51}Cr recovery *in vivo* than did the platelet concentrates prepared with either 4 or 12 nanogram/ml PRP, or the platelet concentrates prepared without PGE(1) (Fig. 3). In studies on J.S., a healthy man, the ^{51}Cr recovery and lifespan of the platelet concentrate prepared with PGE(1) (8 nanogram/ml PRP) and stored at 22°C for 24 hr were similar to those of fresh platelets (Fig. 4).

Autologous platelets prepared from PRP containing 8 nanogram/ml PRP and then frozen with 5% DMSO had ^{51}Cr recovery and lifespan values similar to platelets not treated with PGE(1) (Fig. 5).

Table I shows the relative effectiveness of transfusing fresh and frozen platelets prepared with and without PGE(1). The addition of 8 nanograms of PGE(1) to each ml of whole blood improved the recovery *in vitro* of platelets in the fresh concentrates, but there was no significant effect on the ^{51}Cr recovery *in vivo* or the lifespan. PGE(1) significantly increased the percentage of platelets in the circulation from 40 to 50%, 2 hr after infusion of fresh platelet concentrates.

With the procedure used for freeze-preservation of platelets using DMSO and liquid nitrogen, recovery of about 85% of the platelets was achieved. The presence of PGE(1) in the platelet concentrates did not affect the freeze-thaw-wash recovery of platelets. Two hr after infusion the previously frozen platelets prepared with PGE(1) showed an insignificant increase (5%) in the percentage of ^{51}Cr-labelled platelets in the circulation over the value for platelets prepared without PGE(1) (Table I).

The response to hypotonic stress of platelet concentrates stored at either 4°C or 22°C was similar in units treated with PGE(1) and PGE(2), and those without prostaglandins (Fig. 6). The pH of the platelet concentrates stored at 22°C was significantly decreased at 72 hr compared to that of platelet concentrates stored at 4°C.

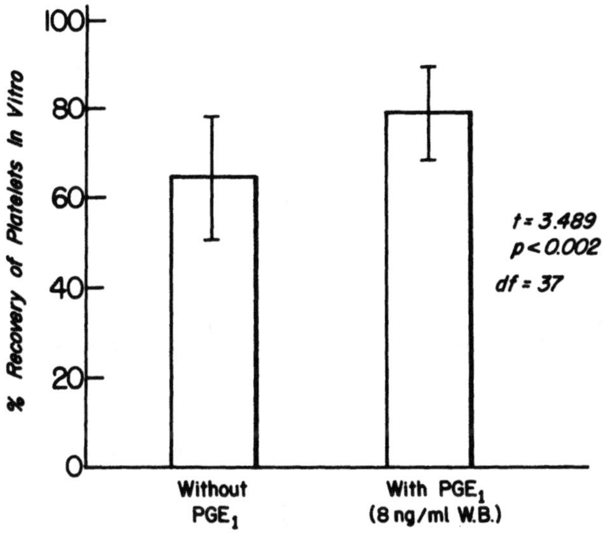

Figure 2. Recovery *in vitro* of platelets from CPD whole blood with and without PGE(1) (final concentration of 8 nanogram/ml whole blood).

Preliminary studies were made on platelet concentrates (stored at room temp for less than 4 hr) prepared from platelet-rich plasma obtained from ACD blood. Six units of fresh platelet concentrates corrected the bleeding time within 2 hr after transfusion into normal volunteers who had been treated with aspirin 24 hr before the transfusion (Handin and Valeri, 1971). This correction occurred with either platelets treated with 8 nanogram/ml PGE(1) platelet concentrate or with untreated platelets; these findings indicate that PGE(1) does not impair the hemostatic effectiveness of fresh platelet concentrates.

Concentrated red cells prepared from CPD whole blood treated with 8 nanogram/ml whole blood and stored at 4°C for one week had 24 hr ^{51}Cr survivals of about 90% with a range from 88 to 95%. Removal of the non-viable red cells during the 30-min posttransfusion period was not associated with hemoglobinemia (as evaluated by plasma hemoglobin elevations and increased plasma radioactivity).

The effects on physical-structural and biochemical measurements were similar whether the concentrated red cells stored at 4°C were treated with PGE(1), PGE(2), or no prostaglandins (Fig. 7). There was no apparent difference in the red cell 2,3 DPG, ATP, sodium ion, or

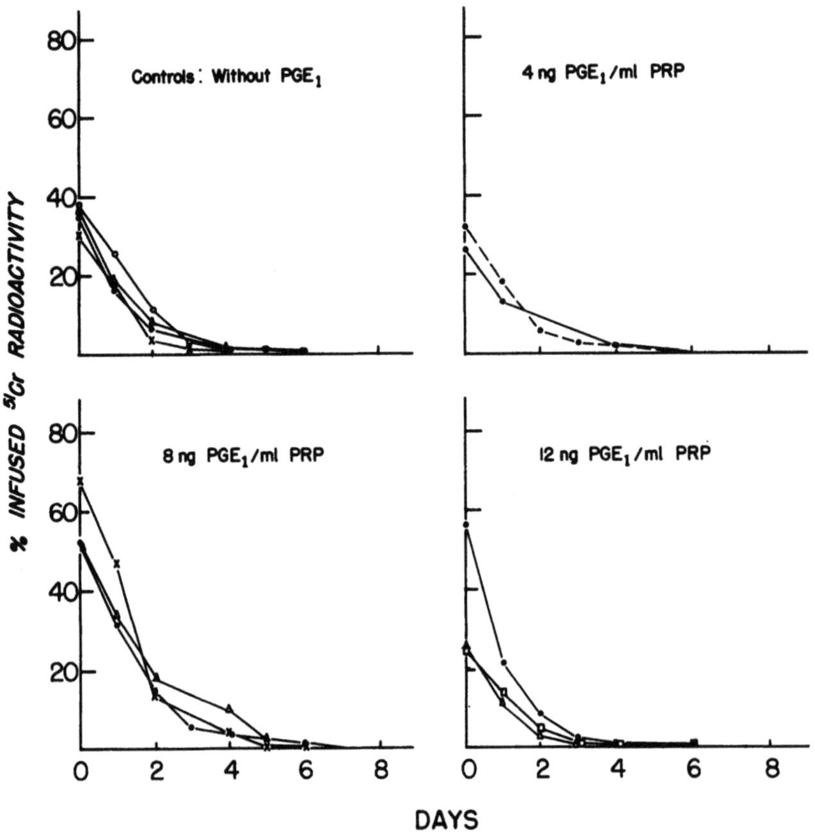

Figure 3. ⁵¹Cr recovery and lifespan of autologous platelets prepared from PRP of healthy volunteers, with PGE(1) (final concentrations 4-12 nanogram/ml PRP), and without PGE(1), after storage as platelet concentrates for 24 hr at 4°C.

potassium ion levels, red cell pH, p50 value of the oxyhemoglobin dissociation curve, red cell osmotic fragility, or red cell filterability whether PGE(1), PGE(2), or no prostaglandin was used.

DISCUSSION

These data demonstrate that the addition of 8 nanograms PGE(1) per ml of CPD-collected whole blood significantly improved platelet recovery *in vitro*. Improved recovery of platelets from blood treated with PGE(1) suggests that red cell concentrates can be prepared from whole blood with minimal platelet contamination. After washing, these red cells should be relatively free

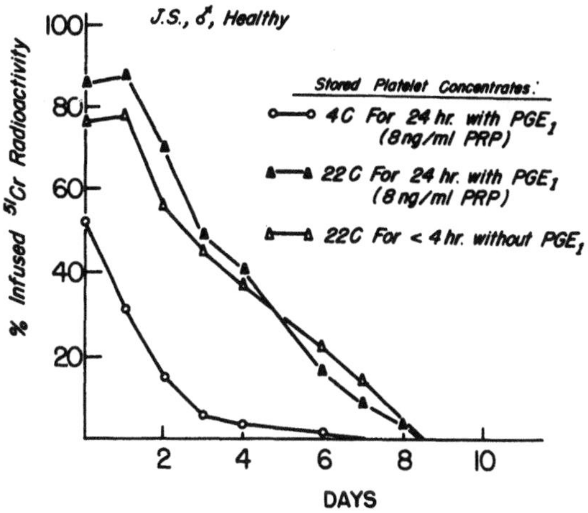

Figure 4. ^{51}Cr recovery and lifespan of autologous platelets in J.S., a healthy man. Effects of the addition of PGE(1) to PRP to a final concentration of 8 nanogram/ml in the preparation of platelet concentrates that were stored at either 4°C or at 22°C. Survival of subject's fresh platelets without PGE(1) is also shown.

of platelets. This is important because patients requiring homotransplantation, having immunologic deficiency states, or with history of febrile, urticarial, or anaphylactoid transfusion reactions, should receive red cells free of white cells and platelets (Valeri, 1970).

Our data confirm the observations of Murphy et al (1970) that it is better to store platelet concentrates at room temp than at 4°C; however, our platelet concentrates were prepared from CPD-collected blood, whereas the studies of Murphy et al (1970) were made on ACD-collected blood. Our findings corroborate and extend the observation of Mourad (1968) that platelet concentrates can be prepared from ACD-collected PRP without acidification if the platelet concentrates are stored at room temp for about 30 min before resuspension. Our data also suggest that with CPD-collected PRP this can be accomplished by storing concentrates at room temp for about 60 min before resuspension. The addition of PGE(1) to whole blood and to PRP permits immediate resuspension of the platelets in the platelet concentrate.

Platelets stored at 4°C for 24 hr with and without PGE(1) had exponential rates of removal from the circulation. Other investigators have also reported an exponential rate of removal of platelet concentrates

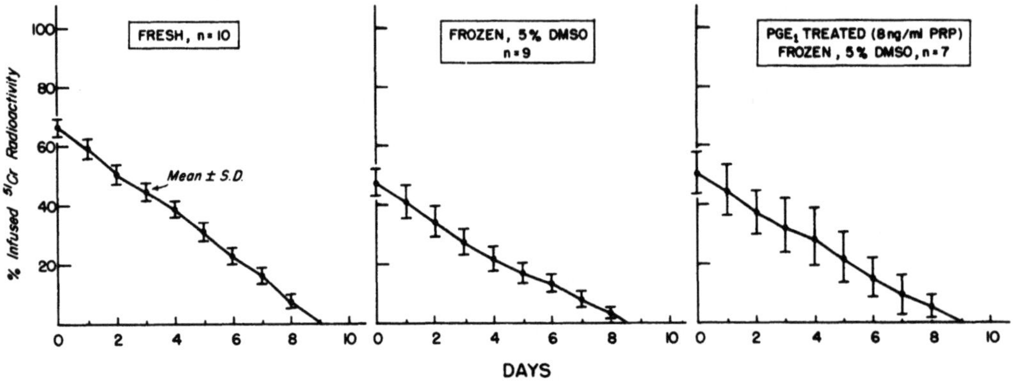

Figure 5. ^{51}Cr recovery and lifespan of autologous platelets prepared from PRP of healthy volunteers with PGE(1) (final concentration of 8 nanogram/ml PRP), and without PGE(1) after freeze-preservation of platelet concentrates.

stored at 4°C for 24 hr (Morrison and Baldini, 1967). On the other hand, a linear rate of removal was observed when ^{51}Cr labelled platelets with and without PGE(1) were stored at 22°C for 24 hr, and when they were frozen with DMSO and liquid nitrogen. Storage at 4°C for 24 hr produced preservation injury manifested by an exponential loss of platelets from the circulation; PGE(1) did not affect this removal pattern.

Hemostatic defects caused by aspirin can be corrected immediately by fresh platelet concentrates. The presence of over 10% circulating fresh platelets shortened the bleeding time and restored secondary aggregation patterns in recipients treated with aspirin (Handin and Valeri, 1971b). Platelets stored at 22°C for 24 hr have adequate survival _in vivo_, but they are not capable of effectively controlling hemostasis, that is, to correct bleeding time _immediately_ after infusion and to aggregate normally _in vitro_ to ADP, epinephrine, and collagen. In a previous study (Handin and Valeri, 1971a) restoration _in vivo_ of platelets that were stored as concentrates for 24 hr at 22°C occurred during the 24 hr posttransfusion period, a finding similar to that of Murphy and Gardner (1971). They also showed restoration _in vivo_ of platelet glycogen. The hemostatic effectiveness of fresh platelet concentrates prepared with PGE(1) has never been studied until now, when PGE(1) was found not to impair the hemostatic effectiveness of fresh platelet concentrates. In addition to inadequacies in hemostatic function after storage of platelet concentrates at 22°C for up to 72 hr,

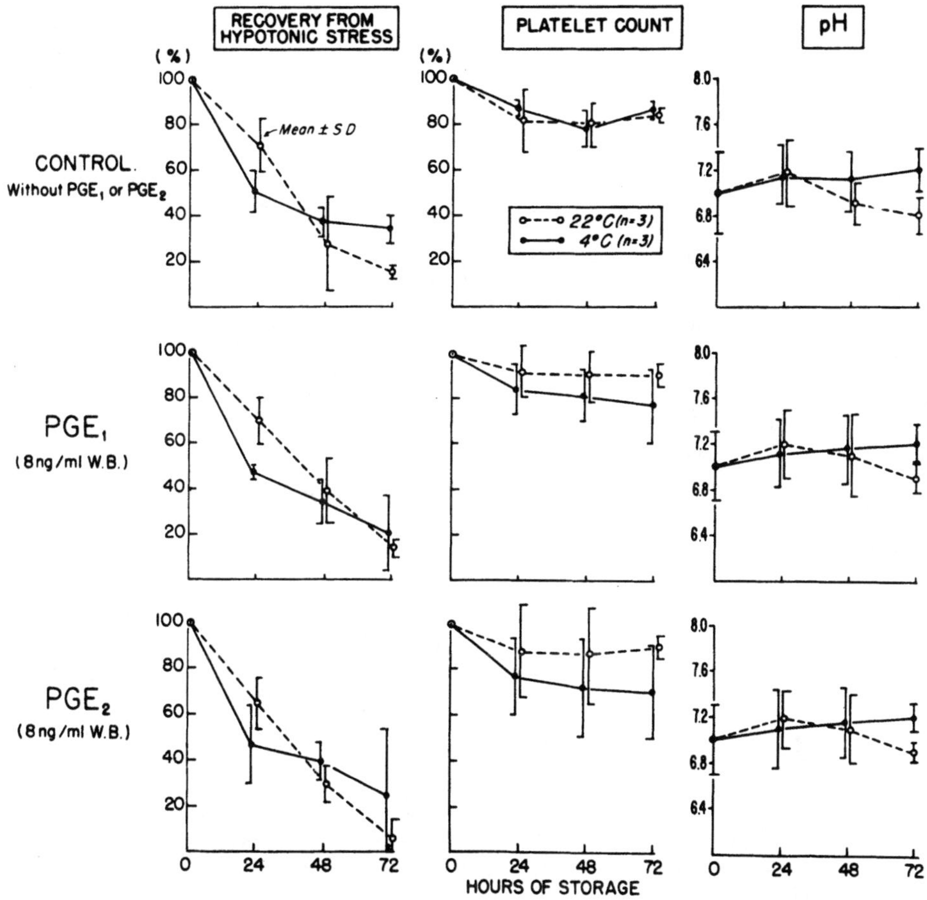

Figure 6. Mean and standard deviation (± S.D) of the following values in three studies: percent recovery of platelets from hypotonic stress, platelet count and pH for platelet concentrates stored at either 4°C or 22°C. Platelet concentrates were prepared from ACD whole blood with PGE(1) (8 nanogram/ml whole blood), with PGE(2) (8 nanogram/ml whole blood), and without prostaglandin.

there have also been reports of bacteriologic contamination (Buchholz et al, 1971).

Platelets frozen with DMSO and liquid nitrogen with and without PGE(1) showed a 15% loss during freezing, thawing, and washing. The hemostatic effectiveness of previously frozen platelets prepared with and without PGE(1) has never been studied.

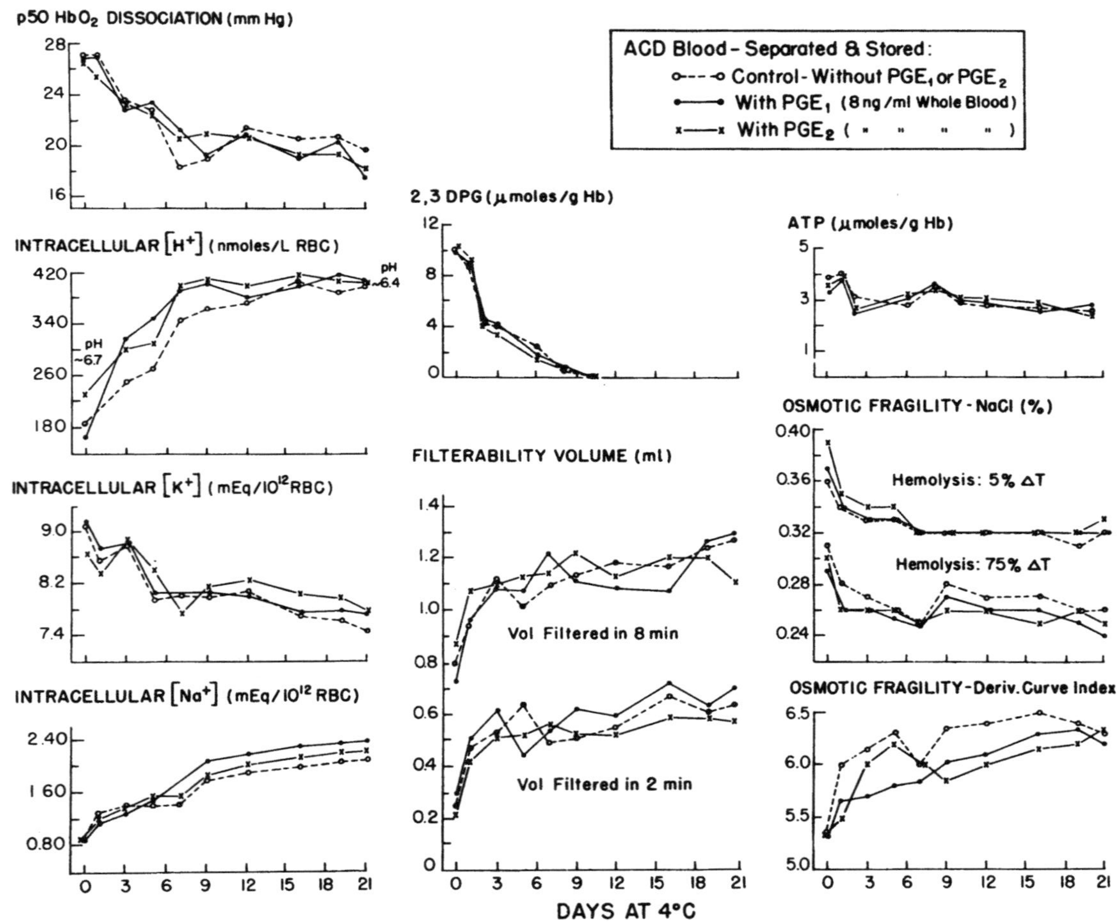

Figure 7. Physical-structural and biochemical effects of PGE(1), PGE(2), and no prostaglandins on stored concentrated red cells prepared from ACD-collected whole blood.

The addition of 8 nanogram/ml PGE(1) to whole blood produced an increase in the number of circulating platelets in fresh platelet concentrates and previously frozen platelets (Table I). This increase reflected only improved recovery of platelets _in vitro_ from whole blood; neither ^{51}Cr recovery _in vivo_ nor lifespan was improved by PGE(1).

The addition of 8 nanogram/ml PGE(1) whole blood did not adversely affect the survival of concentrated red cells stored at 4°C for about one week. The 24 hr survival was similar to that of concentrated red cells stored for the same period without PGE(1) (Valeri et al, in press). The functional state of the stored red cells with PGE(1), PGE(2), and without these prostaglandins was similar, as measured by red cell 2,3 DPG and ATP levels and the p50 values of the oxyhemoglobin dissociation curves (Valeri, 1971).

No adverse clinical effects were observed after the administration of either liquid-stored platelet concentrates, previously frozen platelet concentrates, or the small aliquots of stored concentrated red cells. There were no changes in hematology or chemistry values measured during the week after transfusion of these cellular elements.

Some investigators have reported that with PGE(2) there is an increase in red cell 2,3 DPG (Allen and Rasmussen, 1971; Rorth and Nygaard, 1971), a decrease in red cell filterability, and increased red cell csmotic fragility (Allen and Rasmussen, 1971). In our study, however, no such changes were observed (Fig. 7). The differences in results may be due in part to the final concentration of PGE(2) used. In our study stored red cells were used, whereas other investigators studied the effects of PGE(2) on fresh red cells. Moreover, in the other studies PGE(2) was added directly to the red cells prior to filterability and osmotic fragility measurements, whereas in our study measurements were made at intervals after storage at 4°C.

The therapeutic value of transfused platelets depends on their ability to improve hemostasis as well as to circulate in the recipient. Our major concern then is whether prostaglandin-treated platelets have hemostatic effectiveness (Cronkite, 1966). Studies are in progress to determine the hemostatic effectiveness _in vivo_ of platelets prepared with and without PGE(1). It is considered clinically acceptable to transfuse viable red cells with high affinity for oxygen and the capability of restoration *in vivo*. However, transfusion of stored platelets that must regain hemostatic effectiveness *in vivo* cannot be tolerated in certain clinical situations where immediate hemostatic effectiveness is necessary.

SUMMARY

CPD-collected whole blood treated with 8 nanogram/ml PGE(1) produced a significant increase in recovery of platelets *in vitro*. Platelets with PGE(1) showed recovery *in vitro* of 79.5 ± 11% compared to 64 ± 14% for those without PGE(1).

The presence of PGE(1) improved the recovery of platelets from the whole blood, but it did not significantly affect the ^{51}Cr recovery or lifespan of platelets stored as platelet concentrates at 22°C or 4°C, or after freeze-preservation. The lifespan values of platelet concentrates after storage at 22°C for 24 hr, and after freeze-preservation were similar; platelets were removed from the circulation at a linear rate. Platelet concentrates stored at 4°C for 24 hr, on the other hand, were removed from the circulation at an exponential rate. Storage of platelet concentrates at 4°C produced a preservation injury manifested by an exponential loss of platelets, and PGE(1) did not affect this removal pattern.

PGE did not impair the hemostatic effectiveness of fresh platelet concentrates.

TABLE I

RELATIVE EFFECTIVENESS OF TRANSFUSING FRESH AND FROZEN PLATELETS PREPARED WITH AND WITHOUT PGE(1)

			^{51}Cr Recovery of Circulating Platelets At 2 Hr	% Collected Platelets in Circulation At 2 Hr
Platelets in 450 ml of CPD Whole Blood	With PGE(1) (8 ng/ml)	Platelets In Fresh Concentrates 80%	65%	50%
		Platelets In Frozen-Thawed-Washed Concentrates 68%	45%	30%
	Without PGE(1)	Platelets In Fresh Concentrates 65%	65%	40%
		Platelets In Frozen-Thawed-Washed Concentrates 55%	45%	25%

ACKNOWLEDGEMENTS

The authors appreciate the excellent technical assistance of HM2 Lloyd Adams, USN, Hollace Feingold, and Linda Pivacek. These studies were made possible through Dr. Peter Ramwell and the Alza Corporation, who kindly provided the prostaglandins.

This work was supported by the U. S. Navy. The opinions or assertions contained herein are those of the authors, and are not to be construed as official or reflecting the views of the Navy Department or Naval Service at large.

REFERENCES

Allen, J. E., and Rasmussen, H., 1971, Human red blood cells: prostaglandin E(2), epinephrine, and isoproterenol alter deformability, Science 174:512.

Bellingham, A. J., and Huehns, E. R., 1968, Compensation in haemolytic anaemias caused by abnormal haemoglobins, Nature 218:924.

Brecher, G., and Cronkite, E. P., 1950, Morphology and enumeration of human blood platelets, J. Appl. Physiol. 3:365.

Bucholz, D. H., Young, V. M., Friedman, N. R., Reilly, J. A., and Mardiney, M. R. Jr., 1971, Bacterial proliferation in platelet products stored at room temperature, New Eng. J. Med. 285:429.

Bull, B. S., Schneiderman, M. A., and Brecher, G., 1965, Platelet counts with the Coulter counter, Am. J. Clin. Path. 44:678.

Cronkite, E. P., 1966, Measurement of the effectiveness of platelet transfusion, Transfusion 6:18.

Crosby, W. H., and Furth, F. W., 1956, A modification of the benzidine method for measurement of hemoglobin in plasma and urine, Blood 11:380.

Emmons, P. R., Hampton, J. R., Harrison, M. J. G., Honour, A. J., and Mitchell, J. R. A., 1967, Effect of prostaglandin E(1) on platelet behavior *in vitro* and *in vivo*, Brit. Med. J. 2:468.

Graw, Jr., R. G., Herzig, G. P., Eiser, R. J., and Perry, S., 1971, Leukocyte and platelet collection from normal donors with the continuous-flow blood cell separator, Transfusion 11:94.

Handin, R. I., and Valeri, C. R., 1971a, Improved viability of frozen platelets (Abstract), AABB 24th Annual Meeting, Chicago, Illinois (September 12-16), p. 91.

Handin, R. I., and Valeri, C. R., 1971b, Hemostatic effectiveness of platelets stored at 22°C, New Eng. J. Med. 285:538.

Handin, R. I., Fortier, N. L., and Valeri, C. R., 1970, Platelet response to hypotonic stress after storage at 4°C or 22°C, Transfusion 10:305.

Hays, W. L., 1963, Statistics for Psychologists, pp. 333-335, Holt, Rinehart and Winston, New York.

Hilpert, P., Fleischmann, R. G., Kempe, D., and Bartels, H., 1963, The Bohr effect related to blood and erythrocyte pH, Am. J. Physiol. 205:337.

Jones, A. L., 1968, Continuous-flow blood cell separation, Transfusion 8:94.

Kloeze, J., 1967, Influence of prostaglandins on platelet adhesiveness and platelet aggregation, Nobel Symposium 2, Stockholm, June 1966, Prostaglandins. Eds. S. Bergstrom and B. Samuelsson, p. 241, Almquist and Wiksell, Stockholm.

Morrison, F. S., and Baldini, M., 1967, The favorable effect of ACD on the viability of fresh and stored human platelets, Vox Sang. 12:90.

Mourad, N., 1968, A simple method for obtaining platelet concentrates free of aggregates, Transfusion 8:48.

Murphy, S., and Gardner, F. H., 1969, Platelet preservation: effect of storage temperature on maintenance of platelet viability - deleterious effect of refrigerated storage. New Eng. J. Med. 280:1094.

Murphy, S., and Gardner, F. H., 1971, Platelet storage at 22°C: metabolic, morphologic, and functional studies, J. Clin. Invest. 50:370.

Murphy, S., Sayar, S. N., and Gardner, F. H., 1970, Storage of platelet concentrates at 22°C, Blood 35:549.

Rorth, M., and Nygaard, S. F., 1971, Phosphate metabolism of the red cell during exposure to high altitude. in Alfred Benzon Symposium IV. Oxygen Affinity of Hemoglobin and Red Cell Acid-Base Status. Munksgaard, Copenhagen.

Shio, H., Plasse, A. M., and Ramwell, P. W., 1970, Platelet swelling and prostaglandins, Microvasc. Res. 2:294.

Strumia, M. M., Colwell, L. S., and Dugan, A., 1958, The measure of erythropoiesis in anemias. I. The mixing time and the immediate posttransfusion disappearance of T-1824 dye and of ^{51}Cr tagged erythrocytes in relation to blood volume determination, Blood 13:145.

Teitel, P., 1967, Le test de la filtrabilite erythrocytaire (TFE) une methode simple d'etude de certaines proprietes microrheologigue des globules rouges, Nouv. Rev. Fr. Hematol. 7:195.

Tullis, J. S., Tinch, R. J., Gibson, J. G. II, and Baudanza, P., 1967, A simplified centrifuge for the separation and processing of blood cells, Transfusion 7:232.

Tullis, J. L., Eberle, W. G., II, Baudanza, P., and Tinch, R., 1968, Platelet pheresis description of a new technique, Transfusion 8:154.

Valeri, C. R., 1968, Observations on the chromium labelling of ACD-stored and previously frozen red cells, Transfusion 8:210.

Valeri, C. R., 1970, Recent advances in techniques for freezing red cells. Crit. Rev. Clin. Lab. Sci. 1:381.

Valeri, C. R., 1971, Viability and function of preserved red cells, New Eng. J. Med. 284:81.

Valeri, C. R., and Fortier, N. L., 1969, Red-cell-mass deficits and erythrocyte 2,3 DPG levels, Forsvarsmedicin 5:212.

Valeri, C. R., and Runck, A. H., 1969, Long-term frozen storage of human red blood cells: studies in vivo and in vitro of autologous red blood cells preserved up to six years with high concentrations of glycerol, Transfusion 9:5.

Valeri, C. R., McCallum, L. E., and Danon, D., 1966, Relationships between *in vivo* survival and (1) density distribution, (2) osmotic fragility of previously frozen, autologous, agglomerated, deglycerolized erythrocytes, Transfusion 6:554.

Valeri, C. R., Runck, A. H., and Sampson, W. T., 1969, Effects of agglomeration on human red blood cells, Transfusion 9:120.

Valeri, C. R., Szymanski, I. O., and Pivacek, L. E., (in press)a, Effects of the host on transfused preserved red blood cells, J. Med.

Valeri, C. R., Szymanski, I. O., and Zaroulis, C. G., (in press)b, 24-Hour survival of ACD and CPD stored red cells. I. Evaluation of nonwashed and washed stored red cells, Vox Sang.

Weed, R. I., La Celle, P. L., and Merrill, E. W., 1969, Metabolic dependence of red cell deformability, J. Clin. Invest. 48:795.

SOME EFFECTS OF VASOACTIVE HORMONES

ON THE MAMMALIAN RED BLOOD CELL

J. E. Allen and H. Rasmussen

Department of Biochemistry
University of Pennsylvania
Philadelphia, Penn.

INTRODUCTION

Krogh first observed the marked deformation which red cells undergo in their passage through the capillaries (Krogh, 1922). This deformation is imposed upon red cells since non-distensible capillaries have diameters ranging from 3-12 microns whereas that of the red cell averages 8 microns. In order to traverse the smaller capillaries the red cell is deformed into a sausage shape with a cylindrical diameter approximating that of the capillary, but with a length well in excess of its original diameter. This remarkable change in shape occurs without any significant change in surface area (LaCelle, 1970). Red cell membranes are resistant to stretch but quite readily change shape in response to a deforming pressure (Rand and Burton, 1964a, 1964b). In addition to the shape change which takes place in the capillaries, the red cells normally undergo a change in shape in the small arterioles. This deformation occurs in response to the pattern of forces exerted by laminar flow in the arteriolar circulation (Goldsmith, 1970). The ability of these cells to readily undergo deformation is evident since a negative pressure of only 8 mm H20 is sufficient to draw a normal red cell through a 3 micron microcapillary <u>in vitro</u> (LaCelle, 1969).

Nevertheless, only the properties of the red cells themselves, or more particularly the intrinsic hardness or deformability of their membranes, are not thought of great importance in the moment to moment control of circulatory dynamics: the presently accepted views are that the contraction of the arterioles and pre-capillary sphincters are the prime determinants of flow under a constant pressure head. Recently, however, alterations of red cell deformability have been observed _in vitro_ (Weed, et al, 1969), and a decreased deformability of cells obtained from patients with a variety of hemolytic anemias has been described (Teitel, 1967; LaCelle, 1970). These workers have suggested that deformability may be an important determinant of red cell survival.

The _in vitro_ studies of Weed, et al (1969) have shown that red cells undergo a metabolically-determined, and reversible change in membrane deformability. When cellular ATP concentrations fall _in vitro_ there is an associated increase in cell calcium and a decrease in cell deformability. Weed et al (1969) have proposed that membrane hardness or deformability is determined either by a calcium-dependent contractile element on the inner surface of the red cell membrane as first proposed by Wins and Schoffeniels (1966), or by a sol-gel transformation at the inner surface of this membrane as proposed by Ponder (1948). In either case, Burton (1965) and Weed (1970) have suggested that these changes in membrane hardness, if they occur _in vivo_, may be of importance in the control of the microcirculation. However, the magnitude of the changes in cell calcium and cell ATP content apparently required _in vitro_ to bring about significant changes in membrane deformability, are not normally seen _in vivo_. In addition, the times required to bring about these changes _in vitro_ are measured in hours, whereas circulation times in the whole organism are of the order of minutes, and in the microcirculation of the order of seconds. Clearly, if changes in membrane deformability are to be of importance in the moment to moment control of the microcirculation, then rapidly inducible changes in membrane conformation are a prerequisite.

It is precisely such rapid changes in red cell deformability which we wish to describe for the first time. We have found that the deformability of their cell membranes is a highly dynamic property of red cell populations. Red blood cells taken from normal human donors exhibit changes in deformability depending upon the time of day in which the sample is collected, and by the sex of the donor. Furthermore, in females, deformability is a function of the menstrual cycle. In addition to these changes with a time parameter of hours and days, this property of the cell membrane has been found to undergo short period (about 15 sec) sustained oscillations _in vitro_, as measured by susceptibility of a cell population to osmotic hemolysis. Finally, and perhaps of greatest immediate interest, deformability of red cells *in vitro* changes rapidly in response to hormones which affect acutely the status of the circulation *in vivo*.

All of these data demonstrate that deformability is a highly dynamic property of the red blood cell membrane. This in turn means that the red blood cells must normally be an important factor in the control of the microcirculation. These cells can no longer be considered simply passive participants in the control of blood flow, but must now be considered as active determinants in this control. These conclusions have profound implications in the analysis of the pathophysiology of circulatory disorders.

METHODS AND RESULTS

The Effects of Vasoactive Compounds Upon Red Cell Deformability

Although there are several methods by which erythrocyte deformability can be measured, one of the simplest and most sensitive involves measurement of the rate of flow of a suspension of red cells at high hematocrit through a standard paper filter (Teitel, 1967; Weed et al, 1969; LaCelle, 1970). The method is based on the fact that at hematocrits greater than 60%, red cells are deformed by their close packing (Burton, 1965). Blood flow at these high hematocrits is proportional to the deformability of the cells: for one cell to move relative to its neighbors, changes in shape are necessary. By measuring the flow rate through a paper filter (Schleicher and Schuell #589, white ribbon) deformability can be assayed. The data are expressed in terms of the time taken for half of the total volume of the added blood to pass through the filters - t(1/2) - which is thus inversely proportional to deformability.

To carry out these studies, blood was drawn from adult male and female volunteers between the ages of 20 and 35, without a tourniquet whenever possible. EDTA (0.5 mg/ml) or heparin (10 unit/ml) was used as anticoagulant. The blood was centrifuged, the plasma and buffy coat drawn off, and the cells washed three times with a buffer (in mM, NaCl, 145; KCl, 5; $MgSO_4$, 1; Na_2HPO_4 + NaH_2PO_4, 5; glucose, 10; pH 7.0) If EDTA were used as anticoagulant, $CaCl_2$ was deleted from the buffer for the first wash. Otherwise, the standard buffer was used for all incubations and treatments except in some experiments the pH was adjusted to 7.4 with NaOH. The packed red cells were suspended after the final wash at an hematocrit of 70%. The suspension was divided into equal aliquots; 2 ml being placed in each of eight disposable culture tubes. The blood was then incubated for 10 min at 37°C in a rotatory shaker (stroke 1.5 inch, rpm 300-400). Hormone or carrier was added and the incubation proceeded for a further 10 min. At this time, the aliquots of blood suspension were added to filters and the flow rate determined. The time for 1 ml (0.5 the volume added) to flow through the filter was obtained for each of the eight samples. This quantity, t(1/2), is directly proportional to deformability.

Blood from male donors drawn at 8-10 a.m. was selected for these experiments. A single lot of filter papers was employed since lots varied.

The addition of PGE(2) to washed red cells <u>in vitro</u> gave rise to a significant hardening of red cells (Fig. 1) with a maximum response found at 10^{-10} M. The dose response curve falls off sharply above this concentration so that at 10^{-9} M and above there is no significant effect of the hormone. The erythrocyte is, thus, somewhat sensitive to this compound. This sensitivity may be better appreciated by expressing these data in terms of the number of molecules per cell. At 10^{-11} M PGE(2), there is a significant effect; this concentration corresponds to one molecule of PGE(2) for every 10 cells. At the peak response, 10^{-10} M, there is one molecule per cell. If one goes above this value, to 10^{-9} M, the effect is lost; this difference in sensitivity corresponds to a change in the ratio of molecules to cell from 1 to only 10.

Similar experiments were performed using PGE(1). This compound does not act like PGE(2) but brings about a softening of the cells (t(1/2) control = 3.2; PGE(1) (10^{-10} M) = 2.9, n = 5, P = 0.01, two-tailed Students t-test for matched pairs).

Effects similar to that of PGE(2) are induced by certain catecholamines. Fig. 2 shows that cells are hardened by epinephrine at 10^{-9} M. Concentrations above this value lead to a softening of the cells relative to the peak response. DL-isoproterenol also induces hardening at 10^{-7} M with decreased response as the concentration is increased.

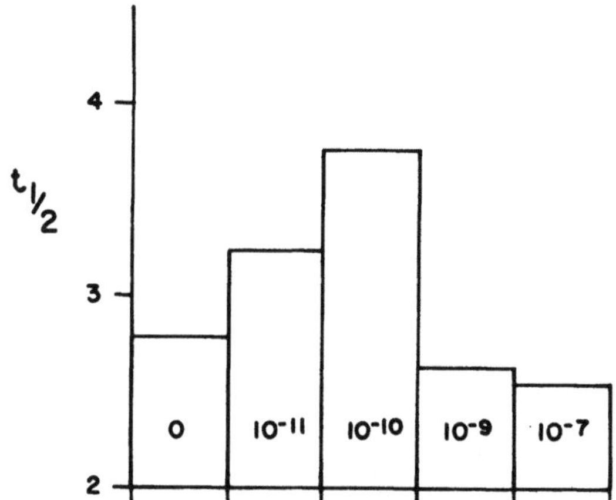

	0	10^{-11}	10^{-10}	10^{-9}	10^{-7}
$\bar{X}$	2.80	3.24	3.77	2.74	2.67
S.D.	0.78	0.81	1.20	0.63	0.44
n	60	9	17	9	4
P	–	<.05	<.005	NS	NS

Figure 1. The effect of PGE(2) upon filterability. t(1/2) = time taken for 1/2 volume of blood to pass filters. Final molar concentrations of PGE(2) in RBC suspensions of 70% hematocrit given in bars. Probability (P) obtained by Students t-test for matched pairs. See text for details of procedures.

Hormonal Control of Hypotonic Hemolysis

Although the filtration technique offers many advantages for determining cell deformability, we felt that such measurement should be supported by data using a different method. We, therefore, examined the alteration of the cells response to hormones by hypotonic hemolysis.

In this technique, washed cells were suspended at an hematocrit of 20%, and incubated at 37°C in a rotatory shaker. Aliquots (50 microliters) were taken serially at intervals and added to 2 ml of vigorously agitated NaCl solution. The content of these salt solutions varied from 0.40 to 0.45 g% depending upon the susceptibility of the particular cell population to hemolysis. The percent hemolysis was determined by measuring the optical density of the supernatant following centrifugation to sediment the unhemolyzed cells; an equal aliquot was added to

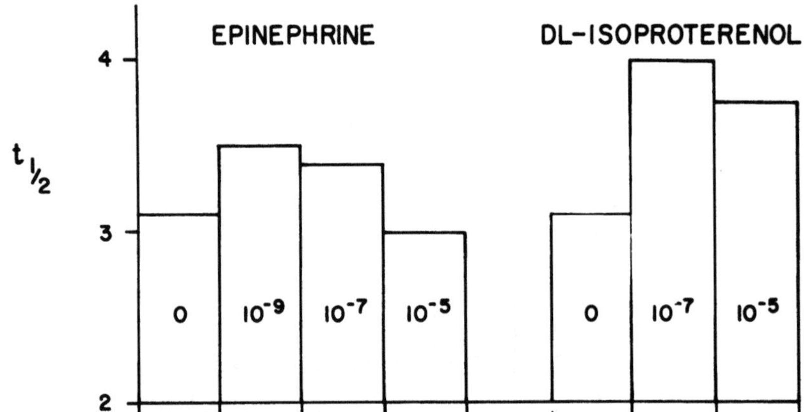

	Epinephrine 0	Epinephrine 10^{-9}	Epinephrine 10^{-7}	Epinephrine 10^{-5}	DL-Isoproterenol 0	DL-Isoproterenol 10^{-7}	DL-Isoproterenol 10^{-5}
$\bar{x}$	3.10	3.51	3.42	3.01	3.10	4.00	3.76
S.D.	0.86	0.72	0.55	0.43	0.86	0.71	1.25
n	35	8	8	8	35	9	13
P	–	<.05	<.05	NS	–	<.001	<.05

Figure 2. The effect of catecholamines upon filterability. See text and Fig. 1 for techniques and legend.

distilled water to obtain the optical density corresponding to 100% hemolysis. The error of the method as determined by serial determinations of 100% hemolysis was found to be < 1%.

The variability of the cell population was found to exceed the error of the method. Serial samples taken at shorter intervals revealed that oscillations of susceptibility to hypotonic hemolysis were occurring and Fig. 3 demonstrates this phenomenon. The samples upon which these curves (fitted by eye) are based were taken at 7 sec intervals. The period of the oscillations was found to be approximately 15 sec. The amplitude, as well as the frequency, was irregular, however. The oscillations persisted *in vitro* for 4 hr if cells were resuspended in fresh buffer every hr.

The technique was automated by using a proportioning pump (Technicon, Autoanalyzer) to continuously sample both the blood suspension and the hemolyzing salt solution. These were mixed, separated into 1 sec aliquots by air bubbles

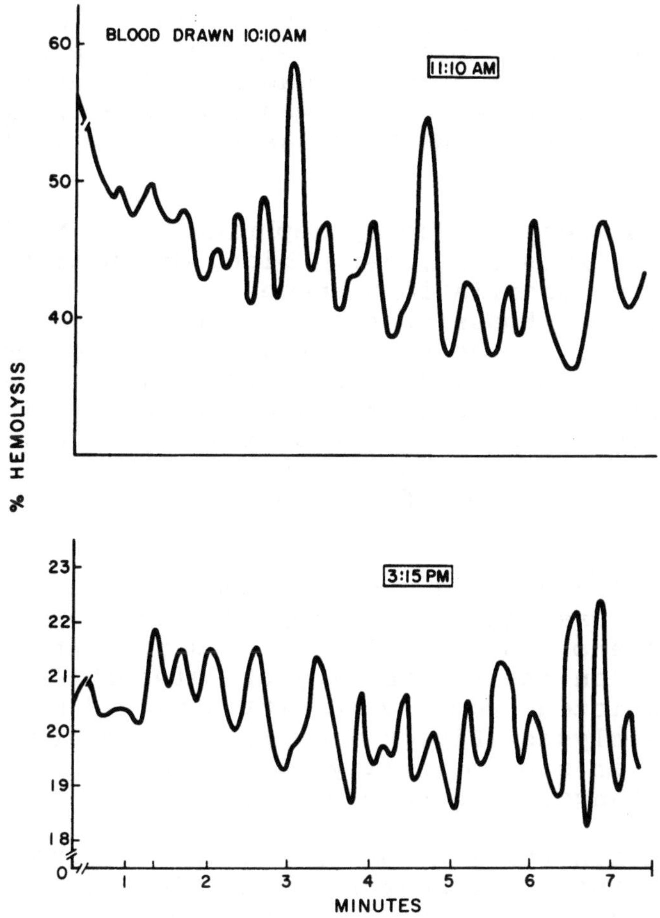

Figure 3. Hypotonic hemolysis: Oscillations. Aliquots of RBC suspension were added serially to hypotonic saline and percent hemolysis determined. NaCl (0.43 g%) used througnout. Cells resuspended in fresh buffer once every hr between the two experiments depicted on the figure.

by the proportioning pump and passed through a 3 min delay coil. This coil both agitated the blood cells in the hemolyzing solution and allowed sufficient time for hemolysis to occur. The solution was debubbled and the suspension led into a flow-through cuvette. The absorbance was measured at 500 nanometers. Absorbance at this wavelength is inversely proportional to the percent hemolysis and a continuous trace of hemolysis with a discrimination time of about 1 sec was recorded.

In Fig. 4 the results of one such experiment are shown. The oscillations in this case had a period of about 15 sec, and the irregularity of amplitude and frequency were again observed. When PGE(2) was added (final conc 2 x 10^{-11} M) there was a small increase in the hemolysis (about 3%) without a major change in the nature of the oscillations. The time taken for complete response was 20-30 sec.

A typical response to epinephrine (10^{-9} M) is shown in Fig. 5. In this case, using blood drawn from another individual, the oscillations showed a greater irregularity in frequency than those of the previous figure. The period was still in the range of 15-20 sec, however. Epinephrine brought about a characteristic increase in hemolysis and a damping of the oscillations. Prior to the damping, however, one cycle of oscillations occurred of amplitude comparable to those seen before hormone addition. The addition of isoproterenol produced similar results.

In Fig. 1 and 2 one can also see that the variance in filterability was reduced by treatment with epinephrine and isoproterenol, but that PGE(2) has no such effect. This effect upon the variability of deformability is what would be predicted from the results obtained measuring hypotonic hemolysis (Fig. 4 and 5). Presumably the oscillations in susceptibility to hypotonic hemolysis are related to changes in deformability.

Interaction between PGE(2) and the catecholamines was indicated by the experiment shown in Fig. 6; PGE(2) (10^{-11} M) was added 10 min prior to addition of isoproterenol (10^{-7} M) and the usual response to PGE(2) was obtained. The addition of isoproterenol led to an unexpected effect, in that a marked increase (about 7%) in hemolysis was followed by a marked decrease. After about 3 min, a similar phenomenon recurred: a marked decrease in hemolysis followed by an increase, and a further swing after about 1 min. After another 3 min (not shown on the figure) another marked decrease and increase in hemolysis occurred. A second smaller peak did not follow, however, and this phenomenon did not recur again. In contrast to similar experiments done in the absence of PGE(2), isoproterenol did not diminish the amplitude of the oscillations. Epinephrine induced similar effects. A further unique aspect of the interaction was found: concentrations of PGE(2) which by themselves had no effect on deformability or hemolysis, i.e. 10^{-9} and 10^{-7} M, were capable of permitting this marked response to catecholamines.

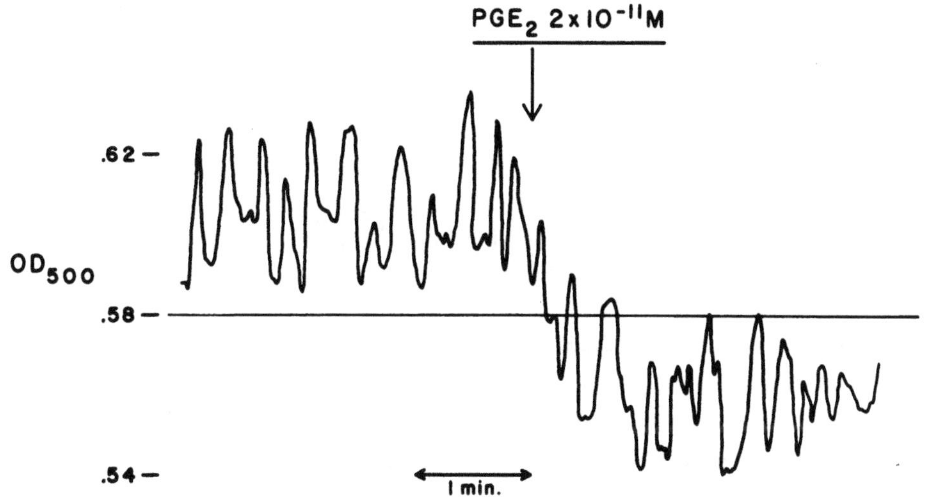

Figure 4. Hypotonic hemolysis: Oscillations and the effects of PGE(2). An automated method (see text) was used. Increasing hemolysis is indicated by a fall in O.D.500.

Hormones and Cell Calcium

Previous models of the mechanism underlying changes in erythrocyte deformability (Weed, et al, 1969), entailed an increase in cellular calcium which then altered the conformation of a protein or proteins in the red cell. This protein interacted with the cell membrane in some way such that deformability was altered. Inadequacy of the usual techniques for measuring cellular calcium led us to develop a new technique for measuring red cell calcium.

Lanthanum at low concentrations has been shown to block the passive flux of calcium across several types of membranes (Van Breemen, 1969; Mela, 1968). It was reasoned that cells incubated in the presence of ^{45}Ca and subsequently added to ice cold buffer solutions containing LaCl3 would lose little of the ^{45}Ca that had already been taken up by the cells, and that further entry would, in addition, be blocked. Other techniques for measurement of red blood cell calcium involve multiple washings in calcium-free media following an initial centrifugation to separate the cells from the incubation medium (Harrison and Long, 1968). Such treatment almost certainly perturbs the intracellular calcium concentration.

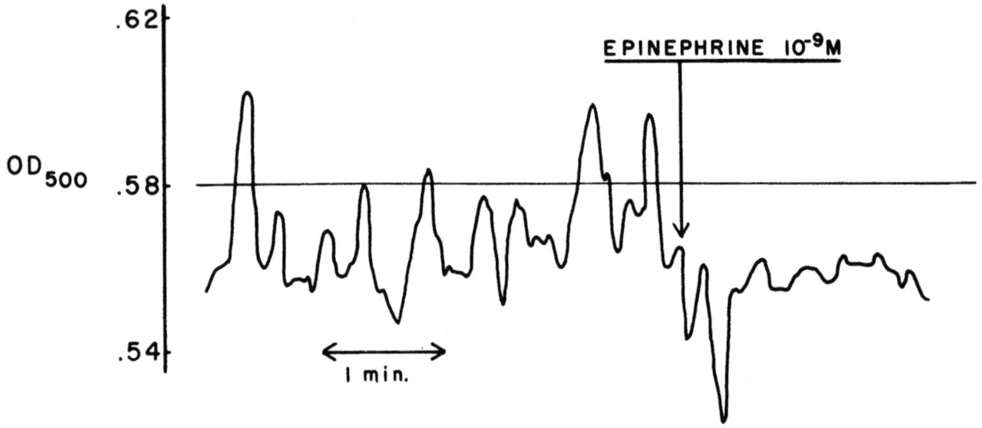

Figure 5. Hypotonic hemolysis: Oscillations and the effects of epinephrine. As in Fig. 4.

The experiments were performed by adding ^{45}Ca to a suspension of red blood cells incubating at 37°C in a rotatory shaker bath. At appropriate intervals, 200 microliter aliquots of red blood cell suspensions (hematocrit 20%) were removed and added to 10 ml of ice cold buffer of the composition given before, at pH 7.0 and 4°C, but containing 2 x 10^{-5} M LaCl3. This suspension was centrifuged within 20 min, and the cells resuspended and washed twice before counting the ^{45}Ca in the red cell pellet.

The ^{45}Ca contained within the cells reached an average maximum value within one min of exposure of the cells to the ^{45}Ca in the original incubation. No further increase in cellular ^{45}Ca was seen after treatments with the isotope of up to 2 hr. Assuming, then, complete equilibration of ^{45}Ca with this pool, i.e. specific activity identical to that in the bathing medium, the mean size of this pool ranges from 3.5 to 4 microM/liter cells which seems constant between individuals. A total calcium pool of 1.6 x 10^{-5} M was found by Harrison and Long (1968) using atomic absorption spectrophotometry on washed human erythrocytes. Assuming this value to be correct, then the exchangeable pool we measure amounts to approximately 20% of the total cell calcium pool.

The size of the calcium pool also exhibited short period oscillations with a waveform similar to that seen in oscillations of susceptibility to hypotonic hemolysis. Therefore, the relationship between hemolysis and calcium was studied. Samples were taken from a red cell

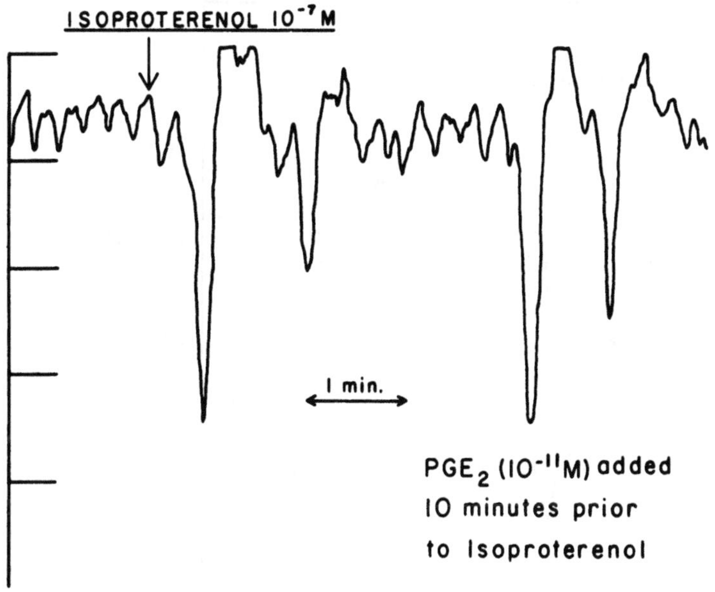

Figure 6. Hypotonic hemolysis: Combined effects of PGE(2) and isoproterenol. As in Fig. 4.

suspension at nearly identical times (± 2 sec); one 200 microliter aliquot was added to LaCl3-containing buffer for ^{45}Ca determinations, another 50 microliter sample was taken for determination of hypotonic hemolysis. Fig. 7 shows an experiment using red blood cells taken, in this instance, from rats. There was obviously a correlation between oscillations of the calcium pool and hemolysis. The frequency was similar and the two parameters appeared in phase.

Fig. 8 shows the response of the calcium pool in human cells to the administration of isoproterenol (10^{-7} M). In these experiments simultaneous samples were taken for the determination of ^{45}Ca and hemolysis, as before, but at 1 min intervals. Isoproterenol increased mean hemolysis, and significantly decreased the variance of the hemolysis. On the other hand, the size of the calcium pool decreased following isoproterenol, and the variance of this measurement also decreased. The decreased variance following isoproterenol agrees with data obtained in experiments similar to that in Fig. 7. The oscillations in intracellular calcium were abolished following isoproterenol in those experiments. In the final period following hormone addition, i.e. 12-18 min, the intracellular calcium pool increased significantly, and the degree of hemolysis returned to near normal levels. The variance did not return to pre-hormone

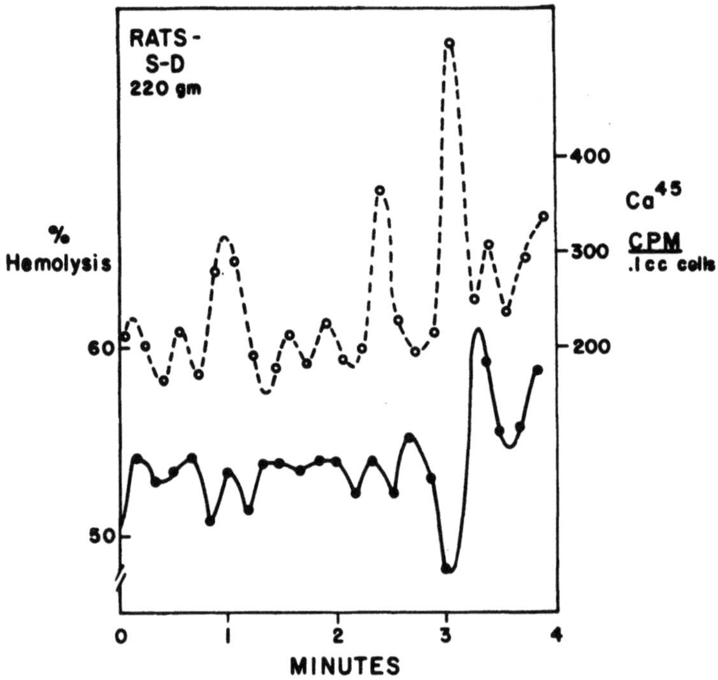

Figure 7. ^{45}Ca incorporation and hypotonic hemolysis. Simultaneous aliquots taken for intracellular ^{45}Ca determinations (lanthanum method) and for determination of percent hemolysis. See text for details.

levels, however, indicating that the oscillations were still suppressed.

Experiments were also performed in which ATP was measured in the erythrocytes following isoproterenol (10^{-7} M). By adding aliquots of cells to distilled water at 4°C, and then sedimenting the red cell ghosts, it was possible to determine the amount of ATP in the cytoplasm, and to obtain a rough measure of the ATP bound to the membranes by acid extraction of the ATP in the sediment containing the membranes. Cytoplasmic ATP fell (1.30 ± 0.12 to 1.12 ± 0.11 mM/liter cells, n = 12) in the first 6 min following isoproterenol treatment, and membrane-associated ATP showed no change.

Fig. 9 shows the results of an experiment employing PGE(2) (10^{-11} M). Although the degree of hemolysis was not determined, it is normally increased by PGE(2) under these conditions. The results with ^{45}Ca and cytoplasmic ATP were essentially the same as those obtained using

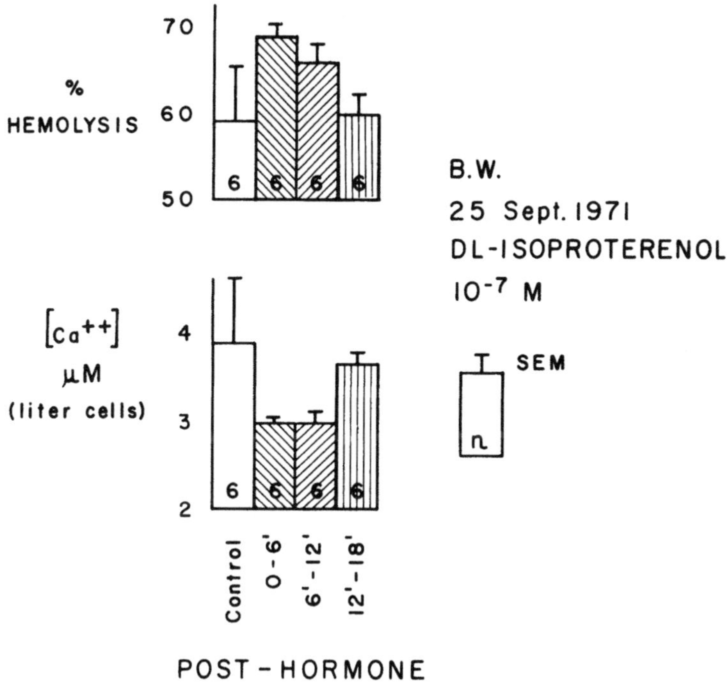

Figure 8. Effects of isoproterenol on RBC calcium and percent hemolysis. Aliquots taken at 1 min intervals from same blood suspension.

isoproterenol. However, the membrane-associated ATP levels were significantly increased by PGE(2). This occurred despite the fact that cytoplasmic ATP levels fell. The variance of the control values of calcium was low in this particular experiment, indicating a smaller amplitude of oscillations.

In spite of the fact that catecholamines and prostaglandins both cause a fall in cytoplasmic calcium and ATP, only PGE(2) causes simultaneous increase in membrane bound ATP. This indicates that these compounds, even though they yield similar results in terms of hemolysis and deformability changes, do not have identical mechanisms of action. The failure of PGE(2) to suppress oscillations supports this concept, as does the interaction of the catecholamines and PGE(2) as shown in Fig. 6.

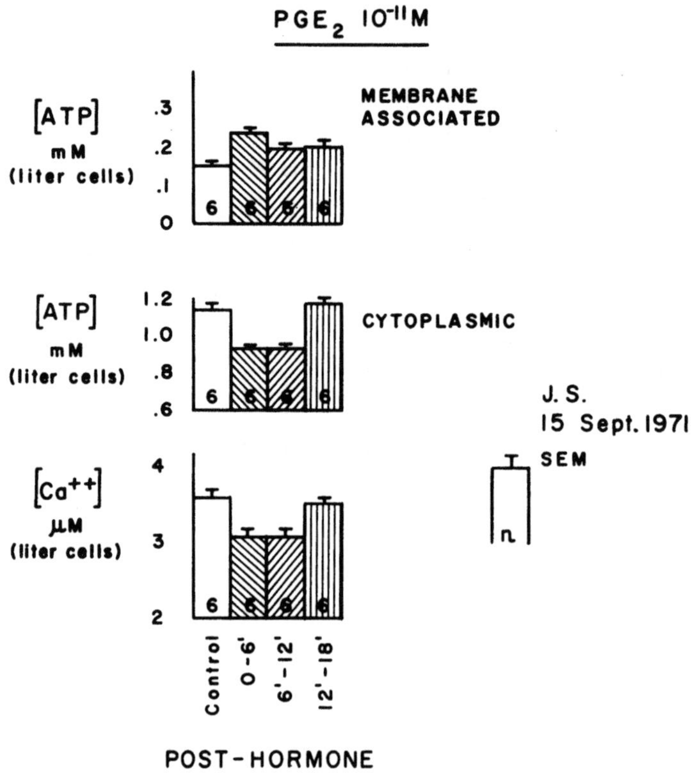

Figure 9. Effects of PGE(2) on both RBC ATP distribution and calcium. As in Fig. 8. Details in text.

Variability of the Donor Population by Sex and Time of Day

In addition to the short period oscillations observed *in vitro*, we have discovered a variability in red cell properties which reflect a longer periodicity and which take place *in vivo*. These results further illustrate the complexity of the erythrocyte response to various physiological stimuli.

The filterability data from a large number of experiments were analyzed according to the sex of the donor, and to the time of day the blood was collected. Both factors were found to influence red cell deformability. Women on the average, yielded cells which were less deformable than those of men (Fig. 10). Furthermore, an analysis of variance using Fisher's F test showed that deformability of cells drawn from women at this time of day showed significantly greater variability than cells drawn from men.

The possibility that this variability was related to the menstrual cycle was then checked by retrospective analysis of data obtained from one female who, fortunately, kept records of her menses. On the first day of her menses (Fig. 11) her cells were hard, relative to the normal mean for women. By the third day, her cells were softer than the normal mean and from that point hardened progressively. Unfortunately, no points were available during the latter half of the subject's cycle, since she left on vacation. Further studies are now in progress.

We had noticed in another series of experiments with rats, that in late pregnancy the deformability of the red cells was greatly decreased. Since oral contraceptives give rise to a hormonal condition similar to pregnancy, we obtained data from a subject on and off such medication. As seen in Fig. 12, cells of this patient, when taking medication, were considerably harder than normal, but returned to normal within a week following cessation of the drug.

Since these initial studies, red cells obtained from other women being treated with oral contraceptives have been examined. Comparable results have been obtained. The cells obtained from women being treated with these drugs are significantly harder than cells obtained from controls, and cell hardness decreases with cessation of drug administration.

A recent study by Aronson, et al (1971) has shown that the viscosity of whole blood obtained from women treated with birth control steroids is increased. However, in these studies, the component or components in whole blood responsible for these changes was not identified. Taken in conjunction with the present observations, it seems quite likely that changes in the deformability of the red cell membrane may be an important part of this change in blood viscosity. These findings raise the question of the role that these changes in red cell properties may have in the thrombo-embolic phenomena seen in this group of patients.

Additional data showed that the deformability of red cells depended upon the time of day at which the blood was drawn. Fig. 13 shows the deformability of red cells drawn from men in the morning and the afternoon. It is apparent

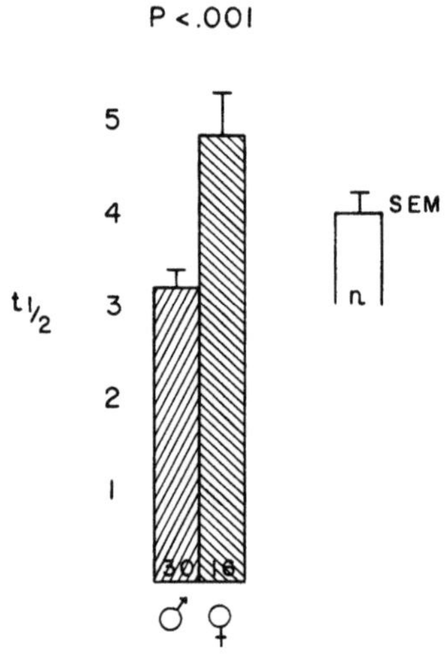

Figure 10. Filterability: Men versus women, morning samples. As in Fig. 1.

that there was a hardening as the day wears on. However, in women, (Fig. 14) there was, on the average, a softening. Analysis of variance in this case showed that the morning samples had a (highly significant) greater variability when compared to the afternoon cells. This, in turn, indicates that one should find a diurnal variation in estrogen and/or progesterone levels, if our analysis is correct. In view of the limited population size we may have a preponderance of cells collected from women in one phase of the menstrual cycle in the morning or afternoon. None of the women included in these data were taking oral contraceptives.

This diurnal variation in red cell deformability is even more dramatically illustrated by studies in experimental animals. Blood was drawn from rats both in the morning and afternoon and determinations made of filterability and susceptibility to hypotonic hemolysis. As shown in Fig. 15, the t(1/2) increased from 2.2 at 8 a.m. to 3.0 at 5:30 p.m. Susceptibility to hemolysis also increased from about 5% in the sample drawn at 7 a.m. to 75% at 2 p.m.

This finding of a diurnal rhythm in a property of the red cell is not new. Price-Jones reported in 1920, that the size of his own red blood cells exhibited a regular daily rhythm with a minimum found at 8-10 a.m. This finding was subsequently disputed by Ponder (1948) who examined both wet mounts and dry mounts of blood cells

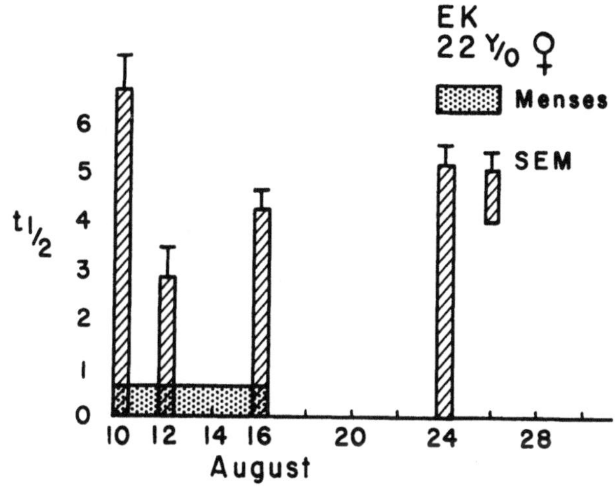

Figure 11. Filterability as a function of the stage of the menstrual cycle. Determinations done in morning on day indicated. As in Fig. 1.

taken at various times of day, and found no difference. Price-Jones' data were based on dry mounts done in a different way than those of Ponder, and it would seem, in retrospect, that the figures obtained by Price-Jones were a reflection of alterations not in the size, per se, but in the way the cells responded to drying in his preparation.

Deformability Changes Induced by PGF(2-alpha)

The complexity of the red cell's response to factors affecting deformability is well illustrated by the effect of PGF(2-alpha) added <u>in vitro</u>.

In cells obtained from men, we have been unable to detect any deformability change in response to PGF(2-alpha) (10^{-9} M). Women, also, generally fail to respond. However, Fig. 16 shows a clear and significant effect of this compound on deformability. PGF(2-alpha) brought about a decrease in t(1/2) on cells taken from two women; one was taking oral contraceptives, the other was at the mid-point of her menstrual cycle, a time of normally high plasma estrogen levels (Baird and Guevara, 1969). One other female on oral contraceptives (DB, Fig. 12) showed a slight, but statistically insignificant fall in t(1/2). Estradiol added <u>in vitro</u> (10 microgram/liter) to cells

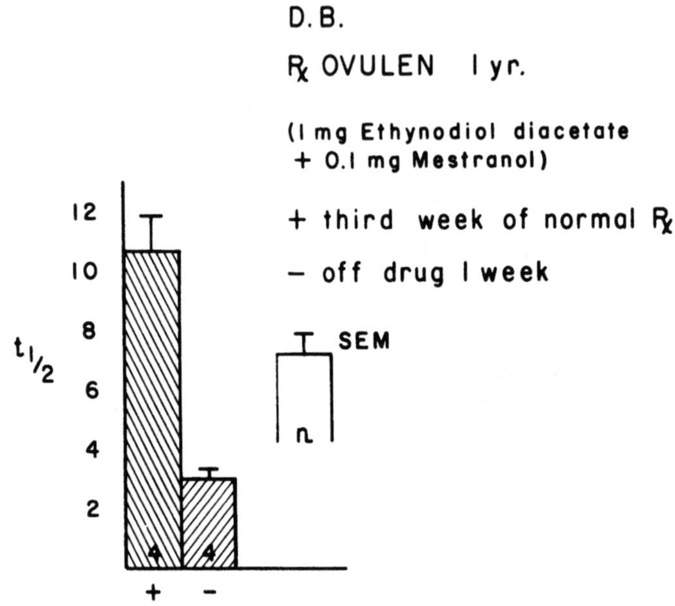

Figure 12. Filterability: Effects of oral contraceptives. As in Fig. 1.

obtained from males did not increase deformability in response to PGF(2-alpha).

These data would indicate that some factor associated with the hormonal status of women plays a part in the response of the red cell to PGF(2-alpha).

Clinical Significance: Shock

The data presented so far indicate that erythrocyte deformability may play a role in the control of the circulation. If that is true, then it is conceivable that the red cell may play a role in the pathophysiology of the poorly understood syndrome of shock. The many aspects of shock can be stated simply: in shock there is ineffective perfusion of the capillary bed (Thal, 1971).

Braasch and Rogausch (1971) found decreased deformability of the red cells in burn shock and proposed that this change might play a pathogenic role. Our data confirm this. In collaboration with Dr. Leonard Miller of the University of Pennsylvania, we have examined the deformability of the red cells of a patient in septic shock. The data are presented in Fig. 17. The patient, a

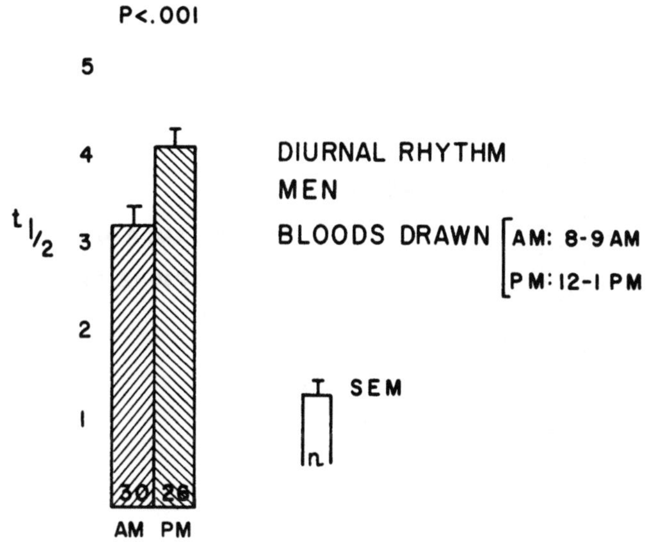

Figure 13. Filterability: Men, morning versus afternoon. As in Fig. 1.

42 year old white female, developed signs and symptoms of severe septicemia. On the day of admission to the Shock and Trauma Unit, her red cells were significantly less deformable than those of a control population as measured by the filterability technique (i.e. t(1/2) increased) and her red cells showed no response to isoproterenol at 10^{-7} M. That evening she developed oliguria which became progressively worse. On the following morning, her cells were much less deformable, and the response to isoproterenol in vitro was the reverse of that seen in normals: her cells became softer. Following that determination of deformability, treatment of the patient with isoproterenol was begun. Renal function improved markedly. Two hours after beginning treatment, another blood sample was taken. The cells had not changed from the morning, and were made more deformable by isoproterenol (endogenous isoproterenol in the drawn blood was removed by washing the cells prior to determining filterability). The patient improved rapidly with antibiotic treatment and isoproterenol infusion, and by the following day she appeared much better. Skin temperature was normal, she was increasingly alert, and had good renal function. At midnight on the 13th of July, isoproterenol infusion was stopped. On the 14th of July, her red cell deformability had returned to the level seen

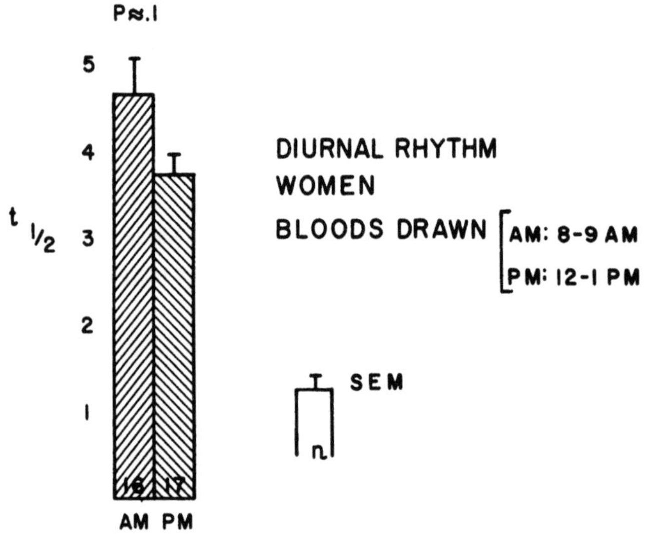

Figure 14. Filterability: Women, morning versus afternoon. As in Fig. 1.

on the first hospital day, and again there was no response to isoproterenol. The patient continued to improve, and was transferred from the Unit on the 15th of July. One week later, another blood sample was drawn. The patient's cells were now almost normal (although still significantly different from controls), and the response to isoproterenol was normal: her cells became less deformable. The patient left the hospital the following week in good condition.

This study does not prove a causal relationship between erythrocyte deformability and shock, but the implications are great. First, deformability of the cells - or rather a lack of it - correlated closely with the patient's condition. Second, the same therapy which in the patient seemed to result in marked inprovement, led *in vitro* to a softening of the cells at the time the therapy was instituted. Finally when the patient was improved and in the recovery phase of her illness, her red cells were found to respond normally to isoproterenol.

These data indicate that the status of a patient in shock may be monitored in the way described here. The efficacy of various drugs which may be used to attempt to combat the shock state may also be assessed *in vitro*.

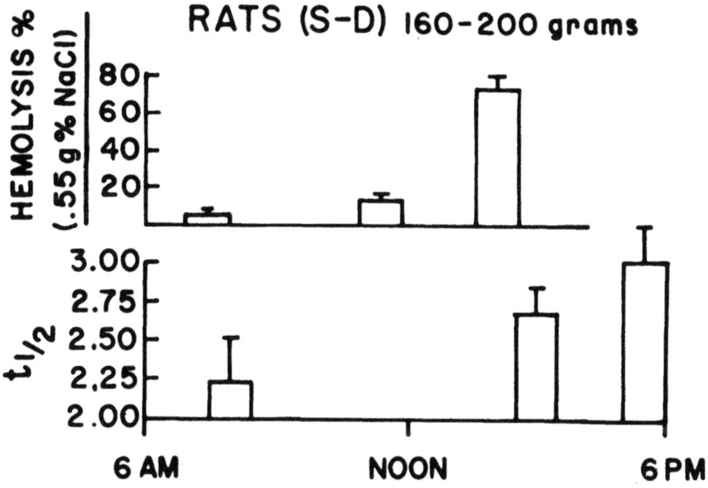

Figure 15. Diurnal cycle in rats: Filterability and susceptibility to hemolysis. Samples obtained by cardiac puncture of ether-anesthetized rats. Colony in light 8 a.m. to 4:30 p.m., balance of day, dark.

Perhaps most important, the data indicate a role of the red cell in the pathogenesis of this state. The poor capillary perfusion may be a result of the decreased capacity of the red cell to deform and hence traverse the smaller capillaries.

DISCUSSION

Our studies have revealed three new phenomena of potential importance in circulatory dynamics: (1) red cells _in vitro_ undergo significant changes in their deformability in response to vasoactive substances at concentrations known to occur _in vivo_; (2) the response of red cells _in vitro_ to osmotic shock (hypotonic hemolysis) exhibit short period oscillations; and (3) red cell deformability changes periodically _in vivo_, exhibiting both a diurnal rhythm, and in women, a monthly rhythm in conjunction with their menstrual cycle. In addition, evidence has been found for a change in this red cell property in response to drugs and disease. All of these facts argue that deformability of red cells _in vivo_ is controlled by a complex set of factors, and that changes in this red cell property are of physiological importance.

The implications of the effects of vasoactive compounds on deformability are, given the importance of

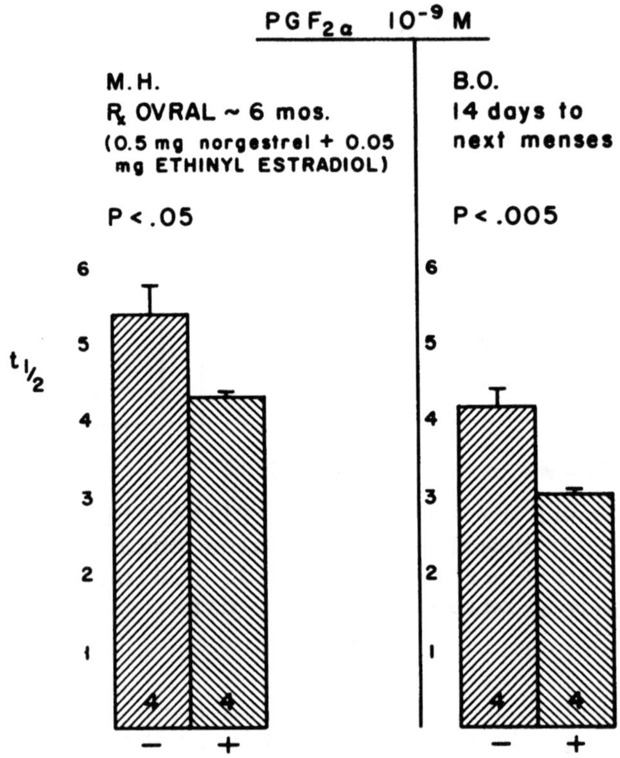

Figure 16. Filterability: Effects of PGF(2-alpha).

deformability in the normal functioning of the microcirculation, obvious: we must reexamine our concepts of the control of the microcirculation. One other point should, perhaps, be made. Prostaglandins are released from almost all tissues under a variety of conditions (Ramwell and Shaw, 1970). One of the functions of this release may be to inform the red cells in the microcirculation of the status of the tissue, i.e. red blood cells passing through a tissue may be one of the major target organs of these agents.

The short term oscillations observed are interesting for at lease two reasons. First, they have a periodicity approximating that of vasomotion and may play some role in this phenomenon. Second, they indicate that red cells communicate in some way with one another: in the absence of any synchronizing signal the oscillations of the individual cells would rapidly cancel out. This raises the immediate question: what is the nature of this synchronizing signal? We have as yet no answer. It also

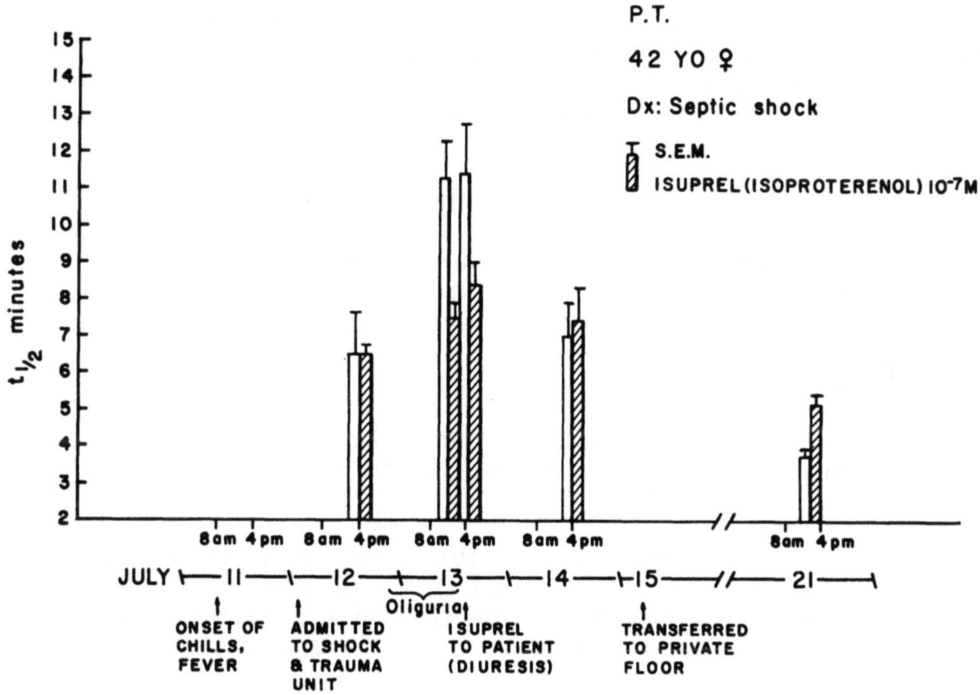

Figure 17. Filterability: Septic shock and the influence of isoproterenol. Details in text.

raised the intriguing possibility that the venous end of the microcirculation may be able, through the continuity of the vascular bed, to inform incoming blood of the metabolic conditions within the tissues. Whatever the signalling compound, it is not unique to red cells and it would not seem unreasonable to suppose that cells in the tissues perfused also participate in these oscillations.

The data obtained showing a diurnal and a menstrual rhythm in cell deformability indicate that the properties of the red cell reflect certain long term *in vivo* processes and presumably play some role in translating these processes via the circulation to the whole organism. Indeed, the red cell may play a role in the daily sleeping-waking cycle, and other aspects of fatigue.

If the data on the patient in septic shock are generally true in the complex shock syndrome, then new modalities of therapy can be devised, and a way of monitoring the process provided. The role of the red cell in other unsolved problems of the microcirculation, e.g. essential hypertension, remains to be explored. It is our present belief that changes in red cell deformability will prove to be of importance in the regulation of capillary blood flow, and thus influence a variety of processes and be involved in the pathogenesis of circulatory disorders.

ACKNOWLEDGEMENTS

The work described herein was supported by a grant from the Office of Naval Research.

REFERENCES

Aronson, H. B., Magora, F., and Schenker, J. G., 1971, Effect of oral contraceptives on blood viscosity, Am. J. Obstet. Gynec. 110:997.

Baird, D. T., and Guevara, A., 1969, Concentration of unconjugated estrone and estradiol in peripheral plasma in nonpregnant women throughout the menstrual cycle, castrate and postmenopausal women and in men, J. Clin. Endocrinol. Metab. 29:149.

Braasch, D., and Rogausch, H., 1971, Decreased red-cell deformability after severe burns, determined with the chlorpromazine test, Pfluegers Arch. 323:41.

Burton, A. C., 1965, "Physiology and Biophysics of the Circulation," Year Book Medical Publishers, Chicago.

Goldsmith, H. L., 1970, Motion of particles in a flowing system, Thromb. Diath. Haemmorrh. Suppl. 40:91.

Harrison, D. G., and Long, C., 1968, The calcium content of human erythrocytes, J. Physiol. 199:367.

Krogh, A., 1922, "The Anatomy and Physiology of Capillaries," Yale University Press, New Haven.

LaCelle, P. L., 1969, Alteration of deformability of the erythrocyte membrane in stored blood, Transf. 9:238.

LaCelle, P. L., 1970, Alteration of membrane deformability in hemolytic anemias, Semin. Hematol. 7:355.

Mela, L., 1968, Interactions of La^{3+} and local anesthetic drugs with mitochondrial calcium and Mn^{++} uptake, Arch. Biochem. Biophys. 123:286.

Ponder, E., 1948, "Hemolysis and Related Phenomena," Grune and Stratton, Inc., New York.

Price-Jones, C., 1920, The diurnal variation in the sizes of red blood cells, J. Path. 23:371.

Ramwell, P. W., and Shaw, J. E., 1970, Biological significance of the prostaglandins, Rec. Prog. Horm. Res. 26:139.

Rand, R. P., and Burton, A. C., 1964a, Mechanical properties of the red cell membrane. I. Membrane stiffness and intracellular pressure, Biophys. J. 4:115.

Rand, R.P., and Burton, A. C., 1964b, Mechanical properties of the red cell membrane. II. Viscoelastic breakdown of the membrane, Biophys. J. 4:303.

Teitel, P., 1967, Le test de la filtrabilite erythrocytaire (TFE) une methode simple d'etude de certaines proprietes microrheologigue des globules rouges, Nouv. Rev. Fr. Hematol. 7:195.

Thal, A. P., 1971, "Shock: A Physiologic Basis for Treatment," Year Book Medical Publishers, Chicago.

Van Breemen, C., 1969, Blockade of membrane calcium fluxes by lanthanum in relation to vascular smooth muscle contractility, Arch. Int. Physiol. 77:710.

Weed, R. I., 1970, The importance of erythrocyte deformability (editorial), Am. J. Med. 49:147.

Weed, R. I., LaCelle, P. L., and Merrill, E. W., 1969, Metabolic dependence of red cell deformability, J. Clin. Invest. 48:795.

Wins, P., and Schoffeniels, E., 1966, ATP and calcium-linked contraction of red cell ghosts, Arch. Int. Physiol. 74:812.

DISCUSSION

RASMUSSEN: I would like to confine my remarks to a whole area of red cells that Dr. Allen didn't have time to talk about, particularly the problem of adenylate cyclase, and what role cyclic AMP may play in phenomena related to red cells. We have gone to considerable trouble to determine if there is an adenylate cyclase in the human red cell and in the rat red cell. We believe that Shepherd and his co-authors are correct; there is an adenylate cyclase in the rat erythrocyte which responds both to epinephrine or norepinephrine and to PGE(2). On the other hand, we can find no evidence of a similar adenylate cyclase in human erythrocytes, and this has been looked at in every conceivable way, from using red cell ghosts, and looking at an adenylate cyclase, to measuring cyclic AMP levels in intact erythrocytes. Now, what I have to add is that Dr. Allen has been talking about human cells, but the same type of responses are observable in rat cells. In terms of the changes in deformability, we can see no qualitative differences between the responses of the rat cell and the human cell, even though there is a clear difference in the adenylate cyclase system.

Finally, I'd like to take the chairman to task for a moment or two concerning one of his slides which showed that 10^{-4} M dibutyryl-cyclic AMP had an effect on platelet function. This kind of evidence has been used as one of the criteria with which to implicate cyclic AMP as a second messenger in hormone action. I want to indicate that this is a pharmacological response and as such it may not teach us as much as we think it does. Let me tell you of some experiments in a completely different system, the fly salivary gland, that illustrate the subtle differences one may be able to observe between the effect of the hormone and added cyclic AMP merely as a reminder that probably in no system do they act identically. The fly salivary gland is a very simple organ which responds to serotonin. The response is to secrete essentially isotonic potassium chloride. If you add 10^{-8} M serotonin you get maximal stimulation. If you add 10^{-2} M cyclic AMP you get an identical physiological response in terms of the rate of secretion and composition of the fluid secreted. However, if you now measure the electrical response by observing the transepithelial potential by having one electrode in the lumen of the gland and the other in the serosal medium, you will find that the addition of serotonin causes a more negative transepithelial potential, whereas the addition of cyclic

AMP causes a more positive transepithelial potential. Hence at that level of analysis the two agents are clearly acting in a different way.

SALZMAN: I suppose I ought to reply to that. There really is no way to refute completely what Dr. Rasmussen has said about high concentrations of dibutyryl-cyclic AMP, but there is some collateral evidence suggesting that it may be operating as intracellular cyclic AMP. If one studies the platelet uptake of labeled dibutyryl-cyclic AMP, the amount of label that ends up in the platelet can be calculated to be roughly of the same order of magnitude as the basal cyclic AMP content of the platelet. So the amount of uptake is consistent with the idea that it is cyclic AMP. Secondly, although cyclic AMP can be shown to duplicate the effects of dibutyryl-cyclic AMP on platelet aggregation, it takes 100 times higher concentration. This would not be expected if it were purely a pharmacological effect.

I wonder if Dr. Allen would say something about the interesting suggestion in his abstract about the protein kinase in the red cell.

ALLEN: That data will be presented at the end of November, but briefly the story is that both human and rat red cells possess a cyclic AMP stimulatable protein kinase associated with the membrane. That is, there is an enzyme, or a series of enzymes, which bring about a phosphorylation of a membrane protein in response to cyclic AMP added to red cell ghosts *in vitro* at 10^{-7} M. That's interesting in the light of what Dr. Rasmussen just said about the fact that the human red cell lacks an adenylate cyclase but has a cyclic AMP stimulatable protein kinase.

AMER: I just want to ask a question of Dr. Rasmussen. If you increase the cyclic AMP intracellularly, supposedly you are having a gradient from intracellular to extracellular for cyclic AMP. But when you perfuse with cyclic AMP, you have a gradient in the reverse direction. Now, would you really expect to have the same effect on the transmembrane potential?

RASMUSSEN: It turns out that it has nothing to do with the cyclic AMP gradient. Serotonin has a clearly distinct and separate effect on the gland that cyclic AMP doesn't produce. The effects involve a calcium dependent effect on the membranes. In the absence of calcium, in fact, serotonin stimulates cyclic AMP production, and then the electrical change is identical to that produced by cyclic AMP.

WEISSMANN: I'd like to confirm your data on effects of hormones on membrane systems, not just for the erythrocyte but for artificial liposomes and lysosomes. PGE(2) at very low concentrations stabilizes both liposomes and lysosomes. What are perhaps more interesting are the hormonal studies. For the last 5 or 6 years we have been studying the effect of low concentrations of 17-beta-estradiol and testosterone on artificial membrane systems. (Bangham, A. D., Standish, M. M., and Weissmann, G., 1965, J. Mol. Biol. 13:253; Sessa, G., and Weissmann, G., 1968, Biochim. Biophys. Acta 150:173; Rita, G. A., and Weissmann, G., The molecular basis of gouty inflammation, II., submitted for publication). As little as 0.001 M of 17-beta-estradiol stabilizes the membranes of artificial liposomes to a significant extent. Testosterone has opposing effects, as does progesterone. Similarly, lysosomes prepared from animals at various levels of hormonal stress of the kind you indicated, namely varying estrogen to progesterone, and varying testosterone - have different properties *in vitro* to the degree that when estrogen is predominant, the organelles are stable (harder). When progesterone is predominant the organelles are weaker and resemble testosterone-treatd cells. Therefore erythrocytes and all other lipid systems that we've studied are stabilized by 17-beta-estradiol, whereas progesterone and testosterone are disruptive.

ALLEN: Do you see any change in the responsiveness of the lysosomes to other substances after they have been drawn from an estradiol treated animal, for example?

WEISSMANN: Yes. For example, we may cite sex-determined differences in response to the agent etiocholanolone, which is amphipathic. As the human female is somewhat resistant to the fever-provoking effects of etiocholanolone relative to the human male, so the estradiol-containing liposome and lysosome are resistant to the membrane disruptive effects of this steroid. The only reason this sort of data may be of interest to a meeting like this is because PGE(2) (as opposed to the other prostaglandins) works in similar fashion at concentrations as low as 10^{-6}M. It had the similar stabilizing effect as the estrogens did, e.g., 17-beta-estradiol.

ALLEN: Fascinating. The reason I asked about the hormonal response is that a prostaglandin, PGF(2-alpha), which in men and most women has no effect on the hardness of the cells, did have a distinctive effect of softening the cells of two women. One of the two was on birth control pills and the other was at exactly the midpoint in her menstrual cycle.

BUCKLES: Concerning the matter of red cell hardness, I am sure that you are aware that a number of other factors affect the flow through a filter bed, and I wonder if you could tell us if you have confirmatory evidence that it is hardness that alters flow rates.

ALLEN: That's why we went to hypotonic hemolysis because we wanted to use an independent method.

LICHTENSTEIN: Your work *in vitro* and *in vivo* seem to fail to go along with one another. In one patient you obviously could soften cells *in vitro*, but when you gave the drug *in vivo* you had no effect. You didn't comment on this.

ALLEN: When the drug was given *in vivo* her clinical status improved. When we measure the effects of isoproterenol after giving the drug *in vivo*, we wash the cells. After we washed the cells free of isoproterenol, we added it back *in vitro*.

LICHTENSTEIN: But the cells were much harder at that point even after being washed.

ALLEN: Yes.

LICHTENSTEIN: In other words, the isoproterenol given had not changed her basic stages.

ALLEN: *In vivo* I would imagine they had but we washed off the effects when we washed the cells *in vitro*.

IBERALL: I think this is a good time for me to get into the argument. I find your work interestingly connected to work that we have done in *in vivo* studies with Dr. Bloch. We have traced the dynamics of red cells by counting their passage in time in capillaries for periods of many hours. Where you find periodicities of the order of 15 sec *in vitro*, we find periodicities of the order of 1-2 min. We are willing to consider that this is likely a simple time scale shift of a related process. We have performed these studies in the frog, mouse, and guinea pig. The corollary to these findings is that we have found the same kind of pulsation showing up in the metabolism of the entire organism. This was a major reason that we went looking for such correlated cycles in the microcirculation. Our initial approach to systems studies in the biological organism was that we could use the overall metabolic heat power produced by the body as an external indicator of internal processes (i.e. the basic method of calorimetry). When we located basic "engine" cycles, then we went inside looking for the control mechanisms. We have done such thermodynamic analysis for a number of other species, too, including man. The reason that I find your work very interesting in demonstrating this kind of temporal gating is that we do not consider the flux of red cells through capillaries to be gated by any elementary hydrodynamic process. We have been trying to persuade hydrodynamicists of this thought for over five years. That is, we do not believe that red cell flow is governed solely by a boundary fluid and correction cells between red cells as they have been trying to develop.

Instead what we propose is yet purely hypothetical. It seems somewhat obvious to me that the red cell form (squirming through say 2-5 micron capillaries in human, rat, mouse, or guinea pig) is not achieved by hydrodynamic stresses, and that basically we must be dealing with another class of forces. Most plausibly these are electrical forces. Thus we posit electrical waves travelling down the capillaries. We propose that these are coupled to the metabolic events of the locally organized cells. For example, with nearby muscle fibers, to illustrate the kind of causal chain we are talking about, we postulate a combustion byproduct, typically lactate, influencing mast cells which could then cross-talk to the capillaries via histamine. It is the biochemical interaction of such substances with the capillary wall, and possibly with internal circulating catecholamines, that we propose may develop the electrical signalling and gating complexes.

What I find absolutely fascinating in what you said was that you were able to demonstrate communicational sensitivity at the order of a single molecule (whether one or a few doesn't really matter). I am not a chemist so I don't care which particular molecule you are talking about. There is a more serious intent than that apparently supercilious (to chemists) remark implies. In the sense that Dr. Allen brought the subject up, of communications, this is a major thesis that exists for pharmacology. We have been stressing it for some time. It is necessary to decide what communicates with what, particularly what communicates with cells or cellular complexes. We are not talking of molecule to molecule interaction, but of molecule to cell communication. In that case you must try to decipher the languages. I am looking at this as a physicist, a "communications" specialist, trying to decide what the interval languages may be. To start with, all chemical languages are gibberish to me. But is is obvious that cells do talk and respond to chemical languages. And the kind of molecules that we consider implicated are things like the catecholemines, histamine, cyclic AMP, possibly prostaglandin, etc. That is why in fact, I am at this conference - to see if I can learn a little more about the prostaglandin language. It is quite basic, if Dr. Allen has been able to show that you can talk to red cells with a "phoneme" sensitivity of a single molecule. This is a much lower sensitivity than we would have suspected. (For further background, see: Iberall, A., and Waddington, C. H., Towards a Theoretical Biology, 2. Sketches, Aldine, 1969; Bloch, E., et al, Introduction to a Biological Systems Science, NASA, CR-1720, Feb. 1971, Nat'l. Tech. Info. Serv., Springfield, Va.; and Iberall, A., Towards a General Science of Viable Systems, McGraw-Hill, 1971).

ALLEN: Yes.

IBERALL: Now my technical question on this is that one of the things about the shift from 15 sec perhaps to our 60 sec (and that Dr. Bloch perhaps did not discuss), is the fact that there is a large difference in deformation if you go through paper pores and the kind of electrical environment that you have in the in vivo capillary. Can you say a bit about how you think those cells go through those holes? What size holes were you using?

ALLEN: The size of the holes in the filter paper is 6-16 microns. The limiting factor in flow isn't the paper itself, but is the viscosity of the blood suspension. The paper is just used to retard flow sufficiently so that we can measure it. We don't know anything about the charge properties of the paper.

SHAW: What effect does changing extracellular calcium have on the response of the red cells to prostaglandins, and on response to isoproterenol? Secondly, did you actually look at the levels of circulating prostaglandins in your shocked patient?

ALLEN: I'll answer your second question first. We have the serums in the freezer at -60°C and we are waiting to get them done. In answer to your first question, in the absence of calcium there is no effect of PGE(2) or catecholamine on deformability. The effects of PGE(2) on calcium fluxes into the cells are very interesting. In order to examine the question, we had to develop a new technique for measuring the calcium in the red cells. What we have done is to use lanthanum chloride at 2 x 10^{-5} M in cold (4°C) buffer as the means of fixing cell calcium. This blocks passive calcium fluxes across smooth muscle membranes and mitochondrial membranes, and now we think the red cell membranes. We add red cell aliquots to this ice cold lanthinum chloride-containing buffer after exposing the cells to ^{45}Ca and hormones. We then measure the ^{45}Ca remaining in the red cell after three washings in this same buffer. We find that there is an extremely dynamic pool of red cell calcium which is decreased by PGE(2) and catecholamines. Previous theories to account for the changes in the red cell membrane deformability have always entailed an increase in intracellular calcium, bringing about a change in the protein associated with the cell membrane. In fact we get this cellular deformability with a decrease in this intracellular calcium pool.

RAMWELL: I think that there are three points that can be made to put all this into perspective. The first point is that prostaglandins appear to be formed by almost all cells of the body, i.e., all tissues contribute to the perfusate. Secondly, the prostaglandins formed are mainly PGE(2) and PGF(2-alpha), due to the ready availability of arachidonic acid. Allen finds that PGE(2) is the most active prostaglandin on red cells and that PGF(2-alpha), in men at least, has no effect. And the third point is that nearly all blood vessels appear to be lined with the metabolizing enzyme 15-dehydrogenase, which therefore possibly limits the prostaglandins to their tissue of origin and so act as local humors perhaps as suggested here, to regulate blood flow.

ALLEN: One piece of data I forgot to include is that we have studied quite intensively the effects of PGE(1) on the red cell. It turns out that if anything, the red cells are softened by PGE(1) and the probability value here is approximately 0.1, so the effect isn't very strong.

LEE: With regard to the interesting probability, that the red cell may have some primary role in control of regional circulation, have you examined the deformability of the red cell in a condition where a regional resistance is deranged without a change in viscosity? I refer particularly to patients with essential hypertension.

Secondly, have you looked at the effects of PGA(1) or PGA(2)?

ALLEN: The answer is no to both questions, although the studies with hypertensive patients are going to be done in conjunction with the Department of Medicine at the University of Pennsylvania.

ELLIS: A rather complex system exists in the red blood cell that is related to radiation damage, aging, and osmotic fragility. This system involves lipid peroxidation and the conversion of sulfhydryl groups to disulfur bridges. Glutathione reductase appears to restore the integrity of the red blood cell membrane by reducing the disulfur bridges to sulfhydryl groups. To what extent do you attribute the changes in the physical characteristics of the red blood cell membranes in your investigations by prostaglandins to physical interactions and how much to changes in sulfhydryl and disulfur groups?

ALLEN: We have absolutely no data with which to answer your question.

IGNARRO: I think perhaps there are agents other than catecholamines or prostaglandins that alter erythrocyte deformability, e.g., it's quite well known that certain non-steroidal anti-inflammatory drugs such as indomethacin, phenylbutazone and flufenamic acid, especially indomethacin, can certainly alter the integrity of the red blood cell membrane. Such drugs markedly stabilize the membrane and protect the cell against hypotonic hemolysis. You can get such effects *in vitro* at 10^{-5} to 10^{-6} M, which is about the concentration that one would find in the blood of a patient suffering from rheumatoid arthritis that is being treated with that particular drug. Now, if one assumes that there is some kind of relationship with regard to the erythrocyte, between membrane stabilization and the deformability that you described, then would one expect that such agents being used in the treatment of rheumatoid arthritis should exhibit an effect on the control of the microcirculation? Do these drugs exert such an effect on the circulation? I don't think so. Is there any evidence for this?

ALLEN: I may make it a general question to the audience. Does anybody know if the therapy for rheumatoid arthritis with high concentration of indomethacin or whatever, ever changes the circulatory dynamics in these people?

RASMUSSEN: May I just make one comment? I think that Dr. Allen did make clear one thing about the calcium data. There is the same oscillation in this pool of labile calcium in the red cell with the same period as the oscillation in the response to hypotonic hemolysis. This is additional evidence that the former is a real phenomenon and not some sort of artifact of one particular kind of method.

ASTER: Have you looked at the effects of these humoral mediators on the induction of red cell sphering by drugs such as vinblastin, colchicine, and sytochalasin B, which appear to affect the contractile and structural proteins of the red cell membranes?

ALLEN: No, we haven't, but what we have done is to look under the microscope to see if there are any gross shape changes induced by these substances themselves. In fact, we could see no change in shape.

USE OF PGE(1) IN PREPARATION AND STORAGE OF PLATELET CONCENTRATES

G. A. Becker, M. K. Chalos,
M. Tuccelli, and R. H. Aster

Milwaukee Blood Center and
Medical College of Wisconsin, Wisc.

INTRODUCTION

With increasing awareness by physicians of the usefulness of platelets in treating hemorrhage due to thrombocytopenia, the demand for platelet concentrates has steadily increased. More intensive use of chemotherapy, radiation, and immunosuppressive therapy in modern medicine and the recognition that bleeding may be due to qualitative as well as quantitative platelet abnormalities make it almost certain that the demand for platelets will increase still further. Yet, except within a few institutions, platelets are often in short supply because of their limited permissible storage time and the high cost in time and materials necessary for preparation of a therapeutic quantity of platelets. The finding that platelets can be stored for several days with only minimal loss of viability if kept at room temperature rather than in the cold as was previously recommended, has helped to increase the quantities of platelets available for transfusion (Murphy and Gardner, 1969; Murphy et al, 1970). Many variables other than temperature which may affect platelet viability during storage require further study however, and questions remain regarding the effectiveness of stored platelets in the treatment of hemorrhage in thrombocytopenic patients.

For the past year we have studied some of these variables, among which are techniques used in centrifuging, resuspending and agitating platelets, temperature of storage, the type of plastic used and characteristics of the donor. As criteria of platelet "viability" we have used the preservation of platelet structure (phase and electron microscopy), metabolic activity, preservation of platelet function (response to various aggregating stimuli) and the recovery and survival times of ^{51}Cr-tagged platelets transfused to normal subjects. We have also sent technologists to local hospitals with platelet concentrates to obtain platelet counts, bleeding times, and measurements of capillary fragility before and after transfusion of platelets stored under various conditions.

Among our findings to date are that after 24 hr of storage at room temp (20°C) platelets deteriorate steadily as indicated by morphological changes, loss of function and reduction of <u>in vivo</u> hemostatic effectiveness. Conversion from "disc" to "sphere" is the first morphological change. When platelets are stored at 4°C, sphering occurs within minutes but function and internal architecture (except for loss of microtubules) are well preserved for 96 hr. At 72 hr, platelets stored in the cold are superior to those kept at room temp for the treatment of thrombocytopenic bleeding.

Because PGE(1) blocks disc-to-sphere transformation of human platelets and inhibits their adhesiveness and capacity to aggregate (Emmons et al, 1967; Shio et al, 1970), we have studied the possible usefulness of PGE(1) as an adjunct to the preparation and short-term preservation of platelets.

METHODS

Crystalline PGE(1) was obtained from Alza Corporation, Palo Alto, California, and stored at -80°C. For use, 1 mg of PGE(1) was dissolved in 0.2 ml absolute ethanol, then diluted with shaking to 20 ml with isotonic saline to produce a working solution of 50 microgram/ml. This was sterilized by filtration through a 0.2 micron filter (Millipore Corp, Bedford, Mass) and stored in a sterile container at 4°C until use. The working solution was stable for 30 days. Its potency was periodically checked by determining its ability to inhibit ADP-induced aggregation of fresh normal platelets.

Platelet morphology was assessed by phase microscopy of wet platelet suspensions and by electron microscopy of platelets fixed with glutaraldehyde and osmium tetroxide, performed by Dr. J. Garancis, Department of Pathology, Medical College of Wisconsin.

Platelet aggregation was studied by a modification of the Born technique (Born and Cross, 1963) using an Aggregometer (Chronolog Corp). ^{51}Cr survival was studied as previously described (Aster and Jandl, 1964; Aster, 1971b).

Blood was collected in Fenwal Triple Packs (FP-297). Platelet-rich plasma (PRP) was prepared by centrifuging whole blood at 2500 rpm (1500 x g) for 6 min in a Sorvall RC3 centrifuge. The plasma was transferred to a satellite bag and centrifuged at 4000 rpm (4000 x g) for 10 min to compact the platelets into a button. All but 30 ml of plasma was expressed into the third pack. The platelet concentrate was then allowed to resuspend without manual manipulation on a slowly moving rotator (Becker and Aster, 1971) and were stored on the rotator which maintained them in a constant state of agitation.

RESULTS

Use of PGE(1) in Preparation of Platelet Concentrates

Platelet concentrates prepared by rapid centrifugation of PRP obtained from blood anticoagulated with acid citrate dextrose (ACD) in the cold almost invariably clump irreversibly when resuspended by manual manipulation. If plasma is not chilled before centrifugation, resuspension of platelets is somewhat more complete but a degree of clumping sufficient to reduce the clinical effectiveness of the concentrates nearly always occurs (Fig. 1). With citrate phosphate dextrose (CPD) anticoagulant, clumping is even more pronounced than in ACD because the plasma has a higher pH (Aster, 1969, 1971a). Platelet concentrates prepared in ACD plasma will resuspend completely if they remain undisturbed for at least 60 min at room temp before being manipulated (Mourad, 1968) or if they are placed on a slowly moving rotator. CPD platelets require 1 to 2 hr (Becker and Aster, 1971). Five to ten percent of concentrates so prepared never resuspend completely, but remain partially clumped.

In all of 24 studies in which PGE(1) in quantities ranging from 4 to 200 micrograms was added to either whole blood or PRP, clumping of platelets was totally abolished in concentrates prepared either at room temperature or 4°C

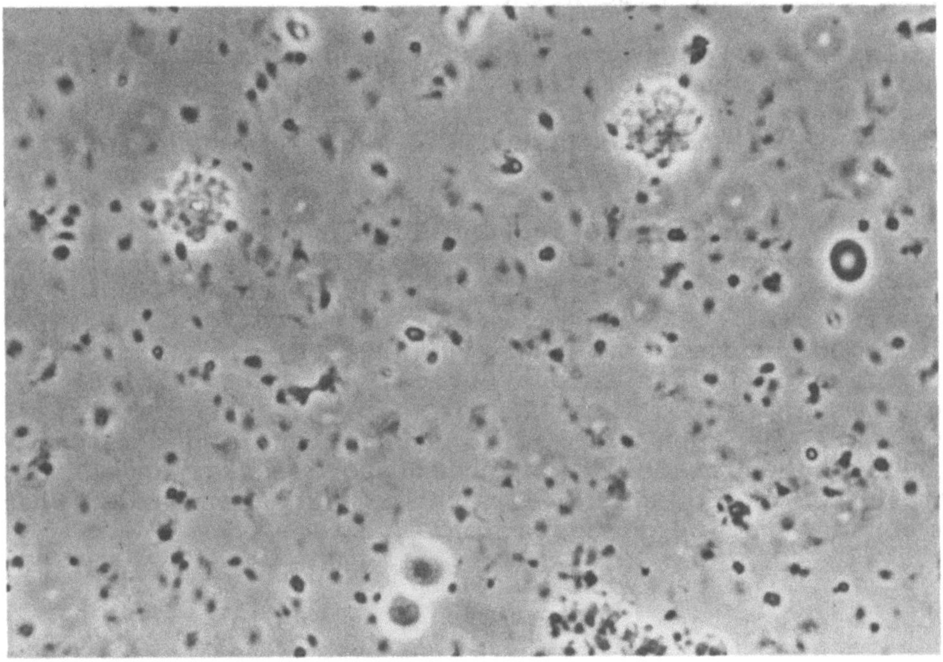

Figure 1. Typical example of platelets concentrated by rapid centrifugation of ACD PRP at room temp and resuspended immediately by manual manipulation. Many clumps of platelets were present which did not disperse completely even after two days of gentle agitation at room temp. About 50% of the total platelets were contained in the aggregates.

using either ACD or CPD anticoagulants (Fig. 2). Concentrates thus prepared could be smoothly resuspended immediately following rapid centrifugation with only a minimum of manipulation. The smallest quantity of PGE(1) used (4 micrograms) resulted in a plasma concentration of about 12 nanogram/ml PGE(1), about one-fourth the amount needed to produce 50 per cent inhibition of aggregation of fresh platelets triggered by 2 x 10^{-6} M ADP.

Platelets from concentrates prepared with PGE(1) were diluted to a final concentration of 300,000 (about 1:10) in ABO-compatible platelet-poor plasma for function studies. Aggregation in response to ADP (2 x 10^{-6} M) collagen, and epinephrine (10^{-5} M) was normal. ^{51}Cr survival times of platelets concentrated and immediately resuspended in PGE(1) were also normal (Fig. 3).

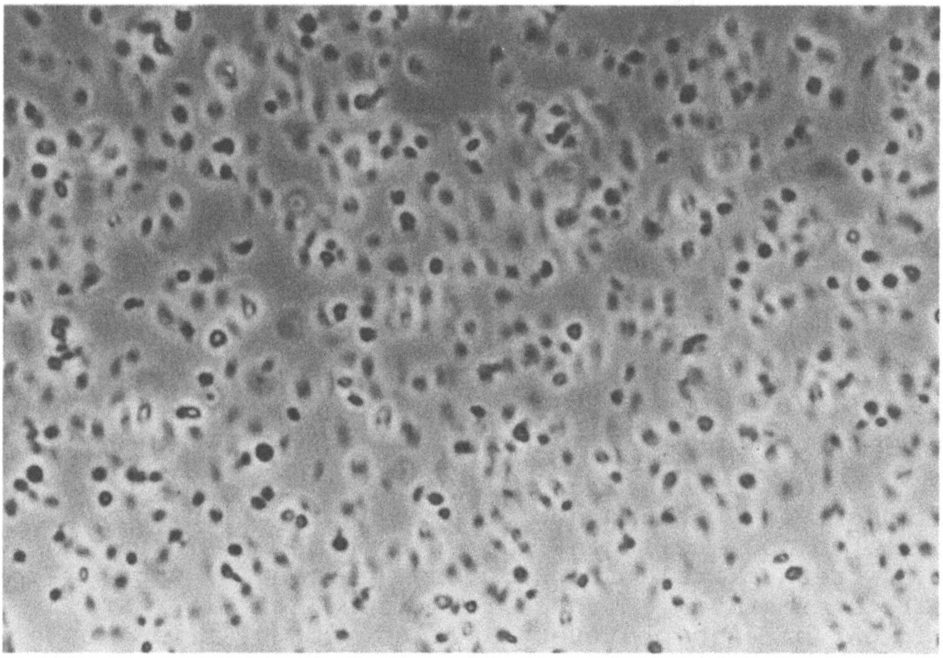

Figure 2. Platelets prepared as in Fig. 1 except for addition of 8 micrograms PGE(1) to PRP prior to rapid centrifugation (final PGE(1) concentration 24 nanogram/ml). A smooth, even suspension of platelets was obtained in 15 to 30 sec.

Effect of PGE(1) on Recovery of Platelets from Stored Whole Blood

Marked clumping of platelets occurs in whole blood stored at 4°C so that after 24 hr an average of only 8 per cent of the platelets originally present can be extracted and concentrated by differential centrifugation (Fig. 4). When PGE(1) in quantities ranging from 4 to 250 micrograms was added to whole blood at the time of collection to produce PGE(1) concentrations between 12 and 750 nanogram/ml, platelet clumping was inhibited and 50 to 100 % of the total platelets could be recovered even after blood was chilled for 24 hr (Fig. 4). With the highest concentrations of PGE(1) used, resuspension of concentrated platelets was almost immediate. With lesser quantities, up to 30 min was required. Platelets thus obtained were spherical in shape but otherwise had normal structure. Upon incubation at 37°C for 1 hr, conversion

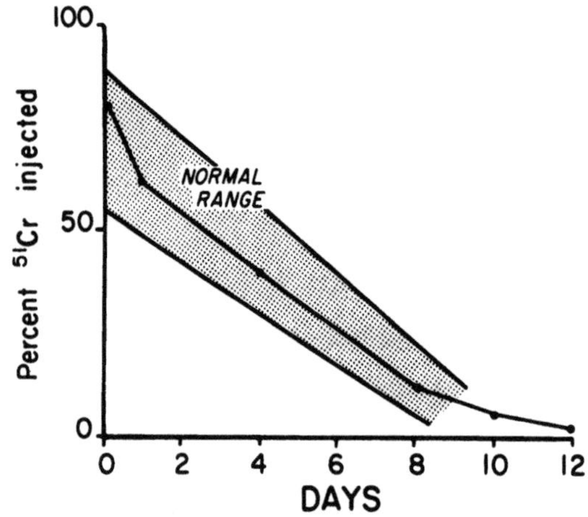

Figure 3. ^{51}Cr survival of platelets concentrated in the presence of PGE(1) 24 nanogram/ml as in Fig. 2.

from sphere to disc did not occur. Aggregation in response to ADP, collagen, and epinephrine was equal to that of fresh platelets. Recovery (percent of total platelets remaining in the circulation 1 hr after injection) of such platelets labeled with ^{51}Cr was normal upon their being transfused (Fig. 5) but their survival time was short (t(1/2)=1.5 days) and comparable to that of platelet concentrates stored for the same period of time at 4°C.

Effect of PGE(1) on Short-Term Preservation of Platelet Concentrates

Our short-term studies of platelet perservation without PGE(1) are referred to above and will be reported in full detail elsewhere. To determine whether PGE(1) might aid in short-term preservation of platelets, a total of 36 studies have been carried out in which the drug was added to platelet concentrates in concentrations ranging from 50 to 700 nanogram/ml. The concentrates were then stored for 24 to 120 hr both at room temp and at 4°C. In the presence of PGE(1), change of platelet shape from disc to sphere was delayed but we have been unable to demonstrate that platelet ultrastructure or function is better preserved with PGE(1) than in the absence of the drug. This was also true of four studies in which 50 micrograms PGE(1) was added daily to the concentrates to compensate for inactivation of PGE(1) during storage. ^{51}Cr survival studies have been performed only on platelets stored at 4°C at a PGE(1) concentration of 250

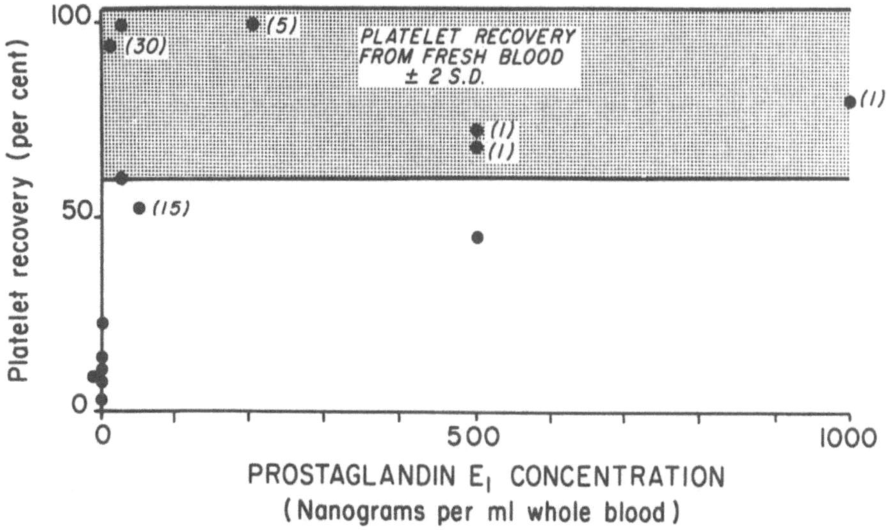

Figure 4. Recovery of platelets by differential centrifugation of whole blood anticoagulated with ACD and stored at 4°C for 24 hr. The ordinate gives the percentage of platelets present in the blood when freshly collected which were recovered in the final platelet concentrate. Figures in parentheses are the number of min required for final resuspension of the concentrates.

nanogram/ml. Survival times of these platelets were uniformly short and were comparable to those of platelets stored without PGE(1) at 4°C for the same period of time (Fig. 6).

Studies of Platelet Freezing

In collaboration with Dr. Donald Greiff, Department of Pathology, Medical College of Wisconsin, we have studied long-term preservation of platelets in the frozen state using present modifications of methods previously reported (Greiff and Mackey, 1970). To date, platelet concentrates prepared using 250 nanogram/ml PGE(1) to facilitate resuspension have not been shown to tolerate freezing in cryoprotective media more readily than untreated platelets.

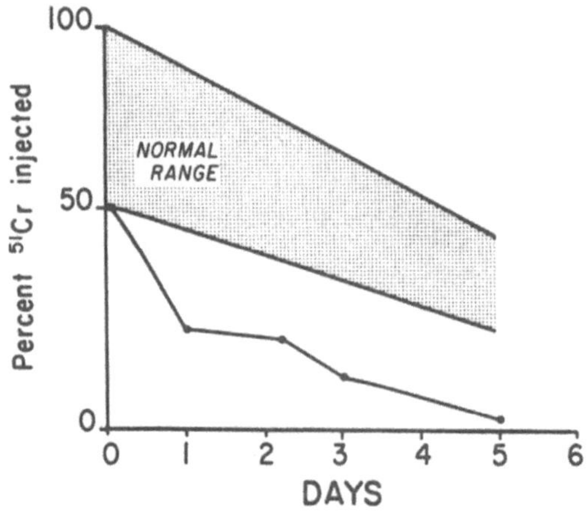

Figure 5. ^{51}Cr survival of platelets harvested from whole blood stored for 24 hr at 4°C and containing PGE(1) 50 nanogram/ml.

DISCUSSION AND CONCLUSIONS

These preliminary studies of the possible usefulness of PGE(1) in preparation and preservation of platelets show that concentrations of PGE(1) as low as 24 nanogram/ml permit platelet buttons to be smoothly resuspended almost immediately after rapid centrifugation without adverse effect on their function or survival. Platelets similarly treated without added PGE(1) clump irreversibly. Thus, addition of small amounts of PGE(1) to whole blood at the time of collection reduces the time required for preparation of platelet concentrates by 1 to 2 hr and may improve the quality of the final product by preventing even microscopic clumping of platelets.

Addition of as little as 4 micrograms PGE(1) to a unit of whole blood at the time of collection permits 50 to 100 % of the platelets initially present to be recovered in platelet concentrates prepared after as long as 24 hr of storage at 4°C. Although the survival time of such platelets is short, they appear to function normally *in vitro*. Clinical studies performed to date indicate that they are effective in producing hemostasis in thrombocytopenic patients. Quantities of PGE(1) (12 nanogram/ml) which are insufficient to inhibit ADP-induced platelet aggregation nonetheless prevent platelet clumping in stored whole blood. The smallest amount of PGE(1) used (4 micrograms) is one one-thousandth the amount safely used to induce abortion in a recent report (Embrey, 1971).

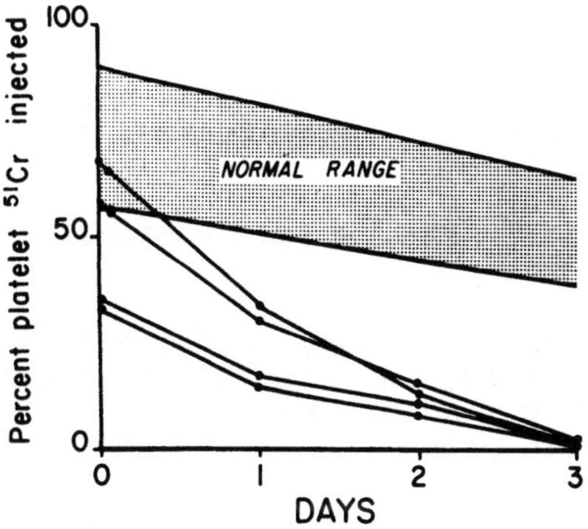

Figure 6. ^{51}Cr survival of platelets stored as concentrates with gentle agitation at 4°C for 72 hr with PGE(1) concentration of 250 nanogram/ml. Recoveries are low compared to those of freshly prepared platelets and survival times are short with t(1/2) of 1 to 2 days. The survival curves were not significantly different from those of similarly prepared platelet concentrates stored without added PGE(1).

One of the factors limiting the availability of platelets for therapeutic purposes has been that much of the blood used for transfusion is collected at mobile sites and must, by regulations of the National Institutes of Health and American Association of Blood Banks, be "chilled immediately" upon collection. By the time this blood has been transported to the processing center, recovery of platelets as concentrates is virtually impossible. Use of PGE(1) may permit recovery of platelets suitable for transfusion from such blood. Further studies are in progress to determine the maximum delay which may be permitted before platelets can be extracted from chilled blood to which PGE(1) has been added without shortening their in vivo survival time.

Our observations to date do not demonstrate that PGE(1) in the concentrations used exert a beneficial effect on platelet viability during short-term storage. These observations are preliminary, however, and additional studies are needed.

REFERENCES

Aster, R. H., 1969, Effect of acidification in enhancing the viability of platelet concentrates: current status, Vox. Sang. 17:23.

Aster, R. H., 1971a, Preparation of platelet concentrates, Vox. Sang., (in press).

Aster, R. H., 1971b, The study of platelet kinetics with ^{51}Cr-labeled platelets, in "Platelet Kinetics", (J. Paulus, R. H. Aster, and M. Breny, eds.). North-Holland Co., Amsterdam, (in press).

Aster, R. H., and Jandl, J. H., 1964. Platelet sequestration in man. I. Methods, J. Clin. Invest. 43:843.

Becker, G. A., and Aster, R. H., 1971, Platelet transfusion therapy, Med. Clin. N. Amer., (in press).

Born, G. V. R., and Cross, M. J., 1963, The aggregation of blood platelets, J. Physiol. 168:178.

Embrey, M. P., 1971, Induction of abortion by prostaglandins E, J. Reprod. Med. 6:15

Emmons, P. R., Hampton, J. R., Harrison, M. J. G., Honour, A. J., and Mitchell, J. R. A., 1967, Effect of prostaglandin E(1) on platelet behaviour in vitro and in vivo, Brit. Med. J. 2:468.

Greiff, D., and Mackey, S., 1970, Cryobiology of platelets. II. Effects of freezing and storage at low temperatures on the survival of isolated human blood platelets as measured by assays for aminopeptidases, Cryobiology 7:9.

Mourad, N., 1968, A simple method for obtaining platelet concentrates free of aggregates, Transfusion 8:48.

Murphy, S., and Gardner, F. H., 1969, Platelet preservation. Effect of storage temperature on maintenance of platelet viability - deleterious effect of refrigerated storage, New Eng. J. Med. 280:1094.

Murphy, S., Sayar, S. N., and Gardner, F. H., 1970, Storage of platelet concentrates at 22°C, Blood 25:549.

Shio, H., Plasse, A. M., and Ramwell, P. W., 1970, Platelet swelling and prostaglandins, Microvasc. Res. 2:294.

DISCUSSION

The following discussion covers the papers by Valeri and by Becker et al. The paper by Becker et al was presented by Dr. Aster.

SALZMAN: It must be very gratifying to someone who treats bleeding patients that basic observations such as those of Shio and Ramwell can be translated into immediate clinical utilization. These last two papers appear to be very important indeed. There is, however, a discrepancy which I would like to discuss. It was my impression that Dr. Valeri did not show any improvement in collection of platelets stored at 4°C with PGE(1), although Dr. Aster did.

ASTER: We do find that recovery of platelets in thrombocytopenic patients and, more importantly, their therapeutic effectiveness, is greatest when platelets are stored at 4°C.

RASMUSSEN: But there was one discrepancy, was there not, on the effect of temperature on platelet storage. The earlier data suggested that storing platelets at 4°C was better that 22°C, in terms of treating thrombocytopenic purpura.

ASTER: The general feeling now is that platelets should be stored at room temp because their survival time as measured with ^{51}Cr is much better at least after 24 hr of storage, and the recovery is better than if you store them at 4°C. On the other hand, as I pointed out, at room temp there is a progressive loss of aggregability, adhesiveness and other measures of platelet function. This loss does not occur at 4°C. The chromium survival time of the cold stored platelets is shorter than that of platelets stored at room temp. At the bedside, though, which is where it counts, we find that the platelets stored in the cold are much more effective in shortening the bleeding time and in stopping hemorrhage in these thrombocytopenic patients. Whether this relates to the better function seen *in vitro* is still unknown. Possibly, their shortened survival time reflects a greater ability to interact with the endothelium of small blood vessels.

VALERI: Some but not all of our results appear to conflict with those of Dr. Aster. Dr. Aster, in reporting on the therapeutic effectiveness of liquid-stored platelets transfused to patients with thrombocytopenia, stated that platelet concentrates stored at 4°C for 3 to 5 days were effective in reducing the bleeding time in these patients. He estimated the survival in vivo of platelet concentrates by the increase in the platelet counts.

We studied the ^{51}Cr survival and the hemostatic effectiveness after transfusion of preserved platelets in normal healthy volunteers treated with aspirin. Our data and those of Dr. Aster on normal volunteers agree, in that ^{51}Cr labelled platelets stored at room temp had a linear rate of removal from the circulation, while ^{51}Cr labelled platelets stored at 4°C had exponential rates of removal. The differences in the survival values of platelet concentrates stored at 4°C as measured on one hand by ^{51}Cr labelling, and on the other hand by the increase in platelet counts, may be due to ^{51}Cr elution.

Although ^{51}Cr survival data showed that platelet concentrates stored at 22°C had better recovery and lifespan than those stored at 4°C, in vitro tests of platelet aggregation showed that platelets stored at 4°C had better aggregation patterns than platelets stored at 22°C.

In our study in healthy volunteers treated with aspirin the prolonged bleeding time was corrected "immediately" by transfusion of fresh platelets prepared with or without PGE(1). The transfusion of platelets that have been stored at room temp for 24 hr did not correct the aspirin induced thrombocytopathy immediately; the bleeding time returned towards normal within 24 hr. We did not transfuse platelet concentrates stored at 4°C to determine whether these platelets can correct the prolonged bleeding time produced by aspirin treatment in healthy volunteers. However, storage of platelet concentrates at 4°C maintained platelet function, and the transfused platelets corrected the bleeding time in patients with thrombocytopenia. Both the circulation and the hemostatic effectiveness of stored platelets are influenced by the physical condition of the recipient. In patients with thrombocytopenia, platelets stored at 4°C may not circulate but may be hemostatically effective. In healthy volunteers, platelets stored at 22°C may circulate but may not be hemostatically effective. It would appear that the circulation and function of preserved platelets may be best evaluated in healthy volunteers.

SHIO: I would like to describe my very first attempt to resuspend platelets in the presence of PGE(1). In this experiment I used ACD platelet rich plasma and plastic culture tubes. Following high speed centrifugation I obtained a very thin film like pellet of platelets in the control samples. There was a macroscopic clumping of cells and the shape of the free cells was spherical. In the experimental tubes I added from 1 to 60 nanogram/ml PGE(1) before centrifugation, and found that about 5-10 nanogram/ml PGE(1) was enough to prevent such an undesired aggregation and sphericalization. Therefore, I took 10 nanogram/ml as a suggested concentration to be used in the blood bag studies of Drs. Aster and Valeri.

EFFECT OF PGE(1) ON HARVESTING OF PLATELETS FROM PRP

PGE(1) Nanogram/ml	Platelet Pellet	Resuspension	Platelet Shape
0	Spread, film-like	Aggr. (++)	Spherical
1	"	Aggr. (+)	"
2	"	Aggr. (+)	"
5	Smooth, round	Aggr. (+)	Flat, Spherical
10	"	No aggr.	Flat
20	"	"	"
60	"	"	"

The appearance of platelet buttons was examined after centrifugation of PRP (1500 x g for 15 min). Platelet aggregates and cell shape were assessed after resuspension in 1/6 original volume of plasma. ++ indicates macroscopic aggregates and + indicates small aggregates consisting of less than 10 cells.

BUCKLES: It seems very strange to me that platelets which are stored at a cold temp seem to be superior biochemically while at the same time they seem to have a shorter biological half life, exhibiting even a different type of curve, suggesting a different mechanism for removal than normal platelets. With regard to the cells that have degraded at 4°C, are they reinfused as well as the good cells? And if they are, is there any possibility that the reinfusion of degenerating platelets could have something to do with the mechanism of restoring normal bleeding times?

ASTER: It could well be that some of the platelets which have degenerated under room temp storage do not take up chromium, and what looks like a relatively good recovery and survival is not as good as one might think on the basis of the isotope data. I mentioned that there is great variability in how well platelets from one donor hold up when stored at room temp. Therefore, there are some donor characteristics which we do not understand. We have not done chromium survival times on platelet concentrates which show marked degeneration.

SALZMAN: I think there is another aspect to Dr. Buckles' question. If platelet A is injured, this may be reflected in a change in function of platelet B. A patient treated with aspirin recovers normal hemostasis when treated with normal platelets, because normal platelets are able to induce aggregation in the aspirin treated patients.

KORNBERG: How is the storage of platelets at 4°C distinguished from that act of chilling the platelets from 37°C to 4°C? If you were to reduce the temp and immediately restore it to 37°C, what would you find? The reason I ask is that with E.coli and B.subtilis it is shocking to the population to be exposed suddenly to temps of about 15°C; many irreversible changes in the membranes and in the enzymes take place.

VALERI: We have no data along these lines.

KORNBERG: Might it be that you are not studying storage, but rather the consequence of reducing the temp?

ASTER: If you chill platelets rapidly to 5°C and infuse these within 1-2 hr, you get a perfectly normal recovery and survival time. Therefore, there must be some storage lesion produced over a period of time.

SALZMAN: On the other hand there is evidence from in vitro studies for a cold induced lesion of platelets which is induced very promptly. It has been shown that if you chill platelets and then immediately rewarm them, they aggregate.

EDELMAN: I have wondered about the possibility that since your infusion contains platelets plus medium, the medium itself might contain materials which would correct the bleeding times.

KORNBERG: Does this indicate that there are different pools of cyclic AMP or different kinds of platelets?

SALZMAN: Unless the platelets are altered, by the adenine or adenosine, it's possible, but this is a trace concentration of 10^{-8} M, so I don't think that's very likely.

ASTER: I am not sure that supernatant from platelet rich plasma has been infused, but there is a great deal of data showing that fragmented platelets are totally ineffective as far as shortening the bleeding time is concerned. Certain clotting assays can be affected by this procedure, probably because of the presence of platelet particles.

EDELMAN: My question specifically referred not to platelet fragments but to their soluble components. The question is, is there a release of some substance into the media from stored platelets which could be biologically active?

ASTER: As far as I know, that point has not been examined.

RAMWELL: Some of the implications of this discussion are not only in therapy for separation of blood components, but also in pathology. During bacterial and traumatic shock, prostaglandins are likely released as a consequence of the ischemia due to decreased tissue perfusion. These prostaglandins, as described earlier, are probably PGE(2) and PGF(2-alpha) due to the ready availability of arachidonic acid, which may also be released as a result of tissue damage releasing acid hydrolases. PGF(2-alpha) appears inactive on platelets and red cells, but PGE(2) aggregates rat platelets in the presence of ADP, and some human platelets are also affected. PGE(2) also hardens red cells, which may lead to reduced capillary transit times. These circumstances seem ideal for thrombus formation and arteriolar plugging. Perhaps an anti-PGE(2) compound may be of therapeutic utility?

PROSTAGLANDIN E(1) AND E(2): QUALITATIVE DIFFERENCE IN PLATELET AGGREGATION

H. Shio and P. W. Ramwell

Institute of Biological Sciences, Alza Corp., Palo Alto, Ca., and Worcester Foundation for Experimental Biology, Shrewsbury, Mass.

INTRODUCTION

Blood platelets exhibit a number of characteristic reactions, i.e. aggregation, adhesion, clot retraction, shape change, etc. Since these reactions are greatly modified by drugs, and since platelets are obtained relatively easily as a single cell preparation, platelets are now considered as a good model system of pharmacological analysis (Shio and Ramwell, 1972a). The aggregation of the cells can be analyzed by means of optical density (OD) change of platelet-rich plasma (PRP) (Born, 1962), and shape change, e.g. by adenosine diphosphate (ADP), is also quantitatively determined with OD recording (Born 1970, Shio et al, 1970).

From a medical point of view, blood is one of the most useful test materials available for clinical analysis. Platelet function has been considered mainly in relation to thrombocytopenic or thrombasthenic disease, but recent investigations indicate the importance of platelet behavior in hypertensive vascular disease (Danta, 1970, Zahavi et al., 1969) and with respect to aspirin ingestion (Sutor et al, 1971). However, it is necessary to pay attention to species differences in platelet reactions, when the platelets are used as a model for medical applications. For example, epinephrine causes platelet aggregation in human PRP, but not in rat PRP. Also the pattern of ADP-induced platelet aggregation is different in these two species; rat platelets aggregate in one phase, whereas human platelets clump in two steps -

with the second phase being far less reversible than the first, and moreover, there is great individual difference in the degree of the second phase aggregation; some 70% of people exhibit two phase aggregation patterns (MacMillan, 1966). Even the difference in content of biogenic amines has not been explained satisfactorily (Markwardt, 1967).

Since the extraordinary high anti-thrombogenic activity of PGE(1) was first reported by Kloeze (1967), the mechanism of action of PGE(1) on platelets has been studied extensively. In vivo inhibition of platelet aggregation has also been established (Emmons et al, 1967; Kloeze, 1970). Of significance in this field is the phenomenon of great specificity of prostaglandin action which is rarely seen in other biological systems. Using rat PRP, Kloeze (1969) studied the structure-activity relation of prostaglandins in aggregation; PGF compounds had no effect on this system and PGE(2) very unexpectedly enhanced aggregation. The structural requirement for PGE(2) potentiation was similar to that for PGE(1) inhibition of aggregation (the presence of 9-keto group, 15-hydroxy group, the length of two carbon chains), the only difference being the delta-5,6 unsaturated bond.

Therefore, it is important to keep these facts in mind when one considers the clinical use or when one analyzes the action of prostaglandins. In this report, we demonstrate the biochemical basis for the qualitative differences in PGE(1) and PGE(2) action in rat platelets, and we also relate those findings to the response of human platelets to prostaglandins. In addition, the significance of the action of prostaglandins on platelets in clinical medicine is considered.

PROSTAGLANDIN ACTION AND CYCLIC AMP IN PLATELETS

As reviewed by Dr. Salzman (see Introductory Remarks), cyclic AMP seems to have a critical role in the regulation of platelet aggregation. Since rat platelet aggregation is clearly potentiated by PGE(2) (Fig. 1a and 1b), the change in cyclic AMP levels in the presence of this prostaglandin was first examined. PGE(2) by itself did not cause any significant effect on cyclic AMP levels of platelets in citrated plasma, but it markedly reduced the PGE(1)-induced cyclic AMP accumulation (Fig. 2). In ADP-induced aggregation, the PGE(1) inhibitory effect was counteracted by PGE(2) (Shio et al, 1972). Therefore, the stimulatory activity of PGE(2) might be related to cyclic AMP reduction in platelets.

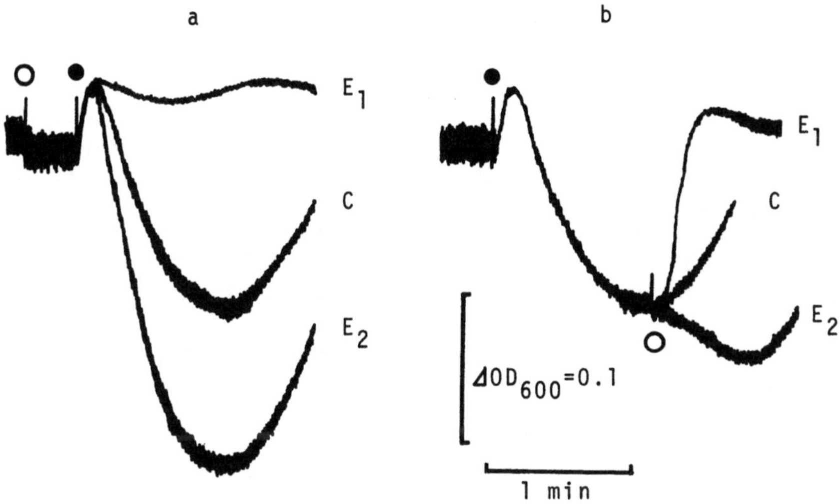

Figure 1. Aggregation curve of rat platelets. Blood was taken into 1/9 vol of sodium citrate (3.8%). PRP was obtained by 375 g x 15 min centrifugation at room temp. Optical density change at 600 nanometers (Born, 1962), was recorded under constant stirring at 37°C. PGE(1) (50 nanogram/ml) or PGE(2) (100 nanogram/ml) was added at °. C indicates control (saline at °). ADP (8×10^{-7} M) was added at •. (a) Effect of pre-incubation with prostaglandins. (b) Effect of prostaglandins added at the maximum point of ADP-aggregation.

However, the dose-response curve of the PGE(2) effect on rat platelet aggregation was biphasic (Kloeze, 1967); at higher concentrations, the potentiation was less than maximum. This raises a question of whether there was PGE(1) contamination in PGE(2) or whether the effect of PGE(2) was really biphasic. Our experiment supported the latter possibility; in the presence of theophylline, totally synthesized PGE(2) (Corey et al, 1970) significantly increases cyclic AMP (Shio et al, 1972) and ^{14}C-cyclic AMP formation from ^{14}C-adenosine in rat platelets (Shio et al, 1971). Therefore, the PGE(2) effect is the summation of at least two qualitatively different effects, and normally the aggregation-enhancing [i.e. anti-PGE(1)] effect is prominent.

Studies of the PGE(2) effect on human platelet aggregation have lead to some confusion. Kloeze (1967) showed that some inhibition occurred at high concentrations. Sekhar (1970) reported that PGE(2) was inhibitory not only on human but also on rat platelet aggregation. We re-examined these phenomena, putting emphasis on the two-phase aggregation pattern, and found that PGE(2) enhances only the second aggretation wave (Fig. 3). When the maximum aggregation in PRP which shows a single phase aggregation, or when initial aggregation rate (in OD) was taken as an index, PGE(2) was clearly inhibitory (Shio and Ramwell, 1972b). Aspirin ingestion or pre-incubation with aspirin *in vitro*, causes a disappearance of the second phase. PGE(2) was inhibitory even on this type of single phase aggregation.

It is interesting to note the relationship between adenylate cyclase stimulation in presence of theophylline (Robison et al, 1969) and inhibition of the first aggregation phase by PGE(2) in human platelets. In both cases, the activity was 1-2% of that of PGE(1). Therefore, it could be interpreted that PGE(2)-inhibition of the first phase is likely dependent on adenylate cyclase stimulation, but the second phase potentiation is due to enhanced cyclic AMP breakdown or other unknown mechanisms - which so far, is similar to the PGE(2) action on rat platelets. Our preliminary experiments indicated that PGE(2) (1 microgram/ml) inhibits PGE(1)-induced cyclic AMP accumulation in human platelets, but the effect of PGE(2) alone is not clear. Moreover, since human platelet aggregation is (in most cases) a sequence of two qualitatively different aggregations, the interpretation of the cyclic AMP experiments is quite difficult. It is very likely that ADP causes a translocation of cell components which in turn cause some shift in enzyme and other drug receptor systems.

Nevertheless, it is clear that PGE(2) enhances the aggregation only when the second phase is present.

PHYSIOLOGICAL AND PATHOLOGICAL EVIDENCE

Thus, the effect of PGE(1) and PGE(2) is interesting from a pharmacological point of view. Clearly, platelet aggregation and adhesiveness are closely related to *in vivo* function of these cells such as in maintenance of vascular integrity, hemostasis, thrombosis and possibly inflammation. Therefore, it is necessary to discuss the possible role of blood prostaglandins in platelet behavior.

First of all, prostaglandins are released from various tissues and organs, and various stimuli (neural, hormonal, ischemic, traumatic, etc.) can evoke release of

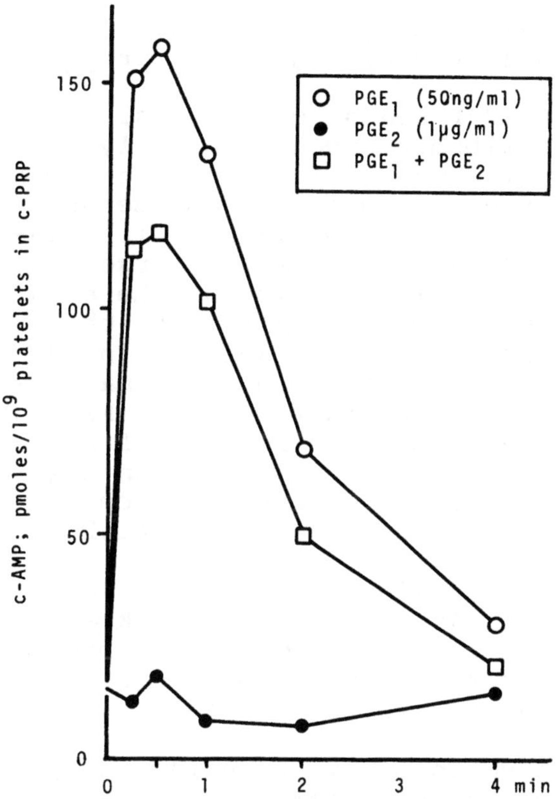

Figure 2. Interaction of PGE(1) and PGE(2) in cyclic AMP changes in rat platelets. PGE(1), PGE(2) or mixture of the two was added to PRP after 10 min pre-incubation. Cyclic AMP was determined with the protein binding method (Gilman, 1970). There was no change in baseline up to 4 min when no prostaglandin was added.

different kinds of prostaglandins (Shio et al, 1971). Under pathological conditions, it is known that a considerable amount of prostaglandins are formed and released, e.g., from lung in anaphylaxis (Piper et al, 1969), and from skin in burns (see Anggard, this Symposium).

Therefore, platelets undergo the continuous exposure to prostaglandin of various concentrations in the blood stream. In vitro experiments indicate that prostaglandins have no significant effect in platelet morphology under physiological conditions, and as reported earlier, prostaglandins are metabolized rapidly when given into the blood stream (Ferreira et al, 1967). However, the rapid response of platelet adenylate cyclase suggests some role of prostaglandin in platelet behavior *in vivo*.

In blood, PGE(1) binds to proteins, and PGE(1) was found to bind to albumin, even when dialyzed (Sekhar, 1971). This binding was confirmed using Sephadex G-200 gel-filtration of citrated plasma and serum of human and rat. When plasma or serum was mixed with 5,6-^{3}H-PGE(1) and applied to a Sephadex column, a large peak of tritium was found in the albumin fraction of the eluate (Fig. 4). Further gel-electrophoretic analysis of this fraction confirmed the PGE(1)-albumin binding. In the gel-filtration experiments, we found some small peaks of tritium prior to the albumin fractions (Fig. 4). The distribution of radioactivity in these fractions was relatively small. Using Cohn's fraction IV(4) of human serum (Pentex), we confirmed the presence of other PGE(1)-binding proteins of higher molecular weight.

However, the albumin-PGE(1) binding does not reduce the activity of PGE(1) on platelets (Sekhar, 1971). This fact indicates the importance of plasma prostaglandins on platelets. The protein binding of PGE(2) and other prostaglandins remains to be studied.

Recent observations have revealed the presence of a prostaglandin formation system in the platelet itself. The prostaglandin content of platelets is low, but a considerable amount of PGE(2) and PGF(2-alpha) was found after thrombin treatment (Smith et al, 1970). This finding suggests that the intracellular regulation of platelet behavior may be through prostaglandin formation in the platelet itself. In this regard, the effect of aspirin is notable. Aspirin selectively inhibits prostaglandin formation in human platelets (Smith et al, 1971). As mentioned before, aspirin inhibits the second aggregation of human platelets (Zucker et al, 1968). Therefore, aspirin is useful to help us to determine the possible role of prostaglandin-forming systems in the regulation of aggregation.

The thrombogenity of long chain saturated fatty acids has been studied extensively (Renaud et al, 1970), but the role of unsaturated fatty acids in platelet function must be reexamined in the light of prostaglandin pharmacology. Also, the regulatory effect of biogenic amines on prostaglandin biosynthesis (Shi et al, 1970) may attract the attention of platelet investigators, since platelets are rich in the biogenic amines which are in a highly localized form.

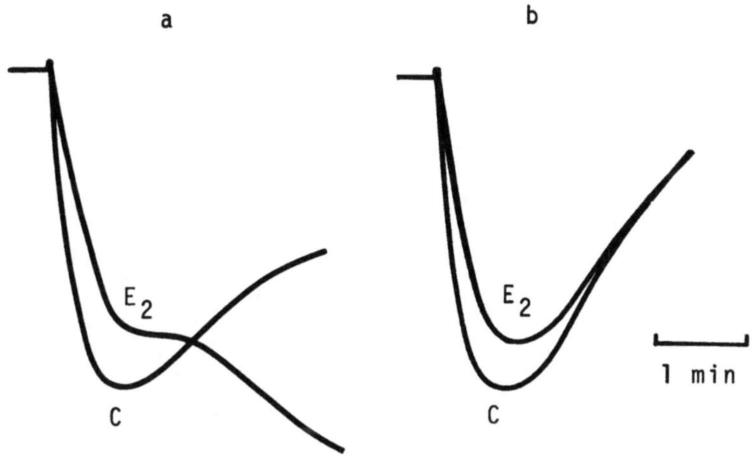

Figure 3. Aggregation of human platelets and effect of PGE(2). Citrated PRP was prepared as in Fig. 1. The OD change at 600 nanometers is shown diagramatically. Aggregation was induced by ADP (2.9×10^{-6} M). C (control), E2 (PGE(2), 100 nanogram/ml, added 30 sec before ADP). (a) Two-phase aggregation. (b) One phase aggregation curve of PRP from the same donor 12 hr after aspirin ingestion.

CONSIDERATION OF CLINICAL APPLICATIONS OF PROSTAGLANDINS

Possible Hazard of PGE(2) Infusion

Among the clinical trials of prostaglandins, therapeutic abortion is the most extensive and it seems to be successful for the second trimester pregnancy. However, the spectrum of prostaglandin actions is too wide for only the induction of the desired effect. A number of side effects have been reported, particularly on the continuous infusion of prostaglandins, mostly PGE(2) and PGF(2-alpha). Gastro-enteric motility and bionetoconstriction are side effects by the PGF compounds. Another finding is the phlebitis which is frequently seen on PGE(2) infusion (Embrey, 1971; Cleft et al, 1971; Hendricks et al, 1971). In a comparison study with PGE(2) and PGF(2-alpha), Hendricks et al (1971) reported the absence of the phlebitis reaction when PGF(2-alpha) is used. They described the reaction as erythema, heat production, pain, tenderness along the vein which received the PGE(2) infusion. Most interestingly, there are differences in the susceptibility of patients to PGE(2) with respect to the phlebitis reaction. Hendricks et al

(1971) found that 60% of patients had phlebitis at 5 microgram/min of PGE(2), while the other patients did not have it even at a 20 microgram/min infusion rate.

The contribution of platelets in this kind of vascular reaction is known (Mustard et al, 1971). Particularly, the release of amines is likely to promote the inflammatory reaction of the vessel wall. PGE(2) does not cause aggregation by itself, but when aggregation is initiated by needle insertion, it may affect platelet behavior, possibly enhancing the second aggregation and release of cell components. Therefore, it is important to examine the PGE(2)-induced phlebitis from this point of view.

Although the phlebitis may stop when the infusion is terminated, and although PGE(2) metabolism is extremely rapid, it must be kept in mind that the degree of PGE(2)-enhancement of platelet aggregation is subject to large individual variations, and moreover, in cardio- and cerebrovascular disease there is a higher tendency for those patients to exhibit two-phase platelet aggregation (Danta, 1970; Zahavi et al, 1969), which is suggestive. Thus PGF(2) infusions may be hazardous in these individuals (see Shio and Ramwell, 1972b).

Antithrombotic Use of PGE(1)

A novel trial in clinical use of PGE(1) has been initiated from our laborabory. Rather than give PGE(1) intravenously to control platelet aggregation, PGE(1) was used in blood collection system to prevent the aggregation during harvesting and storage of platelets for transfusion therapy. Since platelet aggregation is one of the problems in platelet concentrate preparation, PGE(1) was expected to inhibit the centrifuging-induced platelet aggregation. Using polystyrene test tubes as a container for blood, we found about 10 nanograms PGE(1) per ml of PRP is sufficient to prevent aggregation due to cell packing at the bottom of the test tube. This treatment made immediate resuspension of platelets in plasma possible, and moreover, the flat shape of individual cells was maintained during preparation and storage. Based on these investigations (see Shio and RAmwell, 1972c), further studies have been done by Dr. Aster's and Dr. Valeri's groups using blood bags and the more practical methods of the blood bank (see papers in this Proceeding).

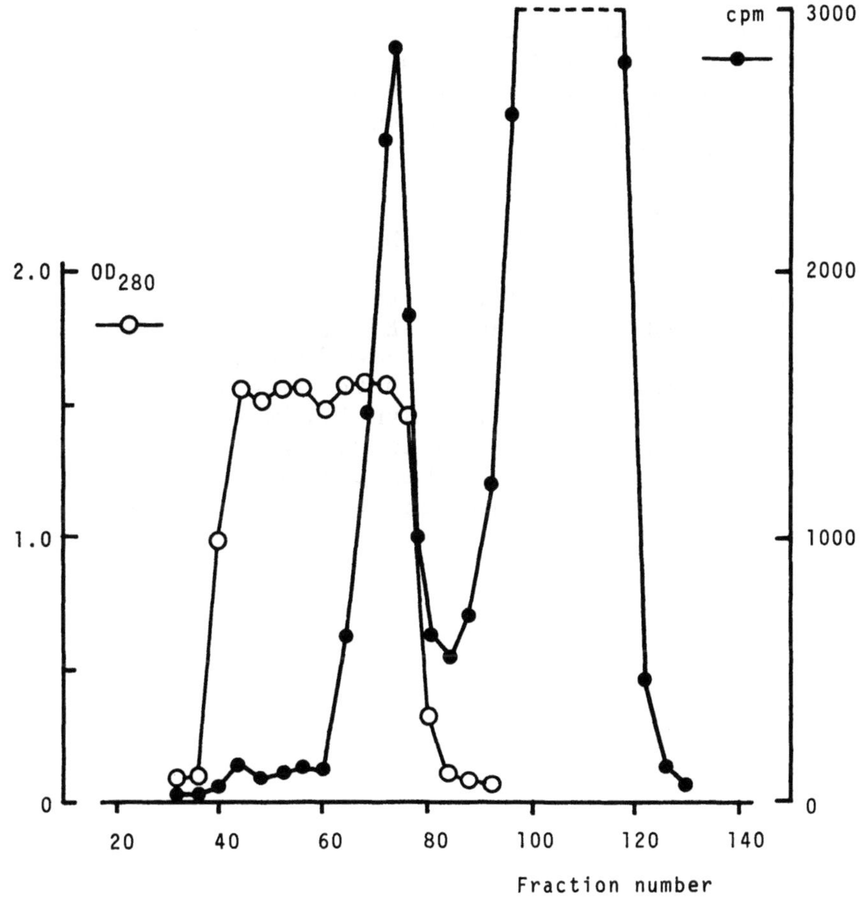

Figure 4. Protein binding of PGE(1) in human serum. 5,6-^{3}H-PGE(1) (2.2 micrograms) was mixed with 3 ml serum and applied on a Sephadex G-200 column. Eluation was performed with 1 M NaCl + 0.1 M tris-HCl pH 8.0. A large peak of radioactivity on right hand side refers to free PGE(1).

The idea of using PGE(1) as an anti-thrombotic agent is a common one, but the wide-range of actions and the extremely rapid metabolism of this prostaglandin discourages <u>in vivo</u> experiment. However, as mentioned before, the platelet-effect of prostaglandins is different from other biological reactions and it may be possible to find a PGE(1) analog or isomer which is specifically effective on platelet reactions. In this regard, the effect of 8-iso PGE(1) (Sekhar et al, 1968) is interesting. Although this compound is a substrate of 15-OH prostaglandindehydrogenase (Nakano et al, 1970), the anti-thrombotic activity is similar to PGE(1) and spasmogenic and vasodepressing effects are only one to ten percent of PGE(1). Also of interest is omega-homo PGE(1), which is 3.8 times as active as PGE(1) in anti-thrombotic effect (Kloeze, 1969).

CONCLUSIONS

This paper shows that even two closely related prostaglandins, as PGE(1) and PGE(2), have qualitatively different actions on single well-defined cells. The consequences of this difference have been discussed with respect to applications in the synthesis of analogs and their ultimate therapeutic application. Such an aim requires detailed study of platelet function to avoid potential side reactions and unnecessary hazards.

ACKNOWLEDGEMENTS

We are grateful to Dr. J. Pike of the Upjohn Co. and Dr. N. Weinshenker of the Alza Corp. for supplying prostaglandins. Also technical assistance of Sheila J. Jessup and Rowena Bray is acknowledged.

This work was supported in part by grants ONR 101-695 and NIH NDS 06444.

REFERENCES

Born, G. V. R., 1962, Nature 194:927.

Born, G. V. R., 1970, J. Physiol. 209:487.

Corey, E. J., Schaaf, T. K., Huber, W., Koelliker, U., and Weinshenker, N. M., 1970, J. Am. Chem. Soc. 92:397.

Craft, I., L., Cullum, A. R., May, D. T. L., Noble, A. D., and Thomas, J. D., 1971, Brit. Med. J. 3:276.

Danta, G., 1970, Thromb. Diath. Haemorrh. 23:159.

Embrey, M., 1971, Ann. N. Y. Acad. Sci. 180:518.

Emmons, P. R., Hampton, J. R., Harrison, M. J. G., Honour, A. J., and Mitchell, J. R. A., 1967, Brit. Med. J. 2:468.

Ferreira, S. H., and Vane, J. R., 1967, Nature 216:868.

Gilman, A. G., 1970, Proc. Nat. Acad. Sci. 67:305.

Hendricks, C. H., Breuner, W. E., Ekbadh, L., Brotanek, V., and Fishburne, J. I., 1971, Am. J. Obst. Gynec. 111:564.

Kloeze, J., 1967, in "Prostaglandins" (S. Bergstrom and B. Samuelsson, eds.), Almqvist and Wiksell, Stockholm, p. 241.

Kloeze, J., 1969, Biochim. Biophys. Acta. 187:285.

Kloeze, J., 1970, Thromb. Diath. Haemorrh. 23:293.

MacMillan, D. C., 1966, Nature 211:140.

Markwardt, F., 1967, in Biochemistry of Blood Platelets (E. Kowalski and S. Niewiarowski, eds.) Academic Press, New York, p. 105.

Mustard, J. F., and Packham, M. A., 1971, in "Inflammation, Immunity and Hypersensitivity" (H. Z. Movat, ed.), Harper and Row, New York, p. 527.

Nakano, J., Angaard and Samuelsson, B., 1969, European J. Biochem. 11:386.

Renaud, S., Kuba, K., Goulet, C., Lemire, Y., and Allard, C., 1870, Circ. Res. 26:553.

Robision, G. A., Arnold, A., and Hartman, R. C., 1969, Pharmacol. Res. Commun. 1:325.

Sekhar, N. C., Weeks, J. R., and Kupiecki, F. P., 1968, Circulation 38: (Suppl. 4):23.

Sekhar, N. C., 1970, J. Med. Chem. 13:39.

Sekhar, N. C., 1971, Thromb. Diath. Haemorrh. Suppl. 42:305.

Shio, H., Plasse, A. M., and Ramwell, P. W., 1970, Microvasc. Res. 2:294.

Shio, H., Shaw, J. E., and Ramwell, P. W., 1971, Ann. N. Y. Acad. Sci. 185:327.

Shio, H., and Ramwell, P. W., 1972a, in "Symposium on Chemistry and Pharmacology of Prostaglandins" (N. Kharasch, ed.) Intra-Science Research Foundation, Santa Monica, in press.

Shio, H., and Ramwell, P. W., 1972b, Nature, in press.

Shio, H., and Ramwell, P. W., 1972c, Science, in press.

Shio, H., Ramwell, P. W., and Jessup, S. J., 1972, Prostaglandins, in press.

Sih, C. J., Takeguchi, C., and Foss, P., 1970, J. Amer. Chem. Soc. 92:6670.

Smith, J. B., and Willis, A. L., 1970, Brit. J. Pharmacol. 40:545.

Smith, J. B., and Willis, A. L., 1971, Nature New Biology 231:235.

Sutor, A. H., Bowie, E. J., and Owen, C. A., 1971, Mayo Clin. Proc. 46:178.

Zahavi, J., and Dreyfuss, F., 1969, Thromb. Diath. Haemorrh. 21:76.

Zucker, M. B., and Peterson, J., 1968, Proc. Soc. Exp. Biol. Med. 127:547.

DISCUSSION

SALZMAN: I would like to show three slides which confirm Dr. Shio's work. We were stimulated by his paper with Dr. Ramwell (1970, Microvasc. Res. 2:294), which came out last summer, and we did some studies with human platelet rich plasma. We found that, in human platelet rich plasma, aggregation induced by ADP was augmented in some instances by PGE(2) at low concentrations. In Fig. 1, for example, at a low concentration of ADP there is primary aggregation but no evidence of a second phase. The same concentration of ADP added one min after the addition of PGE(2) (0.1 microgram/ml) produced a second phase of aggregation. At higher concentrations of PGE(2), the usual response was inhibition of aggregation. We assayed cyclic AMP by Gilman's method.

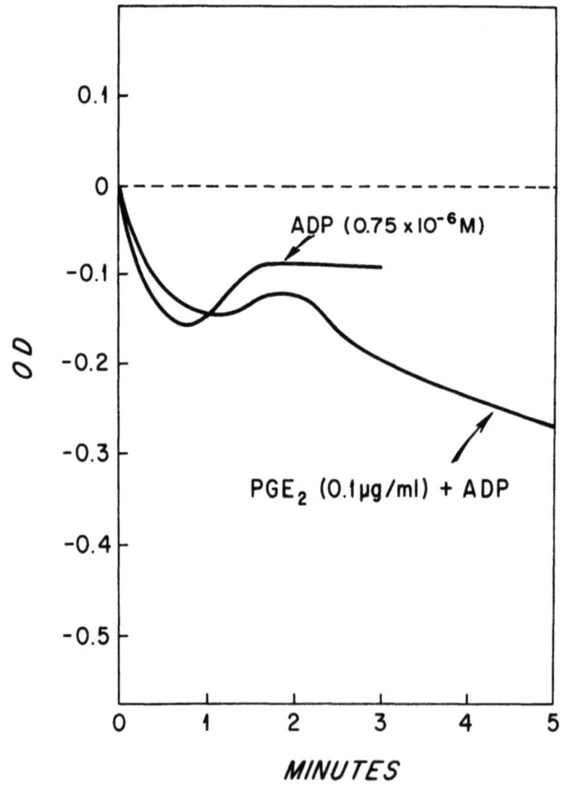

Figure 1. Effect of PGE(2) on platelet aggregation.

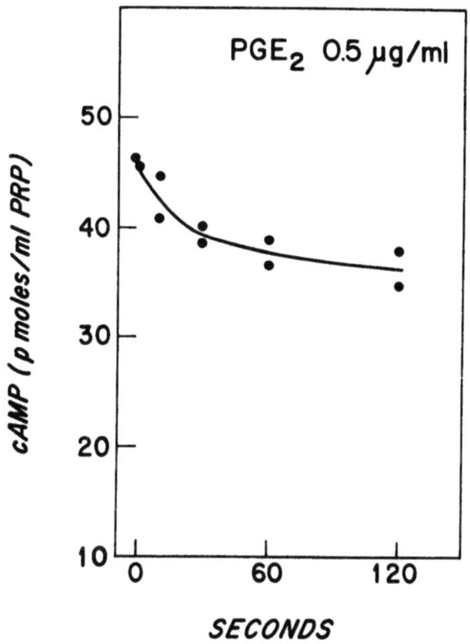

Figure 2. Effect of PGE(2) on cyclic AMP content of platelets.

Fig. 2 shows the cyclic AMP content of platelet rich plasma exposed to PGE(2) for various lengths of time. The response to PGE(2) is very prompt, almost a maximum effect within 30 sec. Fig. 3 shows the effect on cyclic AMP of varying concentrations of PGE(2) compared with PGE(1), measured at 1 min after the addition of prostaglandins to platelet rich plasma. As can be seen, at 1 microgram/ml and less, PGE(2) produced a reduction in cyclic AMP content. At a higher concentration there was an increase in cyclic AMP content. At no concentration of PGE(1) could we find a reduction. I think that these observations confirm Dr. Shio's results.

ZOR: If you incubate PGE(1) and PGE(2) together, does PGE(2) counteract the effect of PGE(1) on this deaggregation? Secondly, does ADP lower cyclic AMP?

SHIO: We found such an interaction between PGE(1) and PGE(2) in rat platelets. PGE(2) inhibits the cyclic AMP response induced by PGE(1).

SALZMAN: ADP does reduce cyclic AMP levels in human platelets. The mechanism of this is quite unknown. We have not been able to show any direct effect of ADP on adenylate cyclase or phosphodiesterase.

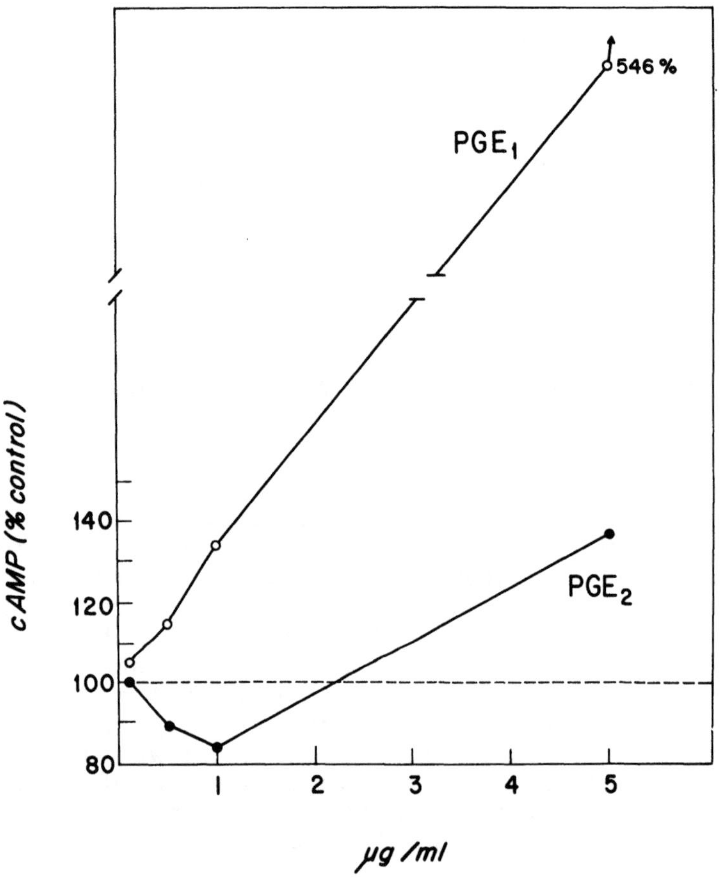

Figure 3. Effect of PGE(1) and PGE(2) on platelet cyclic AMP.

THE EFFECT OF PROSTAGLANDINS, EPINEPHRINE AND ASPIRIN ON CYCLIC AMP PHOSPHODIESTERASE ACTIVITY OF HUMAN BLOOD PLATELETS AND THEIR AGGREGATION

M. S. Amer and N. R. Marquis

Departments of Pharmacology and Biochemistry
Mead Johnson Research Center
Evansville, Ind.

Kloeze (1967) reported that low concentrations of PGE(1) inhibited adenosine diphosphate (ADP)-induced platelet aggregation in plasma of pig, rat and man. Emmons et al (1967) observed that PGE(1) inhibited platelet aggregation induced by other physiological constituents: collagen, thrombin, norepinephrine and serotonin. The extreme potency of PGE(1) in preventing the aggregation phenomenon and its implication with the adenylate cyclic AMP system, as demonstrated by Butcher et al (1967), suggested that the prostaglandin effect might be mediated by cyclic AMP in platelets. Consequently, several laboratories at about the same time reported that PGE(1) increased the synthesis of cyclic AMP by platelet membrane fractions or lysates (Wolfe et al, 1969; Zieve et al, 1969; and Marquis et al, 1969) and by intact platelets (Salzman et al, 1969; Vigdahl et al, 1969).

Cyclic AMP had been observed to be a weak inhibitor of platelet aggregation (Marcus et al, 1965) but the significance of this observation was not fully appreciated until the implication that the adenylate cyclase system might be involved in the regulation of aggregation was considered. Though cyclic AMP has only weak effects on aggregation, the dibutyryl derivative of cyclic AMP is considerably more potent, presumably because of its more lipophilic character (Marquis et al, 1969). Subsequently, Salzman et al (1970, 1971) confirmed that dibutyryl-cyclic AMP was a more potent inhibitor of aggregation than cyclic AMP and that the activity of the dibutyryl derivative could be attributed to its superior uptake by platelets, suggesting an intracellular function for cyclic AMP. Another effect of cyclic AMP is its inhibition of the second phase of platelet aggregation and the release reaction (Salzman et al, 1971).

That cyclic AMP mediates the inhibition of platelet aggregation by PGE(1) is further substantiated by the potency of other prostaglandins, PGA(1) and PGF(1-alpha), relative to PGE(1) on ADP-induced platelet aggregation and cyclic AMP synthesis by platelet membrane fractions (Fig. 1, Marquis et al, 1971). Kloeze (1969) earlier had shown the same relative potency of these prostaglandins. Marquis et al (1970) also demonstrated that the synthesis of cyclic AMP in intact platelets occurred simultaneously with the inhibition of aggregation by PGE(1), further implicating a cyclic AMP mediated mechanism. Finally the observation that caffeine, a known inhibitor of aggregation (Ardlie et al, 1967) and phosphodiesterase (PDE) (Butcher et al, 1962), potentiated PGE(1) inhibition of platelet aggregation fulfilled all the criteria suggested by Sutherland et al (1968) for implicating cyclic AMP mediation of an hormonal effect.

If cyclic AMP were of importance in regulating the aggregability of platelets, then perhaps some of the agents that induce or inhibit aggregation might alter the levels of cyclic AMP in platelets. Indeed, stimulation of adenylate cyclase and/or inhibition of PDE would increase cyclic AMP and lead to an inhibition of aggregation, whereas inhibition of adenylate cyclase and/or stimulation of PDE would decrease cyclic AMP and hence potentiate or even induce aggregation. In fact, numerous agents that have been shown to inhibit platelet aggregation, particularly agents with vasodilator activity are potent inhibitors of platelet PDE (Vigdahl et al, 1971; Mills et al, 1971; and Markwardt, 1967). The stimulation of cyclic AMP synthesis and the increase in platelet intracellular cyclic AMP levels effected by PGE(1) have already been mentioned. In contrast, decreases in cyclic AMP levels in platelets by agents which induce and potentiate platelet aggregation have been observed with ADP, collagen, epinephrine and thrombin (Zieve et al, 1969; Salzman et al, 1969; and Marquis et al, 1970). The mechanism(s) by which some of these agents affect the levels of cyclic AMP and as a result platelet aggregation is(are) the concern of this report.

Two forms of cyclic AMP-PDE have been identified by a number of laboratories (Jard et al, 1970; Kakiuchi et al, 1971; Thompson et al, 1971; and Amer, 1972) in a number of

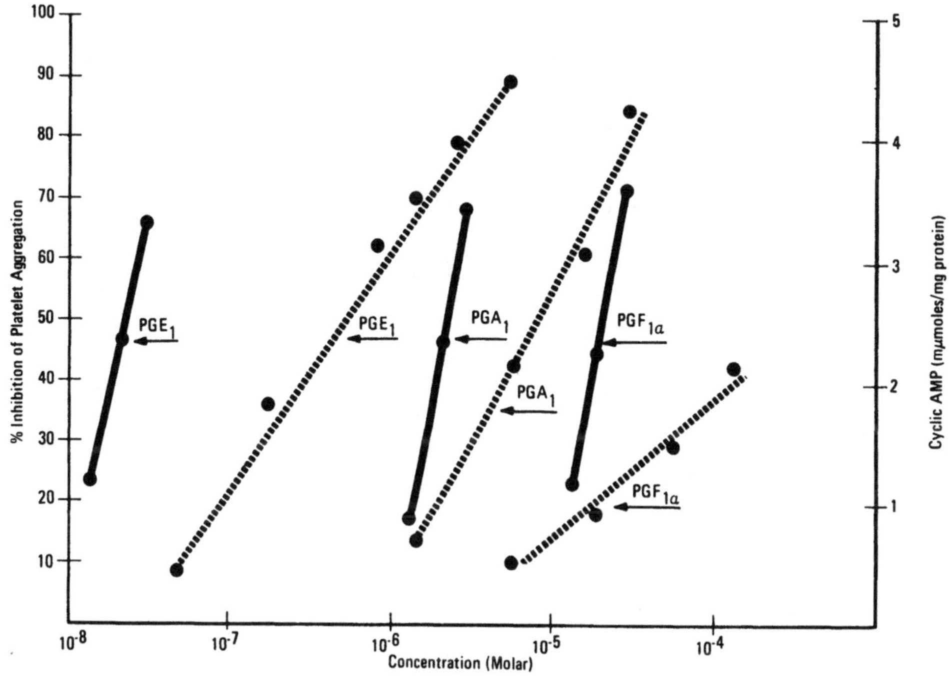

Figure 1: The relative activities of several prostaglandins in inhibiting ADP-induced aggregation and stimulating cyclic AMP synthesis by platelet membrane fractions. Solid lines, platelet aggregation and broken lines, cyclic AMP synthesis. The assays were carried out as described previously (Marquis et al, 1969).

tissues including human blood platelets. The two forms differ significantly in their Km values. In human blood platelets the high Km form or PDE-I has a Km value in the neighborhood of 10^{-4} while the low Km form or PDE-II has a Km of 10^{-6} to 10^{-5} M. In some tissues, notably the brain, the low Km form is at least partly particulate, while the high Km form is mostly soluble (Amer, 1972).

Following the demonstration by Salzman et al (1971) that kaolin and collagen expose PDE activity in platelets which remains membrane bound, it became of interest to see what form of PDE is exposed on release. Results obtained with platelet whole homogenates, membrane fractions and with collagen-released intact platelets is shown in Fig. 2. As can be seen from the Eadie plots, the PDE exposed in the released platelets is of the low Km variety, i.e., PDE-II with Km value (2×10^{-6} M) identical to that in the

12,000 x g membrane fraction and the whole homogenate. This is rather similar to the situation observed in rabbit and monkey brains and supports the hypothesis that the physiologically important PDE is at least partly particulate.

An equilibrium between the two forms of PDE seems to exist in most systems studied (Amer and McKinney, 1970; and Amer, 1971, 1971a, 1972a). Clear demonstration of this equilibrium is difficult in many systems owing to the small proportion of PDE-II activity present, amounting in most cases to less than 5% of the total activity. In human platelets this level is between 1% and 4%. The equilibrium between the two forms can be shifted in either direction with drugs. Epinephrine, for example, seems to shift the equilibrium toward PDE-II, while aspirin seems to affect the reverse process as can be seen in Fig. 3. These effects are more apparent when the proportion of particulate PDE is increased by adding more membrane fraction to the whole homogenate and the percent of total PDE activity present as PDE-II is calculated as shown in Table I. PGE(1) was the most active compound tested thus far while other prostaglandins produce, if any, weak aspirin-like effects. It should be emphasized that only one level of PGE(2) and PGF(2-alpha) were tested in this system. The effects of lower concentrations of PGE(2) on this system are presently being investigated.

The effects of epinephrine and aspirin are also shown in Fig. 4. Upon sucrose gradient fractionation of PDE-II activity, it is clear that PDE-II activity in the heavy fraction is higher with epinephrine and lower with aspirin than the corresponding control values. Fractions two and three had higher activities in the aspirin experiment but since little PDE-II is present in this fraction, this observation does not carry much significance although it does indicate a shift in activity away from the membrane fraction 1. The decrease in PDE-II activity in the first fraction with aspirin is roughly equal to the increase produced in the other fractions. Unfortunately, this is not the case with epinephrine where the increase in activity of fraction 1 was not accompanied by proportionate decreases in the activity assayable in the other fractions.

The effects of PGE(1) in relation to aspirin in the heavy fraction from sucrose gradients is shown in Fig. 5. Here again the percent total activity as PDE-II was reduced by both agents, with PGE(1) being more active. The calculated V-max was increased by both agents and

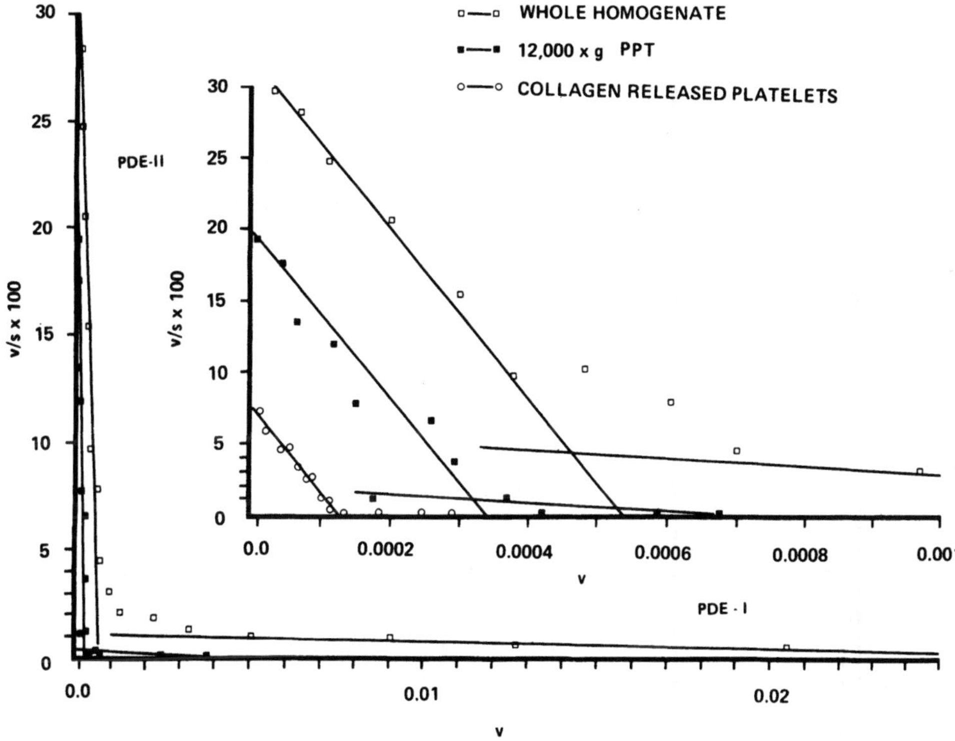

Figure 2: Eadie plots of phosphodiesterase activity from whole homogenate, 12,000 x g precipitate and unhomogenized collagen released platelets. The enzyme assays were carried out according to the method described by Thompson and Appleman (1971a). The substrate levels used were 2.45 x 10^{-7} to 4 x 10^{-3} M. v = initial velocity, s = substrate concentration.

again PGE(1) was more active. The increase in the calculated Vmax is a reflection of the possible shift to PDE-I.

The demonstration of platelet membrane exposed-PDE throws a completely different light on the role that this enzyme might play in the control of platelet cyclic AMP levels and possibly platelet aggregation. It was generally thought that PDE acts as the off switch for the intracellular effects of cyclic AMP. The presence of membrane PDE which is possibly easily accessible on the surface of cells as in the case of platelets permits speculation as to the possible regulatory role of the enzyme on cyclic AMP levels especially since it seems to be responsive to a number of hormones. In addition, this would help explain PDE-effects of large molecular weight

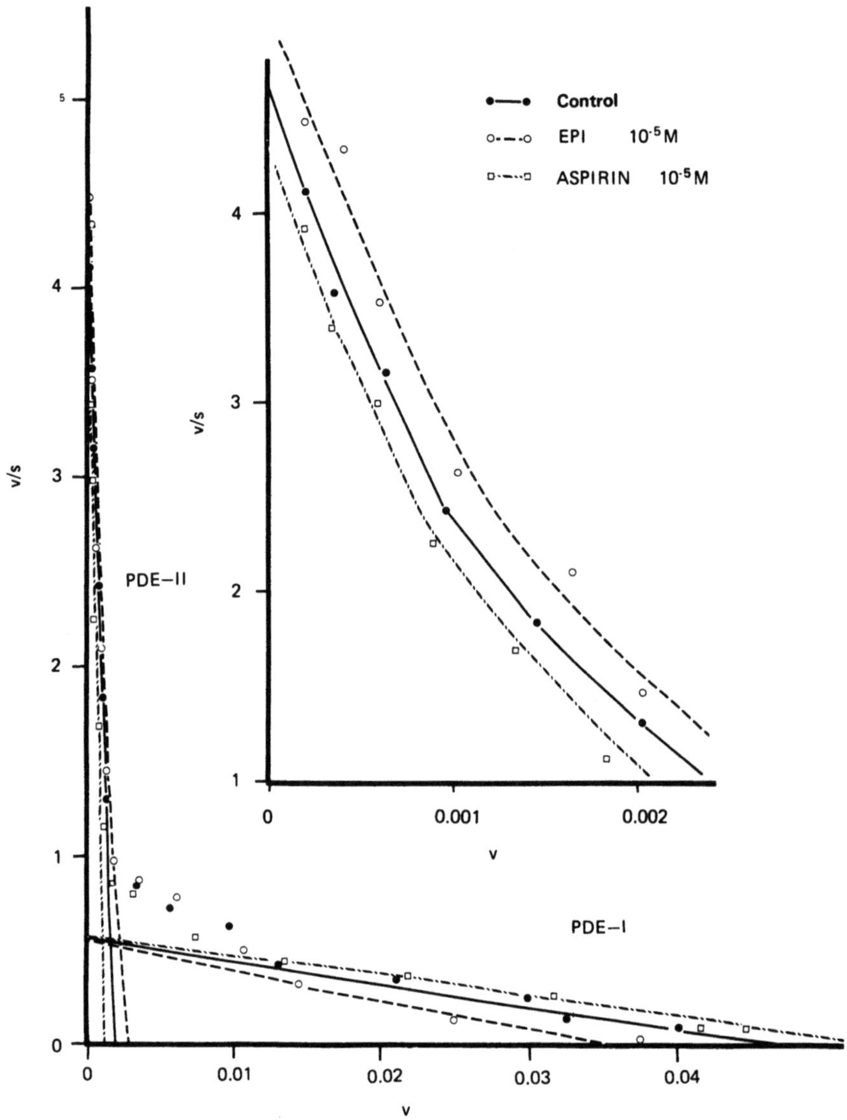

Figure 3: Eadie plots of the hydrolysis of cyclic AMP by fresh human blood platelet 10,000 x g supernates in presence of epinephrine (10^{-5} M) and aspirin (10^{-5} M). The insert is a magnification of the left hand portion of the curve. PDEA-II: Low Km cyclic AMP phosphodiesterase activity. PDEA-I: High Km cyclic AMP phosphodiesterase activity. The assays were done as described for Fig. 2. Note the shift in activity from PDEA-I to PDEA-II with epinephrine and the reverse shift with aspirin.

peptides (e.g., cholecystokinin and other gastrointestinal hormones; Amer, 1972a) and obviate their need to cross the cell membrane.

The effect of PGE(1) in decreasing platelet membrane PDE-II activity might help explain the extreme potency of this compound in elevating cyclic AMP levels since its major effect is the stimulation of adenylate cyclase (Table II) which would be potentiated by the decreased levels of PDE-II. It might also explain the marked synergism observed in the presence of methylxanthines such as caffeine since (1) adenylate cyclase is stimulated by PGE(1), (2) PGE(1) decreases the amount of active PDE, PDE-II and (3) the methylxanthines inhibit the remaining active PDE-II. This could therefore account for the tremendous synergism observed with PGE(1) and the methylxanthines by several investigators on platelet cyclic AMP synthesis (Marquis et al, 1969 and Moskowitz et al, 1971) as well as on the inhibition of aggregation (Mills et al, 1971 and Vigdahl et al, 1971).

In contrast, the effect of epinephrine in decreasing both the basal level and PGE(1)-elevated levels of cyclic AMP may be explained by the action of the catecholamine on increasing the effective amount of PDE-II. It has thus far been difficult explaining why epinephrine, irrespective of the concentration employed, does not effect a greater decrease in cyclic AMP (Table II; Fig. 6). If epinephrine were to affect adenylate cyclase directly it would seem that a more pronounced change would be observed. However, its action on PDE is in keeping with the observations since the levels of cyclic AMP should mirror changes in the Km value of the hydrolyzing enzyme which is decreased by epinephrine (Atkinson, 1969). The effect of caffeine, in opposing the decrease in cyclic AMP induced by epinephrine (Table II), is in support of its effect on PDE. The stimulation of ADP-induced platelet aggregation by epinephrine could also be explained by its effect on PDE. While ADP has been shown to lower cyclic AMP levels in intact platelets (Salzman et al, 1969), ADP has no effect on platelet lysate adenylate cyclase activity. ADP did not affect PDE activity in the present study at a concentration of 10^{-4} M and therefore, as proposed by Salzman et al (1971) and as suggested by the work of Guccione et al (1971), may compete with ATP for the adenylate cyclase of platelets.

Aspirin, a potent inhibitor of the release reaction, secondary aggregation and collagen-induced aggregation, has been shown to decrease the amount of platelet PDE-II

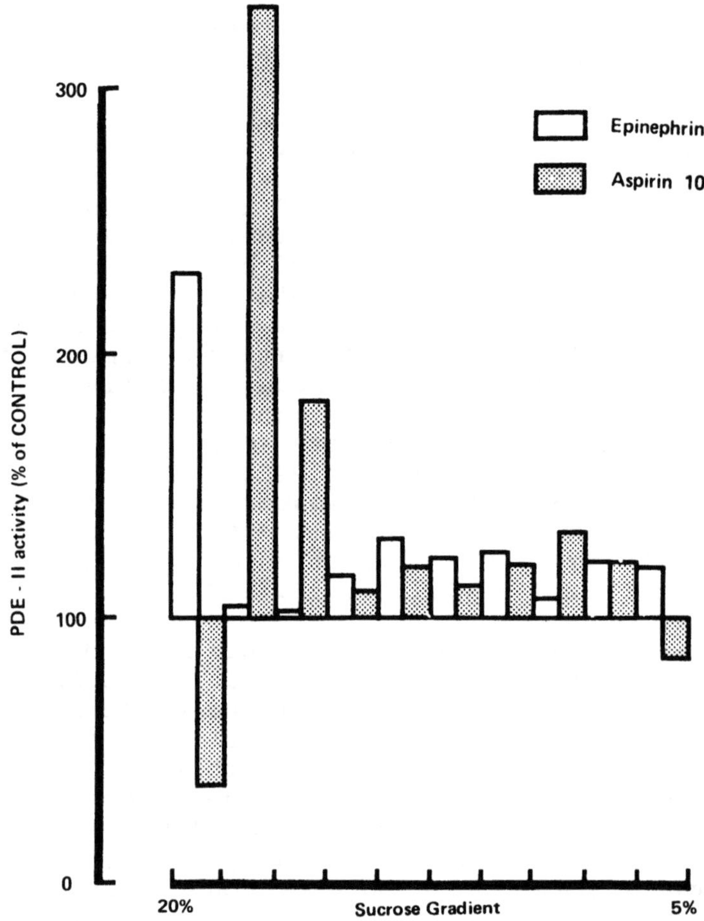

Figure 4: PDE activity in sucrose gradient fractions from human platelet whole homogenate. The heavy fraction (20% sucrose) contained the highest proportion of PDE-II and showed the greatest effects by epinephrine and aspirin. The highest concentration of PDE-I was in fractions 7-9. The PDE assays were done as described for Fig. 2.

activity which in effect would tend to maintain or slightly increase cyclic AMP levels. The effects of aspirin on PDE would result in an effective elevation of the Km value of the enzyme and an accompanying increase in the steady state levels of cyclic AMP. This might not be easy to demonstrate except under conditions where the platelet cyclic AMP levels are significantly reduced. Under these circumstances, aspirin should be expected to effectively antagonize such decreases in cyclic AMP levels. This situation is the reverse of that observed with epinephrine in this system where it was easier to

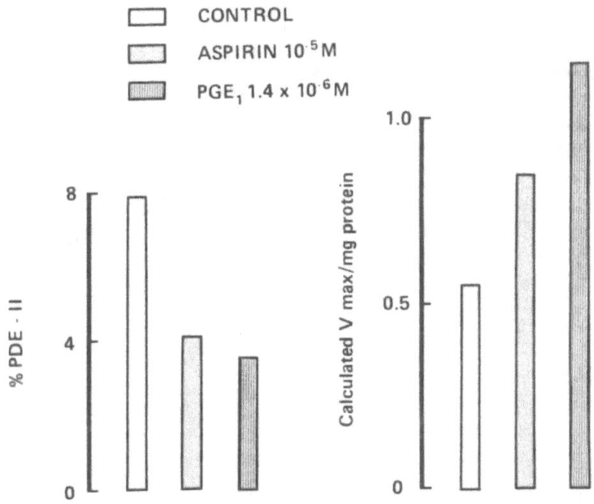

Figure 5: Effects of aspirin and epinephrine added before sucrose gradient fractionation of platelet whole homogenates on the percent PDE-II in and maximal velocity of the heavy fraction. The percent PDE-II was calculated as described for Table I. The maximal velocity was calculated via a computer program developed in this laboratory and based on the method described by Mounter and Turner (1962). The assays were carried out as described for Fig. 2.

demonstrate its lowering effects on cyclic AMP levels previously elevated by PGE(1). In this respect the effects of adenosine are quite interesting. Adenosine inhibits PDE and in the present experiments was found to be an almost selective non-competitive inhibitor of PDE-II with almost no effect on PDE-I (Fig. 7). Theoretically adenosine should affect cyclic AMP levels in a manner similar to aspirin since the effective concentration of PDE-II is reduced in both cases. Adenosine slightly increased basal cyclic AMP levels in platelets and potentiated the effects of other inhibitors on the levels of the cyclic nucleotide (Mills et al, 1971).

The exact role that PDE plays in the control of platelet function is not clear. One can speculate that particulate PDE exposed during release may hydrolyze a membrane-bound pool of cyclic AMP which normally would prevent secondary aggregation. This fits well with the observation previously mentioned of the inhibitory effects of the cyclic nucleotide on secondary aggregation. Though

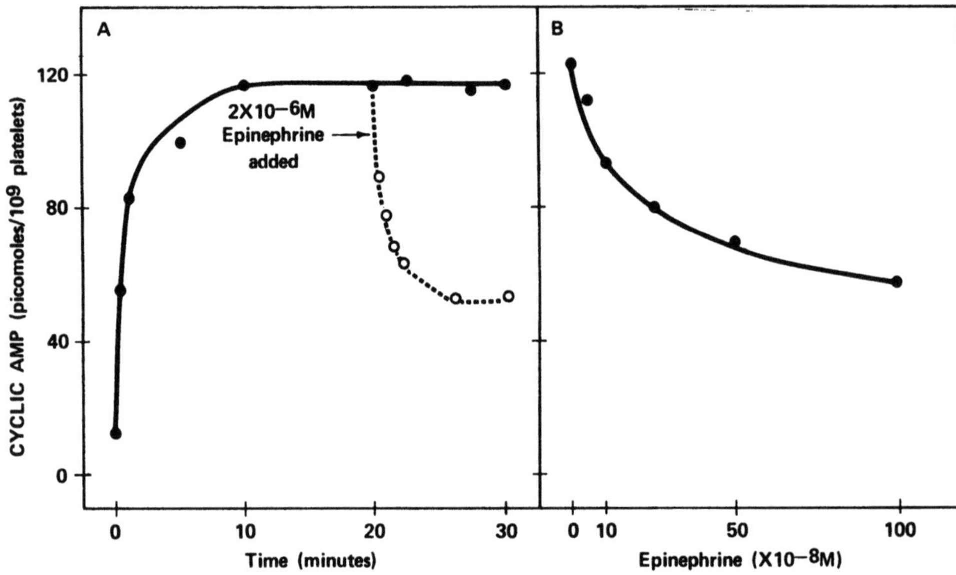

Figure 6: Effects of epinephrine on PGE(1) stimulated cyclic AMP synthesis in intact platelets. Human platelets were isolated from platelet rich plasma and resuspended in a Tris-HCl buffered physiological salt solution containing EDTA and preincubated with adenosine-8-^{14}C to pre-label platelet endogenous ATP. In all experiments, the ^{14}C labeled platelets were preincubated with 1 x 10^{-7} M PGE(1) in order to increase intracellular levels of cyclic AMP before the addition of epinephrine. 1-B) The incubation with varying concentrations of epinephrine were terminated after 20 min.

we have not been able to demonstrate consistent changes in platelet cyclic AMP levels with aspirin in confirmation of the report of Ball et al (1970), its effect on PDE cannot be overlooked.

The recent interesting observations of Smith et al (1971) and Willis et al (this Symposium) that aspirin and other non-steroidal anti-inflammatory agents inhibit the synthesis and release of PGE(2) from platelets when exposed to thrombin, has shed more light on the possible role of PGE(2) as a stimulator of aggregation as originally reported by Kloeze (1967). Shio and Ramwell (this Symposium) confirmed that stimulation of ADP-induced platelet aggregation in rat plasma occurred with PGE(2) as well as observing an inhibitory effect of PGE(2) on PGE(1)-stimulated cyclic AMP synthesis in rat platelets.

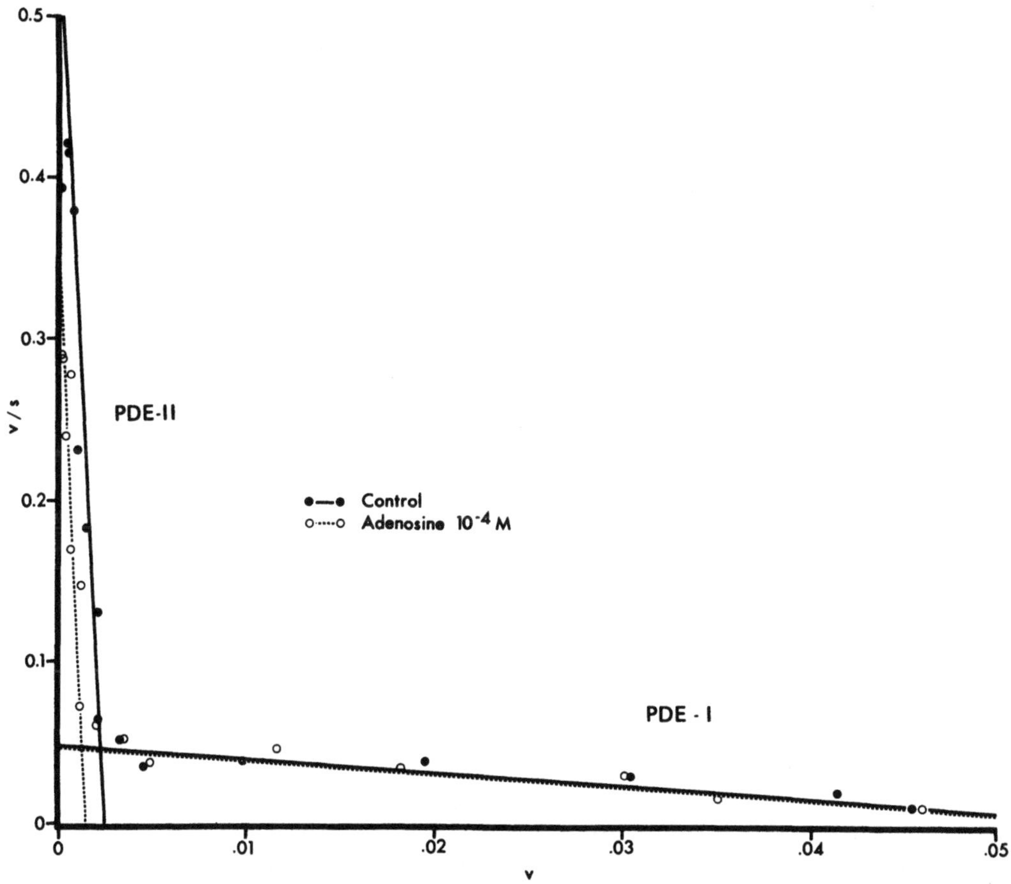

Figure 7: Eadie plots for the effects of adenosine (10^{-4} M) on phosphodiesterase activity from human blood platelet whole homogenates. Conditions are the same as for Fig. 2. Note the selective inhibition by adenosine on PDE-II while no effect is evident on PDE-I.

Salzman (this symposium) in addition showed that PGE(2) decreased basal cyclic AMP levels in intact platelets.

Potentiation of ADP and collagen-induced platelet aggregation by low concentrations of PGE(2) as shown in Fig. 8 further confirms previous observations. Since PGE(2) may become available upon the release of platelet constituents as a result of the aggregation process and since PGE(2) like epiniphrine decreases cyclic AMP levels in the platelet, the prevention of secondary aggregation, as with aspirin and low concentrations of PGE(1), might conceivably prevent this decrease in platelet cyclic AMP. PGE(2), per se, does not induce aggregation and therefore

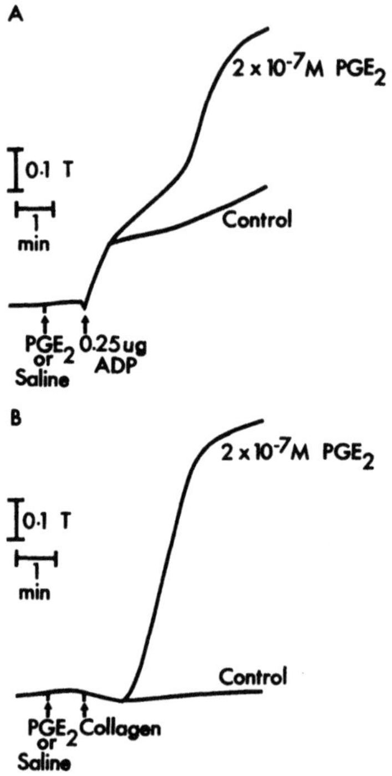

Figure 8: PGE(2)-stimulated human platelet aggregation induced by ADP and collagen. Platelet aggregation was assayed in a Payton Aggregation Module using 0.5 ml platelet rich plasma (PRP). Additions were added in 10 microliter volumes. The PRP was stirred at 1100 RPM and maintained at 37°C throughout the experiment. Data is expressed in transmittance units (T). B) Collagen was used at a concentration which did not by itself produce aggregation.

requires the release or exposure of some factor in order to effect a change in the degree of aggregation. It remains to be shown that inducers of aggregation other than thrombin will initiate the release and synthesis of PGE(2). It is conceivable that (1) aspirin, like other agents, might inhibit the release reaction and PGE(2) release by a cyclic AMP-mediated mechanism and (2) aspirin antagonizes the effect of PGE(2) by lowering the available PDE-II at a membrane site in close proximity to that synthesizing the prostaglandin. The observations of Jaffe et al (this symposium) wherein it was found that dibutyryl-cyclic AMP depressed the release of prostaglandins from a colonic carcinoma culture indicate the possibility that prostaglandin synthesis may be responsive to if not mediated by cyclic AMP.

Since platelet aggregation involves the interaction of membranes of different platelets, adhering in some way to one another, inhibition or induction of this interaction might be expected to occur at or near the cell surface. It is therefore not surprising to find that those enzymes most implicated in this phenomenon are intimately associated with membranes (i.e., adenylate cyclase, PDE, cyclic-AMP-dependent protein kinase, nucleotide diphosphokinase, ATPase and perhaps cthers not yet identified). It is equally tempting to speculate that a pool of cyclic AMP may be membrane-associated, regulated by many of the agents in question, but whose size may be only a small fraction of the total platelet cyclic AMP pool. Hence, changes in the level of cyclic AMP within such a pool would hardly be discernible when assaying total cyclic AMP. Membrane-bound PDE (PDE-II) might play a major role in the regulation of this small membrane-associated pool. Epinephrine, aspirin, adenosine and small concentrations of PGE(1) affect PDE-II and might consequently alter this pool and thus aggregation. Existence of such a pool of cyclic AMP might help explain both the smaller changes seen with the inducers of aggregation and the lack of or small effects shown with inhibitory agents that affect isolated enzyme systems under *in vitro* conditions in the directions expected. It is no coincidence that these agents, with the exception of PGE(1) which also activates adenylate cyclase, produce minimal or no effect on the total cyclic AMP levels. If indeed cyclic AMP plays an important regulatory role in aggregation, either alone and/or via control of PGE(2) release, the effects of potent or even weak inhibitors or inducers of aggregation and their relative effects on the adenylate cyclase-PDE system cannot be easily discarded.

TABLE I

PERCENT PDE-II IN MEMBRANE-SUPPLEMENTED WHOLE HOMOGENATES OF HUMAN BLOOD PLATELETS

Control		7.1
Epinephrine	10^{-5} M	14.3
Aspirin	10^{-5} M	3.8
PGE(1)	1.4×10^{-6} M	2.0
PGF(2-alpha)	2×10^{-4} M	5.5
PGE(2)	2×10^{-4} M	7.0

Effects of a number of compounds on the percent PDE-II activity in membrane supplemented whole homogenate of human blood platelets. The percent PDE-II was calculated via the use of a computer program developed after the treatments of Cleland (1970). The assays were done as described for Fig. 2.

TABLE II

CYCLIC AMP SYNTHESIS BY HUMAN PLATELET MEMBRANES

Additions*	Cyclic AMP (nanoM/mg protein)
none	1.6
Epinephrine (Epi)	1.2
Caffeine (Caff)	2.2
Epi + Caff	2.0
Prostaglandin E(1) [PGE(1)]	5.4
PGE(1) + Caff	24.4
PGE(1) + Epi	1.5
PGE(1) + Epi + Caff	18.4

Agents were present in a final concentration of: epinephrine, 2×10^{-6} M; caffeine, 2×10^{-2} M; PGE(1), 1×10^{-7} M. The methods used were described previously (Marquis et al, 1970).

ACKNOWLEDGEMENTS

The authors wish to thank our colleague, Dr. Robert F. Mayol, for his contribution in providing the sucrose density gradient fractions of the platelets.

REFERENCES

Amer, M. S., 1971, Fed. Proc. 30:220.

Amer, M. S., 1971a, J. Amer. Pharm. Assoc. NS11:118.

Amer, M. S., 1972, In Proc. of the International Conf. on the Physiology and Pharmacology of Cyclic AMP. Ed. by R. Paoletti and G. A. Robison, Raven Press, New York.

Amer, M. S., 1972a, J. Pharm. Exptl. Therap. (in press).

Amer, M. S., and McKinney, G. R., 1970, The Pharmacologist 12:291.

Ardlie, N. G., Glew, G., and Schwartz, C. J., 1967, Thromb. Diath. Haemorrh. 18:670.

Atkinson, D. E., 1969, In Current Topics in Cellular Regulation. Ed. by B. L. Horecker and E. R. Stadtman, Academic Press, New York.

Ball, G., Breneton, G. G., Fulwood, M., Ireland, D. M., and Yates, P., 1970, Biochem. J. 120:709.

Butcher, R. W., Scott, R. E., and Sutherland, E. W., 1967, The Pharmacologist 9:172.

Butcher, R. W. and Sutherland, E. W., 1962, J. Biol. Chem. 237:1244.

Cleland, W. W., 1970, In Enzymes. Ed. by P. D. Boyer, Academic Press, New York.

Emmons, P. R., Hampton, J. R., Harrison, J. J. B., Honour, A. J., and Mitchell, J. R. A., 1967, Brit. Med. J. 2:468.

Guccione, M. A., Packham, M. A., Kinlaugh-Rathbone, R. L., and Mustard, J. F., 1971, Blood 37:542.

Jard, S. and Bernard, M., 1970, Biochem. Biophys. Res. Comm. 41:781-788.

Kakiuchi, S., Yamazaki, R., and Teshima, Y., 1971, Biochem. Biophys. Res. Comm. 42:968-974.

Kloeze, J., 1967, Proceed. of the 2nd Nobel Symposium, Stockholm. Ed. by S. Bergstrom and B. Samuelson, p. 241, Almquist and Wiksell, Stockholm.

Kloeze, J., 1969, Biochem. Biophys. Acta. 187:285.

Marcus, A. J. and Zucker, M. B., 1965, The Physiology of Blood Platelets. Grune and Stratton, Inc., New York.

Markwardt, F. and Hoffman, A., 1967, Biochem. Pharmac. 19:2510.

Marquis, N. R., Becker, J. A., and Vigdahl, R. L., 1970, Biochem. Biophys. Res. Comm. 39:783.

Marquis, N. R., Vigdahl, R. L., and Tavormina, P. A., 1969, Biochem. Biophys. Res. Comm. 36:965.

Marquis, N. R., Vigdahl, R. L., and Tavormina, P. A., 1970, Atherosclerosis: Proc. of the 2nd International Symposium. Ed. by R. J. Jones, Springer-Verlag, New York.

Mills, D. C. B., and Smith, J. B., 1971, Biochem. J. 121:185.

Moskowitz, J., Harwood, J. P., Reid, W. D., and Krishna, G., 1971, Biochem. Biophys. Acta. 230:279.

Mounter, L. A., and Turner, M. E., 1962, Enzymologia 25:225.

Salzman, E. W., Kensler, P., and Levine, L., 1971, New York Academy of Sciences. Conference on Platelets and their Role in Hemostasis (in press).

Salzman, E. W., and Levine, L., 1971, J. of Clin. Invest. 50:131.

Salzman, E. W., and Neri, L. L., 1969, Nature 224:610.

Salzman, E. W., Rubino, E. B., and Sims, R. V., 1970, Series Haematologica 3:100.

Salzman, E. W., and Weisenberger, H., 1971, Proceed. of the International Conference on the Physiology and Pharmacology of Cyclic AMP. Raven Press, New York (in press).

Smith, J. B., and Willis, A. L., 1971, Nature New Biology 231:235.

Sutherland, E. W., Robison, G. A., and Butcher, R. W., 1968, Circulation 37:279.

Thompson, W. J., and Appleman, M. M., 1971, J. Biol. Chem. 246:3145.

Thompson, W. J., and Appleman, M. M., 1971a, Biochem. 10:311.

Vigdahl, R. L., Marquis, N. R., and Tavormina, P. A., 1969, Biochem. Biophys. Res. Comm. 37:409.

Vigdahl, R. L., Mongin Jr., J., and Marquis, N. R., 1971, Biochem. Biophys. Res. Comm. 42:1088.

Weeks, J. R., Chandra Sekhar, N., and Ducharme, D. W., 1968, J. Pharm. Pharmac. 21:103.

Wolfe, S. M., and Shulman, N. R., 1969, Biochem. Biophys. Res. Comm. 35:265.

Zieve, P. D., and Greenaugh, W. B., 1969, Biochem. Biophys. Res. Comm. 35:462.

DISCUSSION

SALZMAN: Dr. Amer's suggestion that there are distinct pools of cyclic AMP in the platelet is supported by other evidence. If one labels platelet cyclic AMP by incubating platelet rich plasma with ^{14}C labeled precursors, the platelet cyclic AMP level behaves differently, depending upon which precursor is used. For example, if one incubates with adenine-^{14}C, the labeled product is not hormone responsive and does not change with epinephrine, ADP, collagen, or thrombin. If one, however, incubates with adenosine-^{14}C, the resultant labeled cyclic AMP is hormone responsive.

KORNBERG: Does this indicate that there are different pools of cyclic AMP or different kinds of platelets?

SALZMAN: If the platelets are altered by the adenine or adenosine, it's possible, but this is with micromolar concentration, so I don't think that's very likely.

LEUKOCYTE CYCLIC AMP: PHARMACOLOGICAL REGULATION AND POSSIBLE PHYSIOLOGICAL IMPLICATIONS

H. R. Bourne

Division of Clinical Pharmacology
University of California School of Medicine
San Francisco, Ca.

INTRODUCTION

Knowledge of both cyclic AMP and the prostaglandins, relative newcomers to biochemical investigation, has burgeoned during the past decade. Cyclic AMP is present in almost every mammalian tissue and its biological role has been well defined in several systems ranging from catabolite repression in *E. coli* to glycogenolysis in mammalian liver. The prostaglandins are equally ubiquitous, though not so well understood. Various prostaglandins exert impressive effects on synthesis of cyclic AMP in many tissues, causing stimulation of synthesis in some and inhibition in others. About two years ago our and other laboratories began investigating possible effects of both cyclic AMP and the prostaglandins in yet another group of cells, mammalian leukocytes.

This review will deal with the effects of prostaglandins and other endogenous compounds or drugs on synthesis and degradation of cyclic AMP in leukocytes. It will attempt to evaluate a number of functional alterations attributed to leukocyte cyclic AMP and/or prostaglandins *in vitro*. Finally, it will discuss the difficult leap from *in vitro* findings, even when the evidence is strong, to physiological or pharmacological significance *in vivo*.

This work was supported by U.S.P.H.S. Grant #HL-09964 and GM-01791. Dr. Bourne is also an established investigator of the American Heart Association.

Extensive study of cells other than the leukocyte has documented the nearly universal role of cyclic AMP as an intracellular "second messenger," relaying a stimulus from the extracellular environment to biochemical machinery within the cell which produces a response useful for survival of the organism (Sutherland et al, 1968). The "first messenger" may be a chemical change in the cell's environment (e.g., glucose deprivation for E.coli) but is represented in more complex organisms by hormones or neurotransmitters which activate adenylate cyclase. The resulting cellular response or "message" is specific for the particular tissue involved: cyclic AMP increases production and release of energy substrates (glucose from liver, free fatty acids from fat cells), contractility in heart muscle, and secretion by endocrine and exocrine glands (e.g., thyroid, beta cells of pancreatic islets, adrenal cortex, and parotid).

Such "stimulatory" effects of cyclic AMP contrast sharply with apparent "inhibitory" effects of the nucleotide in leukocytes, to be detailed below. (A parallel inhibitory effect of cyclic AMP in platelets is discussed by other contributors to this Symposium.) Cyclic AMP in leukocytes (and platelets) in vitro differs from cyclic AMP in other tissues in a second way which may prove more fundamental: in virtually all other biological systems examined so far, both the "first messenger" and the "message" (e.g., epinephrine and hepatic glycogenolysis) were well known before the "second messenger" was discovered; in leukocytes and platelets the sequence has been reversed - the second messenger was known and measured first, and both first messenger and message have been sought by empirical screening of cell functions that can easily be measured in vitro. Such an empirical approach may be unavoidable and can be scientifically respectable; the resulting conclusions as to cause and effect in vivo, however, are often fraught with difficulty. Much of the present literature on prostaglandins or kinins, for example, testifies to the same difficulty.

Nonetheless, the impressive consistency of cyclic AMP's inhibitory effects in formed elements of the blood encourages speculation about the nucleotide's physiological function. Such speculation (presented in the last section of this review) will be warranted if it leads to testable hypotheses, even if they are eventually shown to be naive. Like the fabled blind man, we may find that painstaking examination of a trunk, tusk, or tail leads to a rather unsatisfactory picture of the elephant.

LEUKOCYTE METABOLISM OF CYCLIC AMP

Methodology

In our laboratory, a variety of techniques for measuring metabolism of cyclic AMP have been adapted for leukocytes. Adenylate cyclase was measured in sonicated leukocytes by the method of Krishna et al (1968). Degradation of cyclic AMP by sonicated leukocytes (phosphodiesterase activity, presumably) was assayed by measuring disappearance of radioactive cyclic AMP (Bourne et al, 1971a). De novo synthesis of cyclic AMP by intact leukocytes was measured by a technique adapted from that of Shimizu et al (1969), in which ATP, the intracellular precursor of cyclic AMP, is labelled by pre-incubation of the cells with ^{3}H-adenine (Bourne et al, 1971a; Bourne and Melmon, 1971). In these experiments accumulation of ^{3}H-cyclic AMP could be measured reproducibly only in the presence of theophylline, which prevented degradation of newly synthesized nucleotide. More recently we have assayed cyclic AMP directly according to the sensitive, competition-binding assay of Gilman (1970).

An important reservation regarding most reported experiments with leukocytes is that the cell preparations are heterogeneous. We have been able to separate leukocytes effectively from platelets by differential centrifugation (Bourne et al, 1971a), and erythrocytes, at least in human beings, fortunately contain no adenylate cyclase (Sutherland et al, 1968). We and others have measured cyclic AMP in mixed leukocyte populations, but have also examined cell "functions" peculiar to one or another subset of cells (basophils, neutrophils, lymphocytes). Even purified preparations of lymphocytes are probably heterogeneous, including populations dependent upon either thymus or bursa-equivalent (T and B cells, respectively) and populations of both antigenically committed and uncommitted cells (Craddock et al, 1971).

Synthesis and Degradation

Crude sonicates of human leukocytes catalyze the hydrolysis of ATP to cyclic AMP at rates (per mg protein) comparable to those found in homogenates of other tissues. As in every tissue so far studied, adenylate cyclase in leukocytes requires Mg^{++} and is maximally stimulated by F^- ion (1×10^{-2} M) (Scott, 1970; Bourne et al, 1971a). Subcellular distribution of leukocyte adenylate cyclase has not been studied, but it seems likely that the enzyme is bound to the plasma membrane, as in other cells.

Leukocyte sonicates also degrade cyclic AMP, again at rates comparable to those described in other tissues. Although the degradative enzyme has not been formally characterized as a phosphodiesterase, it is inhibited by the methylxanthines (Bourne et al, 1971a).

Receptor Control of Adenylate Cyclase

Prostaglandins. Prostaglandin E(1) stimulates cyclic AMP production in mixed populations of human leukocytes, purified human lymphocyte preparations, and guinea pig granulocytes (Bourne et al, 1971a; Smith et al, 1971a; Stossel et al, 1970). As shown in Fig. 1, PGE(1) and PGE(2) are almost equally active, followed by PGA(1) and PGF(2-alpha). PGF(1-alpha) is virtually inactive.

Human platelets exhibit a similar spectrum of responses to prostaglandins (E(1) > E(2) > F(1-alpha) (Wolfe and Shulman, 1969). The effects of prostaglandins in leukocytes contrast with the situation described in isolated adipocytes and kidney where PGE(1) strongly inhibits the stimulatory effects of other hormones on adenylate cyclase (Butcher and Baird, 1968; Marumo and Edelman, 1971).

Catecholamines. Adrenergic stimulation of cyclic AMP production in both sonicated and intact leukocytes fulfills most of the requirements for a beta-adrenergic receptor. The order of potency of agonists (isoproterenol and epinephrine are more potent than norepinephrine, which is in turn much more potent than phenylephrine) is characteristic of beta-receptors (Fig. 2), and stimulation by these agents is competitively antagonized by beta-adrenergic blocking agents such as propranolol (Fig. 3) and sotalol, but not by dibenamine or phentolamine (Bourne and Melmon, 1971). Purified preparations of human lymphocytes have a similar beta-adrenergic receptor, which controls synthesis of cyclic AMP and is blocked by propranolol (Smith et al, 1971a).

Beta-receptor control of adenylate cyclase is the rule in tissues where catecholamines have any stimulatory effect (Robison et al, 1967). We have not been able to establish an inhibitory effect of alpha-adrenergic agonists on leukocyte cyclic AMP production, as has been described or inferred in platelets, frog and reptile skin,

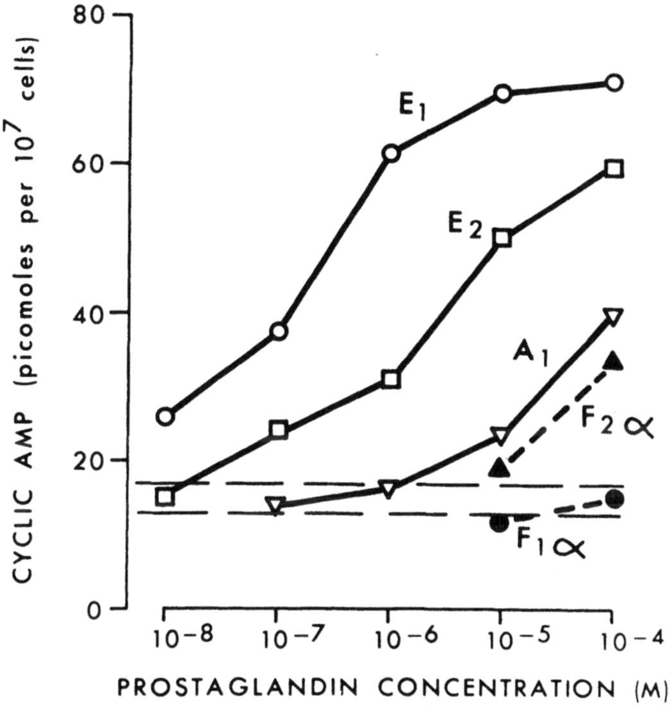

Figure 1. Effect of varying concentrations of five prostaglandins on cyclic AMP content of human leukocytes. Cyclic AMP was measured after a 10 min incubation. Each point represents the mean of duplicates, which differed by not more than 8%. The horizontal dashed lines represent the range of quadruplicate determinations of cyclic AMP content of cells incubated for 10 min in the absence of drug.

and pancreatic islet cells (Marquis et al, 1970; Goldman and Hadley, 1970; Turtle and Kipnis, 1967).

Histamine. Histamine reproducibly stimulates production of cyclic AMP in intact human leukocytes (Bourne et al, 1971b) and in purified mouse splenic lymphocytes (unpublished). The effect in human leukocytes is competitively inhibited by high concentrations of diphenhydramine (Fig. 4), but not by antazolidine, pyribenzamine, or pyrilamine (Bourne et al, 1971c).

Histamine stimulates cyclic AMP production in both brain and myocardium, and causes lipolysis in adipose tissue by a mechanism not yet determined (Nakano and Oliver, 1970; Klein and Levey, 1971). Diphenhydramine blocks the effect of histamine on myocardial adenylate

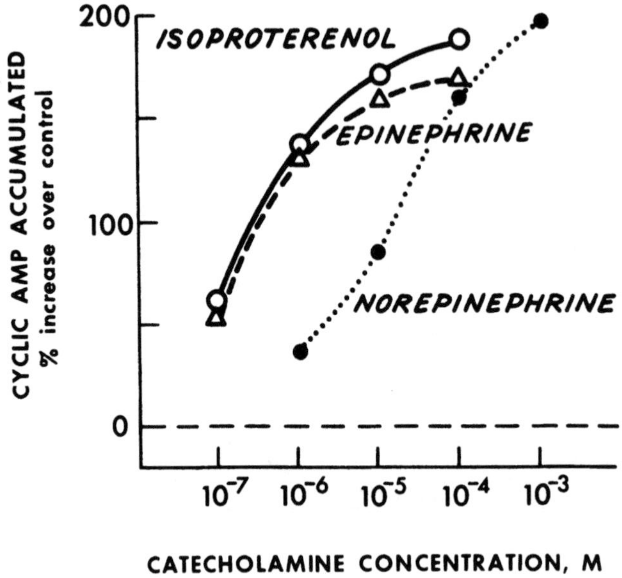

Figure 2. Effect of varying concentrations of catecholamines on accumulation of ^{3}H-cyclic AMP in human leukocytes pre-incubated with ^{3}H-adenine. Theophylline, 1 x 10^{-2} M, was present in all tubes. Each value represents the mean of duplicate determinations in cells exposed to drugs for 10 min. Figure reproduced with permission from The Williams and Wilkins Company, Baltimore: Bourne, H. F. and Melmon, K. L. (1971), Adenylate cyclase in human leukocytes: evidence for activation by separate beta-adrenergic and prostaglandin receptors, J. Pharmacol. Exp. Ther. 178:1.

cyclase at concentrations lower than are required in leukocytes (Klein and Levey, 1971).

Other Potential Agents. Choleragen, the bacterial enterotoxin which produces the profuse diarrhea of cholera, probably does so by activation of adenylate cyclase in the intestinal mucosa (Field, 1971). Purified choleragen, a protein with a molecular weight of 60,000, also causes a dose-dependent increase in leukocyte content of cyclic AMP (Fig. 5). The increase in cyclic AMP does not appear until the leukocytes have been exposed to choleragen for 30-60 min, and is near maximal in 90 min (Fig. 5). This time lag with choleragen, similar to that observed in the gut (Kimberg et al, 1971), contrasts with the almost immediate effect of prostaglandins, catecholamines, and histamine.

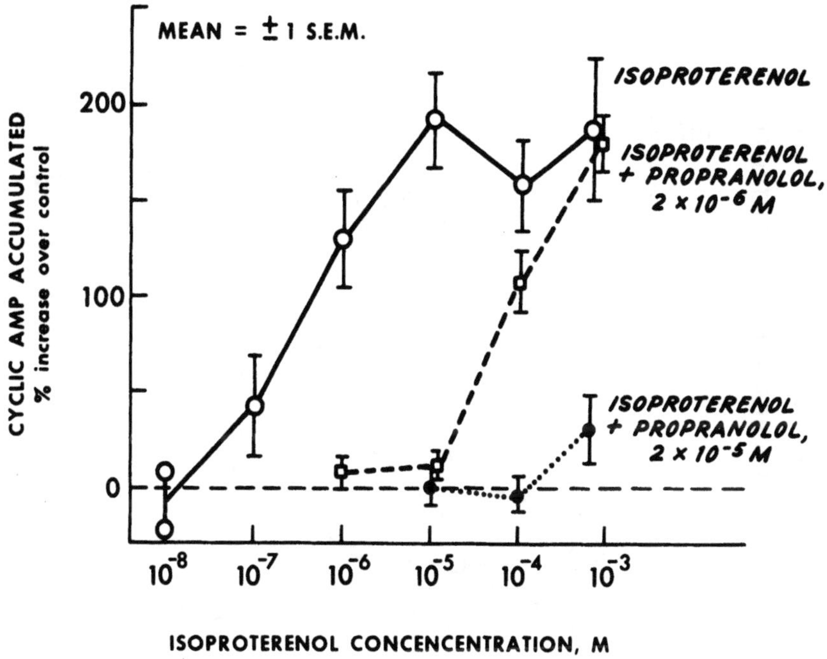

Figure 3. Effect of isoproterenol alone and in combination with two concentrations of propranolol on accumulation of ^{3}H-cyclic AMP in human leukocytes pre-incubated with ^{3}H-adenine. Each value represents the mean ± S.E.M. of four or more determinations in cells exposed to drug for 10 min. Figure reproduced with permission from The Williams and Wilkins Company, Baltimore: Bourne, H. F. and Melmon, K. L. (1971), Adenylate cyclase in human leukocytes: evidence for activation by separate beta-adrenergic and prostaglandin receptors, J. Pharmacol. Exp. Ther. 178:1.

Preliminary investigation of the mechanism reveals no change in phosphodiesterase or NaF-stimulated adenylate cyclase activity in leukocytes incubated with choleragen, although the basal adenylate cyclase activity (in the absence of NaF) is increased. The increase in leukocyte cyclic AMP content is not prevented by cycloheximide, suggesting that it is not dependent on continuing protein synthesis (unpublished data from our laboratory). Similar results in the intestine suggest that choleragen may act on membrane adenylate cyclase by a mechanism analogous to that of NaF (Kimberg et al, 1971).

There is no reason to believe that the effect of choleragen on cyclic AMP production is relevant to the pathogenesis of cholera, except in the gut (Field, 1971).

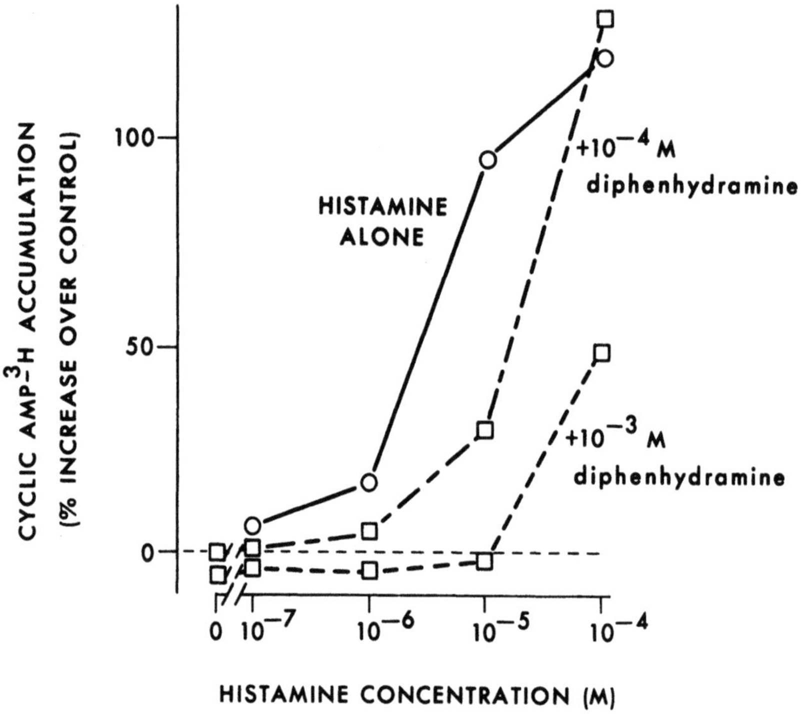

Figure 4. Effect of histamine alone and in combination with two concentrations of diphenhydramine on accumulation of ^{3}H-cyclic AMP in human leukocytes preincubated with ^{3}H-adenine. Each value represents the mean of duplicate determinations in cells exposed to drug for 10 min.

Choleragen may prove a useful tool for study of cyclic AMP in leukocytes, however, because its effect on cyclic AMP production is clearcut (greater than that of PGE(1), for example) and involves a distinctive time lag.

Glucagon and adrenocorticotrophic hormone, peptides which activate adenylate cyclase in other tissues, have no effect on the human leukocyte enzyme (Bourne and Melmon, 1971). Serotonin and bradykinin also fail to increase cyclic AMP in mixed leukocyte populations (unpublished results from our laboratory).

Separate Receptors Controlling Adenylate Cyclase. Abundant evidence has shown that membrane-bound adenylate cyclase in several tissues is controlled by more than one receptor, and that each receptor may combine specifically with a single hormone. The rat epididymal adipocyte contains a single adenylate cyclase which is regulated by

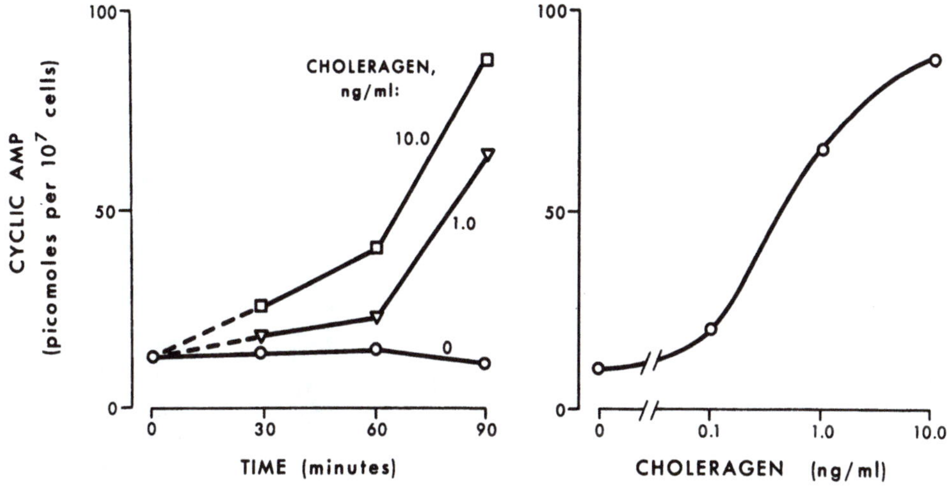

Figure 5. Cyclic AMP content of human leukocytes after exposure to purified cholera toxin: changes with time (left) and with varying concentrations of choleragen (right) at 90 min. Each value represents the mean of duplicate determinations.

three (or more) separate receptors (Birnbaumer and Rodbell, 1969).

Cyclic AMP production in mixed human leukocytes is controlled by at least three separate receptors (Bourne et al, 1971c). As shown in Fig. 6, the stimulatory effect of isoproterenol is specifically blocked by propranolol, which does not prevent the effect of histamine or PGE(1). Similarly, the high concentration of diphenhydramine which blocks the stimulation caused by histamine does not antagonize the effects of isoproterenol or PGE(1).

Because the leukocytes were a heterogeneous population, we do not know whether the separate receptors are associated with the same enzyme, or even whether they are present in the same cell. We have not been able to detect additive effects of maximally effective concentrations of the three compounds; this would suggest that they all stimulate the same enzyme. Such a conclusion is not fully justified, however, since the maximal effects of histamine and isoproterenol are small relative to that of PGE(1), and since truly additive effects could be obscured by experimental variation. When better techniques for separation of subpopulations of leukocytes are developed, this problem can be examined more closely.

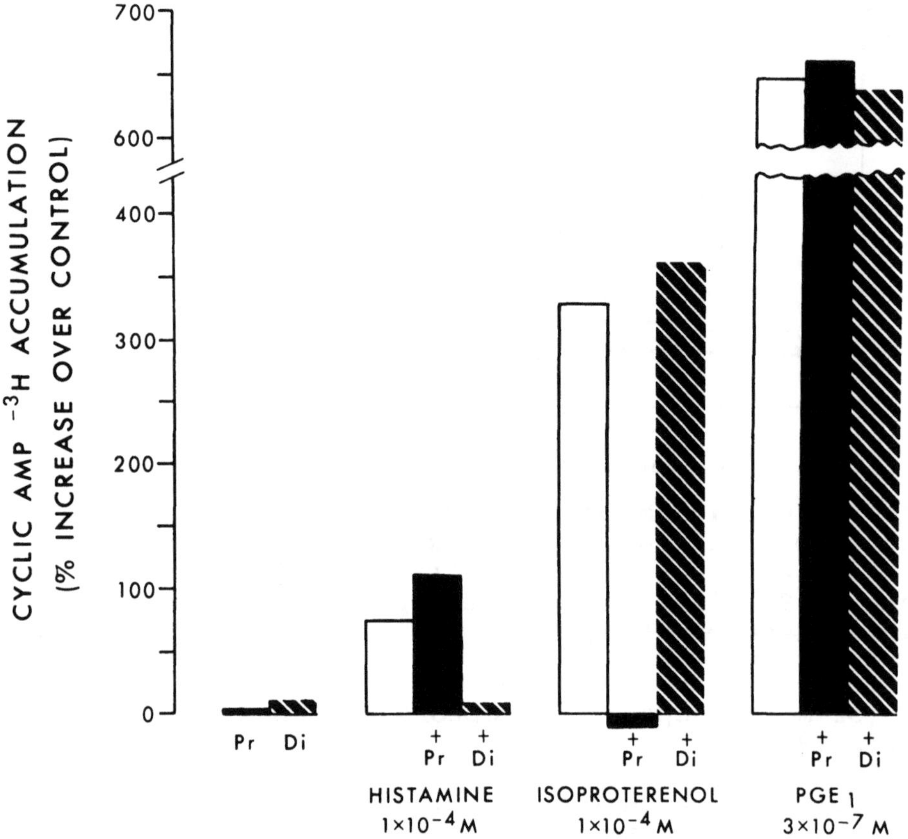

Figure 6. Effect of amine-blocking agents on stimulation of cyclic AMP production in human leukocytes by histamine, isoproterenol, and PGE(1). Each bar represents the mean of duplicate determinations after 10 min exposure to drug(s). Pr = propranolol, 1 x 10^{-4} M; Di = diphenhydramine, 1 x 10^{-3} M.

CYCLIC AMP AND LEUKOCYTE FUNCTION

Topics considered in this section are dealt with in greater detail by other contributors to this Symposium (Weissmann, Parker, Lichtenstein). I will briefly summarize effects attributed to cyclic AMP in neutrophils, lymphocytes, and basophils. The exploratory experiments are of two general types: (1) A leukocyte "function" (transformation of lymphocytes, release of basophil histamine, phagocytosis by neutrophils) is measured in the presence of compounds thought to stimulate cyclic AMP synthesis, block its enzymatic degradation (methylxanthines), or mimic its intracellular effects (cyclic AMP or its dibutyryl derivative). (2) Cellular cyclic AMP is measured in the course of a leukocyte function.

Neutrophils

Circulating neutrophils play an indispensable role in defense of the host by inactivating or removing foreign material and microorganisms from tissues. Performance of this role requires that some or all of the following neutrophil functions be intact (Cline, 1970): (1) migration from blood vessels into injured tissue; (2) phagocytosis of foreign material; (3) killing of micro-organisms; and (4) digestion of dead micro-organisms, foreign materials, or necrotic tissue debris. While performing these functions, (5) the neutrophil may release into tissue microbicidal substances, proteolytic enzymes, chemotactic substances, and chemical mediators of inflammation.

The possible relation of cyclic AMP to these processes has been investigated in several laboratories. The results, while incomplete, suggest that several neutrophil functions are inhibited by exogenous or endogenous cyclic AMP *in vitro*.

Migration of Neutrophils. No formal investigation of the effect of increased intracellular content of cyclic AMP on leukocyte migration has been reported. The ability of caffeine, an inhibitor of phosphodiesterase, to inhibit neutrophil motility at a concentration of 1×10^{-6} M was briefly mentioned in one report (Dimitrov et al, 1969). Another report suggested that cyclic AMP (1×10^{-5}) was actually chemotactic for polymorphonuclear leukocytes in a Boyden chamber, although the variability of response was quite large (Leahy et al, 1970). This area certainly requires further investigation.

Phagocytosis. Two studies of the effect of phagocytosis on cellular content of cyclic AMP presented conflicting results. Incubation of human leukocytes (75%-90% neutrophils) with latex particles for five min, produced a three-fold rise in cellular cyclic AMP (Park et al, 1971). Leukocytes (more than 90% neutrophils) collected from guinea pig peritoneal exudate, on the other hand, showed no change in cyclic AMP content after uptake of latex particles (Stossel et al, 1971). In our laboratory incubation of human leukocytes with C.albicans (1.8 yeasts per neutrophil) similarly produced no change in cyclic AMP. Such inconsistent results may be explained by species differences or by the use of different phagocytic particles. It may be also that the rise in cyclic AMP after exposure to latex particles reported by Park et al (1971) occurred in mononuclear cells rather than neutrophils, as suggested by Mangianello et al (1972).

We have reported slight but consistent inhibition of phagocytosis of C.albicans by pre-incubation of human leukocytes for 30 min with dibutyryl cyclic AMP (3×10^{-3} M), theophylline (3×10^{-3} M), or PGE(1) (1×10^{-5} M) (Bourne et al, 1971a). The burst of oxygen consumption and activation of the hexose-monophosphate shunt following phagocytosis were not significantly affected by the same agents.

Because cyclic AMP regulates hepatic phosphorylase and glycogen synthetase, one might speculate that the nucleotide plays a role in the rapid breakdown of glycogen that accompanies phagocytosis. An impressive study by Stossel et al (1970) showed that this is not the case, at least in guinea pig neutrophils; Cyclic AMP did not change after phagocytosis, and exogenous glucose modified the activity of both phosphorylase and glycogen synthetase without changing cyclic AMP content. Cyclic AMP probably can accelerate glycogenolysis in neutrophils, as suggested by the effects of PGE(1) and NaF in the same study, but does not mediate the effect of phagocytosis.

Microbicidal Activity. In collaboration with Robert I. Lehrer and Martin J. Cline, we have found that the ability of neutrophils to kill C.albicans in vitro is inhibited by agents which increase or mimic leukocyte cyclic AMP (Table I) (Bourne et al, 1971a). The effects of dibutyryl-cyclic AMP, PGE(1), and theophylline were out of proportion to their modest inhibition of phagocytosis. The mechanism by which these drugs decreased neutrophil candidacidal activity is not clear. They did not act by impairment of phagocytosis, post-phagocytic oxygen consumption, or hexose-monophosphate activity, and they did not inhibit leukocyte myeloperoxidase, an enzyme which plays an important part in the killing of C.albicans (Lehrer and Cline, 1969).

The possibility that cyclic AMP mediated the inhibitory effect of PGE(1) and theophylline, by whatever mechanism, cannot be considered proven for several reasons: (1) theophylline was a better inhibitor of candidacidal activity than PGE(1) at the concentrations used, but the reverse was true of the drugs' effects on accumulation of cyclic AMP in leukocytes; (2) although the inhibitory effect of theophylline on neutrophil candidacidal activity was additive to that of PGE(1), a synergistic effect of the two drugs was not demonstrated (Table I), as would be expected if the drugs acted on phosphodiesterase and adenylate cyclase respectively.

Effects of these drugs on the neutrophil's ability to kill other microorganisms have not yet been reported.

Release of Lysosomal Enzymes. After phagocytosis a neutrophil undergoes an important but poorly understood process, "degranulation": the intracellular granules, or lysosomes, are attracted to the phagocytic vacuole; the membranes of granule and vacuole become fused; and the granule's contents are released, some into the phagocytic vacuole, another portion somehow into the extracellular fluid (Cline, 1970). Substances released include cationic proteins with microbicidal activity, and a variety of proteolytic and other enzymes, including myeloperoxidase.

Weissmann, Dukor, and Zurier (1971) have reported that high concentrations (1×10^{-4} M and higher) of dibutyryl-cyclic AMP, theophylline and PGE(1) prevent the appearance of several lysosomal enzymes in the extracellular fluid following phagocytosis by mouse macrophages or human neutrophils (discussed also by Weissmann in this Symposium). They interpret these results as a demonstration that cyclic AMP prevents the fusion of lysosomes with the phagocytic vacuole or with the plasma membrane.

These interesting experiments are suggestive, but not conclusive, for two reasons. (1) The drug effects might well be independent of changes in cellular cyclic AMP content, since high concentrations were necessary and no parallel measurements of the nucleotide were reported. For example, the lowest concentration of PGE(1) (3×10^{-4} M) used in the neutrophil experiments was greater than the maximally effective concentration for stimulation of leukocyte cyclic AMP production in our laboratory (Bourne and Melmon, 1971). (2) Whether or not the drug effects were mediated by cyclic AMP, data reported so far does not establish that degranulation, rather than rate of phagocytosis, was primarily inhibited. Phagocytosis was measured microscopically at the end of a two hr incubation. Since phagocytosis is an exceedingly rapid process, it is possible that the drugs may have caused a considerable inhibition of the initial rate of phagocytosis, as was indeed indicated by our experiments with a different phagocytic particle, C.albicans (Bourne et al, 1971a).

The possibility that cyclic AMP does inhibit degranulation certainly deserves further investigation. Stossel and his co-workers (1971) recently described an elegant method for quantitating phagocytosis and isolating phagocytic vacuoles with their content of lysosomal enzymes. This method should help to resolve the question of cyclic AMP's effect on degranulation.

Lymphocytes

Human Lymphocytes Isolated from Blood. Several laboratories have investigated possible involvement of cyclic AMP in the phenomenon of lymphocyte transformation stimulated by phytohemagglutinin (PHA) (Rigby and Ryan, 1970; Hirschhorn et al, 1970; May et al, 1970; Novogrodsky and Katchalski, 1970). The most complete studies in human lymphocytes are those of Smith and his co-workers (1971a, 1971b), discussed elsewhere in this Symposium by Parker.

These workers presented apparently conflicting data, implicating cyclic AMP as both a mediator and an inhibitor of PHA's mitogenic effect. High concentrations of PHA stimulated adenylate cyclase in broken cells, and produced an early increase, followed by a later decline, in cyclic AMP content of intact lymphocytes. Further data was presented suggesting that cyclic AMP ($6-60 \times 10^{-6}$ M) stimulated lymphocyte DNA synthesis, an effect parallel to, but much smaller than that of PHA. In contrast, isoproterenol, aminophylline, and a series of prostaglandins increased lymphocyte cyclic AMP content and markedly depressed the increased ^{3}H-thymidine incorporation, protein and RNA synthesis caused by PHA. Dibutyryl-cyclic AMP similarly inhibited the response to phytohemagglutinin.

Our laboratory, in collaboration with Dr. Lois Epstein, has confirmed the second set of findings: The prostaglandins, catecholamines, and methylxanthines all stimulated incorporation of ^{3}H-adenine into lymphocyte cyclic AMP and consistently inhibited the mitogenic response of human lymphocytes to PHA. Dibutyryl-cyclic AMP was similarly inhibitory (Bourne et al, 1971d). Using the technique of radioactive adenine incorporation into cyclic AMP, neither we nor other workers (Novogrodsky and

Katchalski, 1970) have been able to detect an effect of PHA on cyclic AMP synthesis in lymphocytes.

These confusing findings have been explained by separation of lymphocyte cyclic AMP into two pools, one responsive to PHA and capable of stimulating mitosis, the other responsive to catecholamines and prostaglandins and capable of inhibiting mitosis (Smith et al, 1971b). Examination of the data, however, suggests a much simpler explanation: it is quite likely that the effect of PHA on lymphocyte cyclic AMP content was nonspecific, since much higher concentrations of PHA were required for this effect than for stimulation of mitosis; in fact, such high concentrations often cause less stimulation of ^{3}H-thymidine incorporation than lower concentrations (Rigas and Tisdale, 1969), which have no effect on cyclic AMP synthesis. The reported stimulatory effect of cyclic AMP on ^{3}H-thymidine incorporation (Smith et al, 1971b) was not statistically significant, was infinitely small in comparison to that of PHA, and has not been confirmed in other laboratories. It is also possible, of course, that cyclic AMP synthesis in response to any of the agents used was actually occurring in a different population of lymphocytes from those which underwent transformation. Purity of the lymphocyte preparations by morphologic criteria certainly does not guarantee homogeneity; immunologic heterogeneity (T and B cells) is in fact quite likely.

In summary, increased intracellular cyclic AMP probably acts to prevent PHA-stimulated mitosis of human lymphocytes isolated from peripheral blood, as suggested by reports from several laboratories. The mechanism of this effect is unknown. The potential complexity of the interactions between catecholamine effects and those due to PHA is indicated by the work of Hadden et al (1971). These workers indicate that catecholamines may actually enhance PHA's stimulation of lymphocyte DNA synthesis, if added at the time of peak PHA effect (68-72 hr). They suggest that both alpha- and beta-adrenergic receptors may be involved, perhaps through modification of glucose and lactate metabolism in the cells.

Thymic Lymphocytes from the Rat. The effect attributed to cyclic AMP in lymphocytes isolated from rat thymus is precisely opposite to that found in human peripheral blood lymphocytes. McManus, Whitfield, and their co-workers have reported increased accumulation of thymic lymphocytes in colchicine metaphase resulting from exposure to a wide variety of compounds, including cyclic

AMP and its dibutyryl derivative, methylxanthines, bradykinin, parathormone, vasopressin, prolactin, growth hormone, cortisol, and epinephrine (McManus and Whitfield, 1969a, 1969b, 1971; Whitfield et al, 1969, 1970a, 1970b, 1970c, 1970d, and 1970e).

The conclusion from these interesting experiments, that cyclic AMP itself mediated the drug effects, is subject to important qualifications. (1) The effect of cyclic AMP was biphasic, in that low concentrations (1×10^{-8} M to 1×10^{-6} M) stimulated, while high concentrations (above 1×10^{-5} M) inhibited mitosis (McManus and Whitfield, 1969a). There is no *a priori* reason to conclude that the lower concentrations produce a more "specific" effect; quite high concentrations of exogenous cyclic AMP are required in some tissues to produce effects, even when the evidence that cyclic AMP mediates those effects is on considerably firmer ground (Sutherland et al, 1968). (2) With the single exception of epinephrine (McManus et al, 1971), the documented mitogenic effect of the hormones studied has not been combined with measurement of cyclic AMP synthesis in the cell population involved. The wide variety of hormones which are said to act through cyclic AMP in this system is in itself disturbing: if the thymic lymphocyte does indeed respond to each of them *in vivo*, it is an exceedingly busy cell.

Lymphoid Tissues from Other Species. Some of the difficulties in defining the relation of cyclic AMP to lymphocyte transformation may be due to species differences, differences in the type of lymphocyte studied, or age-related changes in cyclic AMP metabolism. In lymphocytes isolated from pig blood, cyclic AMP produces as great a mitogenic effect as phytohemagglutinin (Cross and Ord, 1970), for example.

The possibility that cyclic AMP metabolism varies in lymphocytes from different tissues in the same animal is just beginning to be explored. Makman (1971) reported that basal and hormone-responsive adenylate cyclase was greater in lysates of thymocytes than in splenic or mesenteric node cells, in both the mouse and the rat. He also found that thymic adenylate cyclase activity in the mouse declined by about one-half with increasing age of the animals (between five and eight weeks), although the response of hepatic adenylate cyclase increased during the same period.

Cytolytic Activity of Mouse Lymphocytes. The ability of lymphocytes to kill cells which the lymphocytes have been specifically immunized is undergoing extensive investigation, principally because of its possible relevance to allograft immunity. Henney and Lichtenstein (1971) have reported that isoproterenol, methylxanthines, and dibutyryl-cyclic AMP block the cytolytic activity of mouse lymphocytes against allogeneic mast cells in vitro.

These findings have recently been confirmed and extended, in collaboration with our laboratory (discussed in more detail by Dr. Lichtenstein in this Symposium). Four compounds which inhibit lymphocyte cytolytic activity; histamine, PGE(1), isoproterenol and theophylline, also increase the content of cyclic AMP in the same cells. Their relative effects on cyclic AMP content and inhibition of cytolytic activity are similar. Furthermore, propranolol prevents the effect of isoproterenol, but not that of the other agents, on both cyclic AMP and cytolytic activity.

The physiologic significance of these in vitro findings remains to be clarified. The system appears promising, however, because of its relevance to delayed hypersensitivity and allograft immunity.

Basophils and Mast Cells

In vitro models of immediate hypersensitivity have provided the best evidence to date that cyclic AMP may play a regulatory role in formed elements of the blood. A number of investigations stem from the observation of Lichtenstein and Margolis (1968) that catecholamines, methylxanthines and dibutyryl-cyclic AMP inhibit the antigenic release of histamine from leukocytes of allergic donors. This work is discussed by Dr. Lichtenstein in this Symposium and will only be touched upon here.

Antigenic release of histamine from leukocytes of allergic donors is an active secretory process mediated by cell-fixed IgE (Lichtenstein and Osler, 1964). Perhaps the chief advantage of this system over those discussed above lies in its relevance to a relatively well defined disease process: In vitro measurements of histamine release correlate well with the symptomatic course of subjects with ragweed allergy (Lichtenstein et al, 1966).

In collaboration with Lichtenstein we have found an excellent structure-activity correlation between stimulation of cyclic AMP production in leukocytes and inhibition of IgE-mediated histamine release by prostaglandins, catecholamines, and histamine (Bourne et al, 1971c). In both systems PGE(1) and PGE(2) are equally active, while PGF(1-alpha) is not. Similarly, adrenergic agents with predominant beta-activity (isoproterenol, epinephrine) are much more active in both systems than alpha-agonists, such as phenylephrine. Propranolol blocks the effects of catecholamines, but not those of prostaglandins or histamine, in both systems.

Circulating leukocytes (even basophils) probably do not contribute significantly to the release of inflammatory mediators in immediate hypersensitivity. It is important, therefore that similar results attributed to cyclic AMP have been described in human and monkey lung and rat peritoneal mast cells (Assem and Schild, 1969; Koopman et al, 1970; Ishizaka et al, 1971).

A cogent criticism of the leukocyte studies is that histamine is concentrated in basophils, whereas changes in cyclic AMP content were measured in mixed leukocyte populations. The same reservation applies to the other systems. Further work on purified mast cell preparations or with mast cell tumors in tissue culture will be necessary to meet this criticism. More importantly, such experiments will allow investigation of the mechanism by which cyclic AMP produces the inhibitory effect. It may even turn out that the combination of antigen and IgE produces histamine release by decreasing cellular cyclic AMP.

CONCLUSIONS

The biochemical machinery for synthesis and degradation of cyclic AMP is present in leukocytes of many species. Both adenylate cyclase and phosphodiesterase in leukocytes are similar to the enzymes studied in other tissues. The activity of adenylate cyclase in human leukocytes, like that in myocardial or adipose cells, is regulated by more than one specific receptor. We do not know the relative distribution of these receptors among the subclasses of leukocytes, although purified lymphocytes do contain receptors responsive to histamine, catecholamines and prostaglandins.

We also do not know at present which, if any, of these compounds acts as a "first messenger" controlling the activity of leukocyte adenylate cyclase *in vivo*.

Effects in the test tube require higher concentrations of all three compounds than are present in normal plasma. We may eventually find, in fact, that all three receptors are evolutionary rests, analogous to the many receptors in the rat fat cell; they may be present only because in the absence of high plasma concentrations of their respective hormones they confer no selective disadvantage on the organism.

If leukocyte cyclic AMP acts as a second messenger, we have not yet deciphered the physiologic "message". Evidence for several messages has been reported, however, as summarized above: (1) inhibition of degranulation in neutrophils and macrophages (Weissmann et al, 1971); (2) inhibition of neutrophil candidacidal activity (Bourne et al, 1971a); (3) inhibition of cytolytic activity of sensitized lymphocytes (Henney and Lichtenstein, 1971); (4) both inhibition and mediation of lymphocyte transformation (Smith et al, 1971a, 1971b; McManus et al, 1971); (5) inhibition of antigenic release of inflammatory mediators from basophils or mast cells (Lichtenstein and Margolis, 1968; Assem and Schild, 1969; Koopman et al, 1970; Ishizaka et al, 1971; Bourne et al, 1971b, 1971c). In addition, it has been suggested that: (6) phagocytosis itself is a first messenger stimulating production of cyclic AMP, although an actual "message" was not defined (Park et al, 1971).

The contention that cyclic AMP is primarily involved in any of these <u>in vitro</u> systems cannot be considered proven. Although touched upon in the preceding discussion, the important reservations are common to many of the experiments reported, and are worth summarizing here (see Table II).

<u>Measurement of Cyclic AMP</u>. An essential criterion for implicating intracellular cyclic AMP as the mediator of a physiologic or pharmacologic effect is that cyclic AMP itself should increase or decrease appropriately with the functional effect observed (Sutherland et al, 1968). The studies of lysosomal enzyme leak after phagocytosis and many of the papers on lymphocyte transformation do not include measurements of cyclic AMP.

<u>Potential Nonspecificity of Drug Effects</u>. Sutherland et al (1968) suggested that an effect due to intracellular cyclic AMP should (1) be potentiated by methylxanthines, which prevent inactivation of the nucleotide by phosphodiesterase; (2) occur at concentrations of first messenger compounds which also stimulate adenylate cyclase

or accumulation of cyclic AMP; and (3) be reproduced by cyclic AMP itself or its dibutyryl derivative. Some of the studies reported have not examined the effects of methylxanthines in combination with other drugs, while others have shown additive, but not synergistic effects (Bourne et al, 1971a; Weissmann et al, 1971). A number of experiments have involved concentrations of putative first messenger compounds far in excess of those needed to stimulate cyclic AMP production, (Weissmann et al, 1971) or have found a discrepancy between the ability of drugs to elevate cyclic AMP accumulation and their ability to produce the effect observed, as in our study of neutrophil candidacidal activity (Bourne et al, 1971a). A number of studies have shown the appropriate effect of cyclic AMP or dibutyryl-cyclic AMP, but in some systems other nucleotides produce the same effect at comparable concentrations, as with inhibition of lymphocyte transformation (Smith et al, 1971b).

Heterogeneity of Cell Populations. Virtually every leukocyte system so far examined has involved mixed populations of cells. Even when the cells were morphologically similar (e.g., lymphocytes) it is likely that they were functionally heterogeneous. The resulting qualification of conclusions relating changes in cellular cyclic AMP content to functional effects is obvious.

In summary, it is not possible at present to state without reservation that cyclic AMP carries any specific message in leukocytes. The system studied most extensively, by Lichtenstein and workers in other laboratories (including our own), implicates cyclic AMP as an inhibitor of antigenic release of inflammatory mediators from basophils or mast cells. These experiments meet many of the criteria listed in Table II and are internally consistent; species differences (so far) are not prominent.

In addition to the problem of cellular heterogeneity, another difficulty is that many of the test systems are themselves complex and poorly understood; the mechanisms of neutrophil degranulation or candidacidal activity, or of lymphocyte transformation are mysterious enough without considering cyclic AMP at all. Conclusive evidence regarding the role of cyclic AMP in these cells will require study of more fundamental cell functions in homogeneous populations of cells, preferably in cultured cells with a uniform genotype. Such investigations are beginning in several laboratories, including our own. The relative consistency of cyclic AMP's inhibitory effects in a variety of cell types (see Table II) may be a clue that the nucleotide actually modifies a more fundamental cellular process, and that changes in "functions" such as degranulation, histamine release, or candidacidal activity are merely indirect consequences. Candidates for a primary locus of action of leukocyte cyclic AMP include: (1) Activation of a cyclic AMP dependent protein kinase which in turn activates or inhibits diverse enzyme systems (by phosphorylation) to produce the observed changes in cell function; (2) an effect on carbohydrate metabolism which leads to inhibition of cell function; (3) a specific or generalized effect on protein synthesis; (4) a specific change in distribution of Ca^{++} within the cell, by analogy with the theory expounded for other cells by Rasmussen (1970); (5) direct interference with a microtubular system which may eventually be shown to regulate phenomena as diverse as phagocytosis, degranulation, histamine release and mitosis of lymphocytes (Weissmann, this Symposium).

Of these possible actions, only the second has been tested: an attempt was made to correlate the effects of certain drugs, including cyclic AMP, on utilization of glucose with their inhibition of lymphocyte transformation (May et al, 1970). The results were inconclusive, in part because glucose utilization was measured in mixed leukocyte populations, while mitosis in response to PHA occurred only in lymphocytes.

SPECULATIONS: LEUKOCYTE CYCLIC AMP *IN VIVO*

A more basic objection to all of the possible conclusions discussed in the preceding section comes from this fact: the "second messenger" has been investigated empirically, because of its importance in other tissues, but without well defined hypotheses relevant to known messages received by leukocytes *in vivo*. One potentially productive path to such hypotheses will lead deeper into the cell, into investigation of cyclic AMP-dependent protein kinase, carbohydrate metabolism, intracellular distribution of calcium, or effects of cyclic AMP on microtubules, as noted above.

A second approach appears equally necessary: Assume that cyclic AMP actually does act in leukocytes of the intact organism in the same fashion as has been observed in a model system *in vitro*; then ask what the *in vivo* consequences of such an action would be and design experiments to show that these consequences do or do not occur. I propose a speculative scheme which may allow the importance of leukocyte (or mast cell) cyclic AMP to be tested in vivo: Cyclic AMP mediates the effects of an endogenous first messenger system directed at limiting or controlling inflammatory, allergic or immune responses.

The proposed scheme rests at present on two slender lines of evidence. First, if the in vitro models discussed above are not all misleading it is difficult to ignore the pattern that emerges (see Table II). Each study examines a leukocyte function which is thought to be necessary for survival of the organism, but in each system the effect of cyclic AMP is inhibitory. Second, a few pharmacological experiments in vivo suggest that cyclic AMP could mediate an anti-inflammatory effect in leukocytes or mast cells: (1) Isoproterenol and theophylline were reported many years ago to prevent the wheal and flare response to intradermal injection of antigen in allergic subjects (Sheldon et al, 1951). Since isoproterenol did not prevent the cutaneous response to injected histamine, it may have acted on beta-receptors in mast cells to prevent release of histamine. (2) Similarly, isoproterenol prevented dextran-induced inflammation in rats, an effect apparently mediated via beta-adrenergic receptors (McKinney and Lish, 1966). (3) Administration of prostaglandins can inhibit progress of an inflammatory arthritis in rats, an effect interpreted as mediated by cyclic AMP (Weissmann, this Symposium).

Such evidence is certainly inconclusive and lies outside the mainstream of most investigations of inflammation and immunity, which have focused on factors initiating or maintaining inflammatory or immune processes. It appears likely, however, that these processes are subject to inhibitory as well as activating influences, analogous to the intricately balanced effects of thrombin and plasmin in coagulation of blood, or the finely tuned feedback loops controlling secretion of many hormones.

In the in vitro models of immediate hypersensitivity, for example, histamine and PGE(1) prevent antigenic (IgE-mediated release of histamine, probably through increased production of leukocyte cyclic AMP (Lichtenstein, this Symposium). Considerable evidence, however, indicates that both histamine and PGE(1) can reproduce some of the vascular changes of inflammation when injected into the skin of man and other animals and that both are present in the extracellular fluid of inflamed tissue (Spector and Willoughby, 1965; Willis, 1969; Crunkhorn and Willis, 1971).

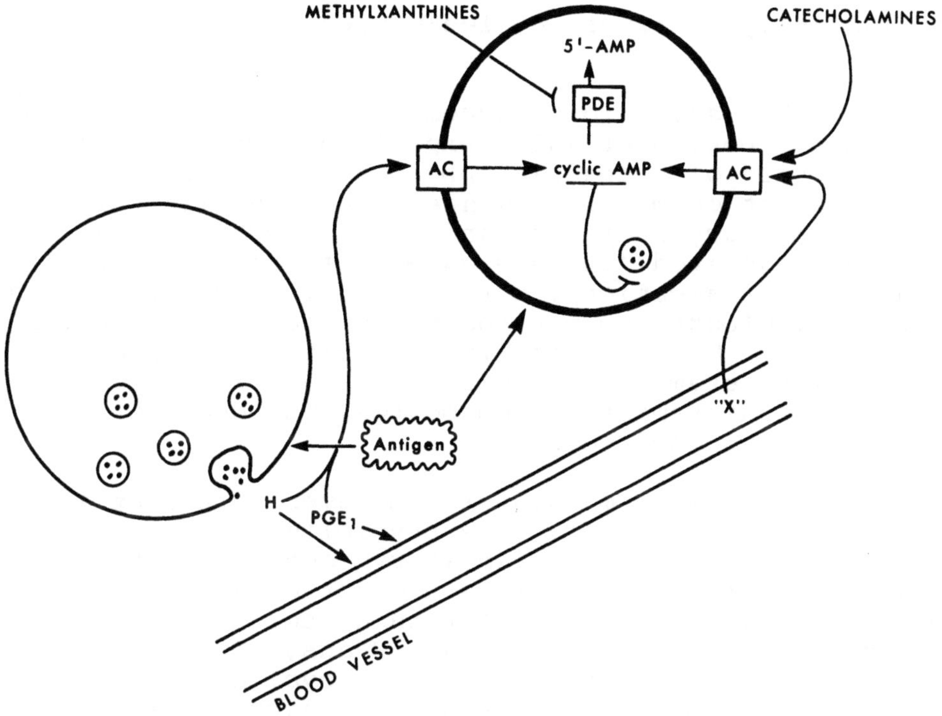

Figure 7. Speculative scenario for possible anti-inflammatory role of leukocyte cyclic AMP. The inflammatory stimulus (antigen) causes release of histamine and (perhaps) PGE(1) from a mast cell (left). These inflammatory mediators contribute to the microvascular changes of inflammation, but are also available to stimulate adenylate cyclase in the cell membrane of a leukocyte newly arrived at the inflammatory site (right). The resulting rise in leukocyte cyclic AMP prevents further release of inflammatory mediators (such as histamine) contained in leukocyte granules. It is also possible that catecholamines released by the sympathetic nerves or humoral agents ("X" in the diagram) could modify inflammation by stimulating cyclic AMP production in leukocytes.

This apparent paradox suggests the possibility, diagrammed in Fig. 7, that histamine and prostaglandins may act not only on blood vessels, but also as links in a physiological feedback loop which could limit the intensity of response to a given inflammatory stimulus. After initial release from mast cells, for example, either compound might act via cyclic AMP to block further release of inflammatory mediators by leukocytes newly arrived at the scene.

For such a scenario to become an hypothesis, it must be testable. Design of valid experiments will not be easy: Detection of a feedback phenomenon requires not only measurements of chemical messengers involved (in this case, inflammatory mediators and cyclic AMP) and the "function" regulated (in this case, the complex vascular response of inflammation), but also an intervention which specifically blocks one portion of the loop while leaving others intact. No such specific inhibitors are presently available (for example, an antihistamine which prevents the effect of histamine on leukocytes, but does not affect the response in blood vessels).

Many homeostatic mechanisms, however, are most easily detectable when they fail. The physiological significance of one endogenous "anti-inflammatory" substance, the C-1-esterase inhibitor of plasma, would not be appreciated if it were not absent in patients with hereditary angioneurotic edema (Donaldson and Evans, 1963). An analogous defect related to leukocyte cyclic AMP is presently being sought in several laboratories, with atopic disorders and asthma the most likely candidates. Hyporesponsiveness of beta-adrenergic receptors has actually been postulated in asthma, although convincing documentation is lacking (Szentivanyi, 1968). The possibility that synthesis of cyclic AMP in leukocytes of asthmatic patients is relatively unresponsive to beta-adrenergic agents is discussed in this Symposium by Parker.

An additional subplot in this scenario can be subjected to experimental investigation: A first messenger aimed at modulation of inflammation or immune responses might arrive via the blood or the autonomic nervous system. An anti-inflammatory role for catecholamines has been suggested, on rather meager evidence (Spector and Willoughby, 1965). Nonetheless, the possibility that the sympathetic nervous system can modify the responses of mast cells or leukocytes, as well as those of blood vessels and glands, should be considered.

Finally, an anti-inflammatory (or immunosuppressive) system mediated by cyclic AMP could have therapeutic implications. It may be, for example, that methylxanthines and catecholamines are useful in asthma, urticaria, and anaphylactic reactions not only because of their effects on bronchi and blood vessels, but also because they act to prevent release of mediators of the inflammatory or allergic process. If this proves to be true, the adenylate cyclase system in leukocytes could be useful for designing and testing possible anti-allergic or anti-inflammatory compounds with greater efficacy or specificity than those presently available.

TABLE I**

INHIBITORY EFFECT OF DRUGS ON LEUKOCYTE CANDIDACIDAL ACTIVITY

Drug	Candidacidal activity	Significance*
	% of control+	P<
Dibutyryl-cyclic 3',5'-AMP, 1 x 10^{-3} M/l	29.7±2.1(6)	0.001
Theophylline		
1 x 10^{-3} M/l	32.8±11.8(5)	0.01
1 x 10^{-4} M/l	93.9±6.9(8)	NS
PGE(1) (in 0.33% ethanol)		
1 x 10^{-5} M/l	75.0±7.1(9)	0.01
1 x 10^{-6} M/l	84.1±7.3(8)	NS
PGE(1), 5 x 10^{-6} M/l + theophylline, 1 x 10^{-4} M/l		
Calculated "additive" effect	78.0±9.2(8)	□
Actual effect in combination	67.3±8.8(8)	□
Ethanol, 0.33%	101.5±6.8(5)	NS

* Paired t test.
+ Mean ±SE (n).
□ Not statistically different by paired t test (0.1 < P < 0.05).

**Bourne, J. R., Lehrer, R. I., Cline, M. J., and Melmon, K. L., 1971, Cyclic 3',5'-adenosine monophosphate in the human leukocyte: synthesis, degradation, and effects on neutrophil candidacidal activity, J. Clin. Invest., 50:927. Reprinted with permission.

TABLE II

EVIDENCE FOR CYCLIC AMP AS A SECOND MESSENGER IN LEUKOCYTES

Cell type	First messenger	"Message"	Cellular content of cyclic AMP	Effect of first messenger plus methylxanthine	Effect of cyclic AMP (or analog)	Homogenous cell population
Neutrophil	PGE(1)	Inhibition of lysosomal enzyme release(a)	Not measured	Additive only	Appropriate	No
"	PGE(1)	Inhibition of candicacidal activity(b)	Discrepancy between relative effects of PGE(1) and theophylline on cyclic AMP and on candidacidal activity	Additive only	Appropriate	No
"	Phagocytic particle	?	contrasting results in different laboratories(c)	Not tested	---	No
Lymphocyte	PGE(1), catecholamines	Inhibition of transformation(d)	Increases appropriately	Not tested	Appropriate (at high concentration)	?

TABLE II (CONTINUED)

Lymphocyte	PHA	Stimulation of transformation(d)	Inappropriately high concentration of PHA for effect on cyclic AMP	Not tested	Marked species differences	?
"	PGE(1), catechol-amines, histamines	Inhibition of cytolytic activity(3)	Appropriate changes with agonists and specific antagonist (propranolol)	Not tested	Appropriate	?
Thymocyte	Epinephrine and many others	Mitogenic effect(f)	Measured with epinephrine only	Caffeine potentiates	Biphasic effect, dependent on concentration	?
Basophil/ Mast cell	Prostaglandins E(1),E(2) Beta adrenergic agents, Histamine	Inhibition of histamine release(g)	Appropriate changes with agonists and specific antagonist (propranolol)	Caffeine + theophylline potentiate	Appropriate	No

a. Weissman et al, 1971.
b. Bourne et al, 1971a.
c. Park et al, 1971; Stossel et al, 1970; Mangianello et al, 1971.
d. Smith et al, 1971a, 1971b; Bourne et al, 1971d; Cross and Ord, 1970.
e. Henney and Lichtenstein, 1971.
f. McManus and Whitfield, 1969a, 1969b; McManus et al, 1971; Whitfield et al, 1969, 1970a, 1970b, 1970c, 1970d, 1970e.
g. Lichtenstein and Margolis, 1968; Bourne et al, 1971b, 1971c.

ACKNOWLEDGEMENTS

The purified choleragen (effect shown in Fig. 5) was kindly supplied by Dr. Richard A. Finkelstein, Dallas, Texas.

REFERENCES

Assem, E. S. K., and Schild, H. O., 1969, Inhibition by sympathomimetic amines of histamine release induced by antigen in passively sensitized human lung, Nature (London) 224:1028.

Birnbaumer, L., and Rodbell, M., 1969, Adenylate cyclase in fat cells. II. Hormone receptors, J. Biol. Chem. 244:3477.

Bourne, H. R., and Melmon, K. L., 1971, Adenylate cyclase in human leukocytes: Evidence for activation by separate beta-adrenergic and prostaglandin receptors, J. Pharmacol. Exp. Ther. 178:1.

Bourne, H. R., Lehrer, R. I., Cline, M. J., and Melmon, K. L., 1971a, Cyclic 3',5'-adenosine monophosphate in the human leukocyte: Synthesis, degradation, and effects on neutrophil candidacidal activity, J. Clin. Invest. 50:920.

Bourne, H. R., Melmon, K. L., and Lichtenstein, L. M., 1971b, Histamine augments leukocyte adenosine 3',5'-monophosphate and blocks antigenic histamine release, Science (Washington) 173:743.

Bourne, H. R., Lichtenstein, L. M., and Melmon, K. L., 1971c, Pharmacologic control of allergic histamine release *in vitro*: Evidence for an inhibitory role of 3',5'-adenosine monophosphate in human leukocytes (submitted for publication).

Bourne, H. R., Epstein, L. B., and Melmon, K. L., 1971d, Lymphocyte cyclic adenosine monophosphate (AMP) synthesis and inhibition of phytohemagglutinin-induced transformation, J. Clin. Invest. 50:10a (abstract).

Butcher, R. W., and Baird, C. E., 1968, Effects of prostaglandins on adenosine 3',5'-monophosphate levels in fat and other tissues, J. Biol. Chem. 243:1713.

Cline, M. J., 1970, Leukocyte function in inflammation: The ingestion, killing, and digestion of microorganisms, Ser. Haematol. 3:3.

Craddock, C. G., Longmire, R., and McMillan, R., 1971, Lymphocytes and the immune response, New Eng. J. Med. 285:324.

Cross, M. E., and Ord, M. G., 1970, The transformation of lymphocytes by adenosine 3',5'-cyclic monophosphate, and consequent changes in histone microstructure, Biochem. J. 120:21.

Crunkhorn, P., and Willis, A. L., 1971, Cutaneous reactions to intradermal prostaglandins, Brit. J. Pharmacol. 41:49.

Dimitrov, N. V., Miller, J., and Ziegra, S. R., 1969, The effects of caffeine on glucose metabolism of polymorphonuclear leukocytes, J. Pharmacol. Exp. Ther. 168:240.

Donaldson, V. H., and Evans, R. R., 1963, A biochemical abnormality in hereditary angioneurotic edema: Absence of serum inhibitor of C'1-esterase, Am. J. Med. 35:37.

Field, M., 1971, Intestinal secretion: Effect of cyclic AMP and its role in cholera, New Eng. J. Med. 284:1137.

Gilman, A. G., 1970, A protein binding assay for adenosine 3',5'-cyclic monophosphate, Proc. Nat. Acad. Sci. 67:305.

Goldman, J. M., and Hadley, M. E., 1970, Cyclic AMP and adrenergic receptors in melanophore responses to methylxanthines, Eur. J. Pharmacol. 12:365.

Hadden, J. W., Hadden, E. M., and Good, R. A., 1971, Adrenergic mechanisms in human lymphocyte metabolism, Biochem. Biophys. Acta. 237:339.

Henney, C. S., and Lichtenstein, L. M., 1971, The role of cyclic AMP in the cytolytic activity of lymphocytes, J. Immunol. 107:610.

Hirschhorn, R., Grossman, J., and Weissmann, G., 1970, Effect of cyclic 3',5'-adenosine monophosphate on lymphocyte transformation, Proc. Soc. Exp. Biol. Med. 133:1361.

Ishizaka, T., Ishizaka, K., Orange, R. P., and Austen, K. F., 1971, The pharmacologic inhibition of the antigen induced release of histamine and slow reacting substance of anaphylaxis (SRS-A) from monkey lung tissues mediated by human IgE, J. Immunol. 106:1267.

Kimberg, D. V., Field, M., Johnson, J., Henderson, A., and Gushon, E., 1971, Stimulation of intestinal mucosal adenylate cyclase by cholera enterotoxin and prostaglandins, J. Clin. Invest. 50:1218.

Klein, I., and Levey, G. S., 1971, Activation of myocardial adenylate cyclase by histamine in guinea pig, cat, and human heart, J. Clin. Invest. 50:1012.

Koopman, W. J., Orange, R. P., and Austen, K. F., 1970, Immunochemical and biologic properties of rat IgE. III. Modulation of the IgE-mediated release of slow-reacting substance of anaphylaxis by agents influencing the level of cyclic 3',5'-adenosine monophosphate, J. Immunol. 105:1096.

Krishna, G., Weiss, B., and Brodie, B. B., 1968, A simple, sensitive method for the assay of adenylate cyclase, J. Pharmacol. Exp. Ther. 163:379.

Leahy, D. R., McLean, E. R., Jr., and Bonner, J. T., 1970, Evidence for cyclic-3',5'-adenosine monophosphate as chemotactic agent for polymorphonuclear leukocytes, Blood 36:52.

Lehrer, R. I., and Cline, M. J., 1969, Leukocyte myeloperoxidase deficiency and disseminated candidiasis: The role of myeloperoxidase in resistance to Candida infection, J. Clin. Invest. 48:1478.

Lichtenstein, L. M., and Margolis, S., 1968, Histamine release *in vitro*: inhibition by catecholamines and methylxanthines, Science (Washington) 161:902.

Lichtenstein, L. M., Norman, P. S., Winkerwerder, W. L., and Osler, A. G., 1966, *In vitro* studies of ragweed allergy: Changes in cellular and humoral activity associated with specific desensitization, J. Clin. Invest. 45:1126.

Lichtenstein, L. M., and Osler, A. G., 1964, Studies of the mechanisms of hypersensitivity phenomena. IX. Histamine release from human leukocytes by ragweed pollen antigen, J. Exp. Med. 120:507.

McKinney, G. R., and Lish, P. M., 1966, Interaction of beta-adrenergic blockade and certain vasodilators in dextian-induced rat paw edema, Proc. Soc. Exp. Biol. Med. 121:494.

McManus, J. P., and Whitfield, J. F., 1969a, Stimulation of DNA synthesis and mitotic activity of thymic lymphocytes by cyclic adenosine 3',5'-monophosphate, Exp. Cell. Res. 58:188.

McManus, J. P., and Whitfield, J. F., 1969b, Mediation of the mitogenic action of growth hormone by adenosine 3',5'-monophosphate (cyclic AMP) Proc. Soc. Exp. Biol. Med. 132:409.

McManus, J. P., Whitfield, J. F., and Yondale, T., 1971, Stimulation by epinephrine of adenylate cyclase activity, cyclic AMP formation, DNA synthesis, and cell proliferation in populations of rat thymic lymphocytes, J. Cell. Physiol. 77:103.

Makman, M. H., 1971, Properties of adenylate cyclase of lymphoid cells, Proc. Nat. Acad. Sci. 68:885.

Mangianello, V., Evans, W. H., Stossel, T. P., Mason, R. J., and Vaughan, M., 1972, The effect of polystyrene beads on cyclic AMP concentration in leukocytes, J. Clin. Invest. (in press).

Marquis, N. R., Becker, J. A., and Vigdahl, R. L., 1970, Platelet aggregation. III. An epinephrine induced decrease in cyclic AMP synthesis, Biochem. Biophys. Res. Comm. 39:783.

Marumo, F., and Edelman, I., 1971, Effects of Ca^{++} and prostaglandin E, on vasopressin activation of renal adenylate cyclase, J. Clin. Invest. 50:1613.

May, C. D., Lyman, M., and Alberto, R., 1970, Effects of compounds which inhibit lymphocyte stimulation on the utilization of glucose by leukocytes, J. Allergy 46:21.

Nakano, J., and Oliver, R. D., 1970, Effect of histamine and its derivatives on lipolysis in isolated rat fat cells, Arch. Int. Pharmacodynam. 186:339.

Novogrodsky, A., and Katchalski, E., 1970, Effect of phytohemagglutinin and prostaglandins on cyclic AMP synthesis in rat lymph node lymphocytes, Biochem. Biophys. Acta. 215:291.

Park, B. H., Good, R. A., Beck, N. P., and Davis, B. B., 1971, Concentration of cyclic adenosine 3',5'-monophosphate in human leukocytes during phagocytosis, Nature New Biol. 229:27.

Rasmussen, H., 1970, Cell communication, calcium ion, and cyclic adenosine monophosphate, Science (Washington) 170:404.

Rigas, D. A., and Tisdale, V. V., 1969, Bio-assay and dose-response of the mitogenic activity of the phytohemagglutinin of Phaseolus vulgaris, Experientia 25:399.

Rigby, P. G., and Ryan, W. L., 1970, The effect of cyclic AMP and related compounds on human lymphocyte transformation (HLT) stimulated by phytohemagglutinin, Rev. Europ. Etudes Clin. et Biol. 15:774.

Robison, G. A., Butcher, R. W., and Sutherland, E. W., 1967, Adenylate cyclase as an adrenergic receptor, Ann. N. Y. Acad. Sci. 139:703.

Scott, R. E., 1970, Effects of prostaglandins, epinephrine, and NaF on human leukocyte, platelet and liver adenylate cyclase, Blood 35:514.

Sheldon, J. M., Husted, J. R., and Lovell, R. G., 1951, Effect of isuprel on antigen-antibody and histamine skin reaction, Ann. of Allergy 9:45.

Shimizu, H., Daly, J. W., and Creveling, C. R., 1969, A radioisotopic method for measuring the formation of adenosine 3',5'-cyclic monophosphate in incubated slices of brain, J. Neurochem. 16:1609.

Smith, J. W., Steiner, A. L., Newberry, W. M., Jr., and Parker, C. W., 1971a, Cyclic adenosine 3',5'-monophosphate in human lymphocytes: Alterations after phytohemagglutinin stimulation, J. Clin. Invest. 50:432.

Smith, J. W., Steiner, A. L., and Parker, C. W., 1971b, Human lymphocyte metabolism: Effects of cyclic and noncyclic nucleotides on stimulation by phytohemagglutinin, J. Clin. Invest. 50:442.

Spector, W. G., and Willoughby, D. A., 1965, Chemical mediators. III. in The Inflammatory Process (Zweifach, B. W., Grant, L., and McCluskey, R. T., eds.), pp. 427-448, Academic Press, New York.

Stossel, T. P., Murad, F., Mason, R. J., and Vaughan, M., 1970, Regulation of glycogen in polymorphonuclear leukocytes, J. Biol. Chem. 245:6228.

Stossel, T. P., Polland, T. D., Mason, R. J., and Vaughan, M., 1971, Isolation and properties of phagocytic vesicles from polymorphonuclear leukocytes, J. Clin. Invest. 50:1745.

Sutherland, E. W., Robison, G. A., and Butcher, R. W., 1968, Some aspects of the biological role of adenosine 3',5'-monophosphate (cyclic AMP), Circulation 37:279.

Szentivanyi, A., 1968, The beta-adrenergic theory of the atopic abnormality in bronchial asthma, J. Allergy 42:203.

Turtle, J. R., and Kipnis, D. M., 1967, An adrenergic receptor mechanism for the control of cyclic 3',5'-adenosine monophosphate synthesis in tissues, Biochem. Biophys. Res. Comm. 28:797.

Weissmann, G., Dukor, P., and Zurier, R. B., 1971, Effect of cyclic AMP on release of lysosomal enzymes from phagocytes, Nature New Biol. 231:131.

Whitfield, J. F., Perris, A. D., and Yondale, T., 1969, The calcium-mediated promotion of mitotic activity in rat thymocyte populations by growth hormone, neurohormones, parathyroid hormone, and prolactin, J. Cell. Physiol. 73:203.

Whitfield, J. F., McManus, J. P., and Gillan, D. J., 1970a, Cyclic AMP mediation of bradykinin-induced stimulation of mitotic activity and DNA synthesis in thymocytes, Proc. Soc. Exp. Biol. Med. 133:1270.

Whitfield, J. F., McManus, J. P., and Rixon, R. H., 1970b, Cyclic AMP-mediated stimulation of thymocyte proliferation by low concentrations of cortisol, Proc. Soc. Exp. Biol. Med. 134:1170.

Whitfield, J. F., McManus, J. P., and Rixon, R. H., 1970c, Potentiation by antidiuretic hormone (vasopressin) of the ability of parathyroid hormone to stimulate the proliferation of rat thymic lymphocytes, Horm. Metab. Res. 2:233.

Whitfield, J. F., McManus, J. P., and Rixon, R. H., 1970d, The possible mediation by cyclic AMP of parathyroid hormone-induced stimulation of mitotic activity and deoxyribonucleic acid synthesis in rat thymic lymphocytes, J. Cell. Physiol. 75:213.

Whitfield, J. F., McManus, J. P., and Gillan, D. J., 1971e, The possible mediation by cyclic AMP of the stimulation of thymocyte proliferation by vasopressin and the inhibition of this mitogenic action by thyrocalcitonin, J. Cell. Physiol. 76:65.

Willis, A. L., 1969, Parallel assay of prostaglandin-like activity in rat inflammatory exudate by means of cascade superfusion, J. Pharm. Pharmacol. 21:126.

Wolfe, S. M., and Shulman, N. R., 1969, Adenylate cyclase activity in human platelets, Biochem. Biophys. Res. Commun. 35:265.

DISCUSSION

YATES: I would like to caution against depending upon changes in cyclic AMP content of tissues as evidence that the cyclic AMP system was activated or inhibited. Not only is there the obvious difficulty that tissue content of any substance is the resultant of at least two processes, but turnover might be changing even if content isn't. Park has been studying the response of perfused rat livers to glucagon, with respect to glucose secretion. That system involves cyclic AMP activation, yet the cyclic AMP content in the liver samples doesn't change appreciably at a time when glucagon is producing its effect. The way you see that cyclic AMP is involved is that it washes out in hepatic venous blood. I realize that it's something of a cliche to call attention to compartmentalization, but it seems to me that at least some of the time you are going to get false negatives if you rely heavily on change in content of cyclic AMP as indication that it is acting as "second messenger". Sometimes the content will not change, but there will be a rearrangement or redistribution within the cell or a change in turnover, and cyclic AMP may then indeed be acting as a second messenger, without your knowing it.

BOURNE: I think it's good to bring ghosts from the closet now and then. This could be a ghost in the closet unless we can test it. We can measure the cyclic AMP in the supernatant fluid and have done so. We find that, in the unstimulated situation, about 10% or the total cyclic AMP in solution will be outside of the cell and 90% in the leukocytes. The more effective the stimulation of adenylate cyclase, the more comes out percentage-wise, so that the experiments with prostaglandins show falsely low concentrations of cyclic AMP by about 30%. The supernatant would have about half as much as the cell in that particular experiment. We have looked at compounds that stimulate cyclic AMP to a much lesser extent than prostaglandins, and there isn't much leak of the cyclic AMP outside the cell. It's also entirely possible that a compound could activate the phosphodiesterase and the cyclase at the same time and could not produce a net change in cyclic AMP. At the moment, aside from looking at phosphodiesterase inhibitors in combination with the adenylate cyclase stimulators, there is no good way of testing this possibility.

GOTH: The long latent period before choleragen works in your system shows up very beautifully also in the inflammatory models, as we have shown in our laboratory. In addition it is blocked by cycloheximide. Have you considered the possibility that in your system cycloheximide might work and the effect of choleragen is on the synthesis of something rather than on the enzyme directly?

BOURNE: Yes. I am glad you asked that. I didn't have time to include the fact that we have been interested in how choleragen could produce this delayed effect on cyclic AMP. In the gut the effect on secretion is blocked by cycloheximide, but it has not been shown that the effect on cyclic AMP is blocked by cycloheximide. Presumably there the cyclic AMP itself may do something to protein synthesis which produces the message. In the leukocytes, cycloheximide does not prevent the effect of choleragen. Interestingly the choleragen appears to stimulate adenylate cyclase activity in a way enterely different from the prostaglandins or other drugs. We have sonicated the cells at the end of 1.5 hr incubation. If they were previously incubated with PGE(1), after washing, these sonicates would not show any increase in adenylate cyclase activity. On the other hand, when the cells have been exposed to choleragen, the basal adenylate cyclase activity is increased tremendously, but the sodium fluoride stimulated activity is not increased. I think that means the choleragen is doing something directly to the cyclase. We have preliminary evidence that it does so very early in the game, even though the effect occurs later. In a preliminary experiment, if we wash the choleragen off after 10 min and let the cells incubate for another 80 min, then cyclic AMP goes up willy nilly. It looks as if choleragen is acting very differently from the other drugs.

EDELMAN: In the case of stimulation of renal cyclic AMP production by vasopressin, cyclic AMP can be detected in the urine. It is prabable that adenylate cyclase is bound to the plasma membrane and faces inward, and that cyclic AMP is produced on the membrane and released into the cell. Thus, the plasma membrane has "sidedness". The rate of cyclic AMP escape from the cell should be determined by the transmembrane concentration gradient. If there is no transport system for cyclic AMP, there should be no sidedness with respect to the relationship between gradient and movement across the cell membrane. The intracellular concentrations of cyclic AMP which have generally been considered to be effective physiologically are about 10^{-7} M. The concentrations of cyclic AMP in isolated systems which are required to produce a discrete physiological response is 10^{-3} to 10^{-2} M. These results suggest that concentration gradients between three and five orders of magnitude are required to get effective intracellular concentrations, and yet in the experiments you cited, cyclic AMP appears in the medium at concentrations that are approximately half those that are estimated to be present in the cell. There appears to be something very peculiar about the ability of cyclic AMP to escape from cells, compared to the high degree of hindrance to entry into cells. One possible explanation for this apparent discrepancy is that cyclic AMP generated inside of cells may change the permeability of the cell membrane to cyclic AMP. If this were the case, addition of cyclic AMP to the outside solution should facilitate the entry of cyclic AMP, i.e. the response should be cooperative. I am curious to hear if there is any explanation for what appears to be the degree of sidedness with respect to escape from or entry of cyclic AMP into cells.

BOURNE: Another way of testing this would be to expose the cells to radioactive cyclic AMP in the presence of a compound which tremendously stimulates endogenous cyclic AMP production. In fat cells the stimulation of endogenous cyclic AMP production facilitates entry of exogenous cyclic AMP into the cells (P. Schonhofer, personal communication). That is a more direct way of looking at the entry.

KORNBERG: Do you have any information about the localization of adenylate cyclase among the membranes of the cell?

BOURNE: We have steered away from that, primarily because of the tremendous difficulty of breaking up leukocytes and making sure that you really have a fraction that means something, other than the lysosomes.

KORNBERG: Have people isolated nuclei, mitochondria and other organelles, and tested among them for the cyclase and diesterases?

BOURNE: That's been done to some extent in lymphocytes and thymocytes, and in general they have been associated with fractions that were called a membrane fraction.

KORNBERG: Plasma membranes?

BOURNE: Plasma membranes, right.

<u>KORNBERG</u>: As discrete from nuclear membranes or lysosomal membranes?

<u>BOURNE</u>: Yes, but I can't speak for whether those are valid studies or not. That's not an area that I feel competent in. Is there somebody here who can?

<u>RAMWELL</u>: Oscar Hechter has made the claim that he found a nuclear adenylate cyclase.

<u>KORNBERG</u>: Well, it seems to me important to know that. With respect to what Dr. Edelman was saying, it doesn't surprise me that the concentration of a given effector required for a pure enzyme would be quite different from that enzyme fixed in some membranous location. Various inhibitors of mitochondrial electron transport or oxidative phosphorylation may not act at all upon an isolated mitochondrial component, whereas their effects are striking when that component is localized within the membrane. It seems to me, then, that much would be learned about many of these cellular responses by studying membrane fractions from a homogeneous population of cells. It would be important to know, then, how the vesicles from various parts of cells respond with respect to the Sutherland postulates.

<u>BOURNE</u>: There is another kind of compartmentalization that may be important, which I think Dr. Weissmann is going to talk about. If you measure cyclic AMP concentration in virtually any cell, and express them per cell water, they come out to be around 10^{-6} M. But most of the protein kinases are far more than fully activated at 10^{-6} M cyclic AMP. This suggests that a good portion of what you actually measure in the basal unstimulated state is not available to the protein kinase, if the protein kinase is important. So that's another thing which eventually is going to have to be looked at. When we can examine specific fractions of cells, we may find that there is some way of sequestering some of that cyclic AMP from its effector within the cell.

EFFECT OF PROSTAGLANDINS UPON ENZYME RELEASE FROM LYSOSOMES AND EXPERIMENTAL ARTHRITIS

R. B. Zurier and G. Weissmann

Departments of Medicine
Cell Biology and Genetics
New York University School of Medicine, N. Y.

EFFECT OF PROSTAGLANDINS ON LYSOSOMAL ENZYME RELEASE FROM HUMAN LEUKOCYTES

Cells engaged in the uptake of particles become the center of acute inflammatory reactions (Metchnikoff, 1905). This phenomenon has been attributed, in part, to the release from phagocytic cells of enzymes previously sequestered within lysosomes (Weissmann, 1967). Exposure of joints to lysosomal extracts results in inflammation and cartilage degradation (Weissmann et al, 1969). Therefore, inhibition of enzyme release from cells which participate in the inflammatory response may modify the inflammation and subsequent tissue injury.

Lysosomes of phagocytic cells constitute part of the vacuolar system and their shuttle and flow has been compared to the kind of directed intracellular traffic observed in secretory cells (deDuve and Wattiaux, 1966). Consequently it appeared reasonable that substances such as cyclic AMP which regulate the flow of stored and exportable proteins in such tissues as pancreas, salivary gland or thyroid might also act to regulate granule flow and merger in phagocytic cells. Cyclic nucleotides, theophylline, and PGE(1) have been implicated in the regulation of flow of subcellular organelles in a variety of systems (Dingle, 1968), and may interact with microtubule protein.

We therefore undertook studies of the mechanisms of enzyme release from phagocytes and investigated the pharmacological control of enzyme release.

We have previously documented the selective release of lysosomal enzymes from human leukocytes exposed to zymosan or to particles formed by complexing rheumatoid factor with heat denatured gamma-globulin (Weissmann et al, 1971a and 1971b). Enzyme release is selective in that substantial proportions of total beta-glucuronidase are discharged to the medium. During this time there is no significant escape from cells of the cytoplasmic enzyme lactic dehydrogenase (LDH). It is clear that cells remain viable since cell death can be monitored readily by striking increases in LDH release such as those observed after 6 hr incubation and by the increase in the number of cells which lose the ability to exclude the dye eosin Y (Fig. 1). Enzyme release under these conditions, then, is not associated with generalized membrane damage. It is possible that enzyme release occurs by a mechanism which has been termed "regurgitation during feeding" (Weissmann and Dukor, 1970), whereby lysosomal enzymes escape from phagocytic vacuoles through channels created by virtue of incomplete fusion of vacuolar membrane with plasma membrane. Electron microscopic images consistent with this hypothesis have, in fact, been presented (Zucker-Franklin and Hirsch, 1964; Karnovsky et al, 1970).

We investigated the effects of agents which increase intracellular levels of cyclic AMP on enzyme release. Human peripheral blood leukocytes were obtained from freshly drawn heparinized blood after sedimentation of the red cells with dextran. The white cells were washed once and resuspended in buffered saline at pH 7.3. Cells were preincubated at 37°C with compounds for one hr prior to exposure to particles.

Cyclic AMP by itself had no effect on zymosan induced extrusion of lysosomal enzymes (Table I). However, in combination with theophylline cyclic AMP caused a considerable reduction of enzyme release, though not of particle uptake. Theophylline alone was less effective. PGE(1) which has been shown to raise the level of intracellular cyclic AMP in human peripheral blood leukocytes to 15 nanoM/10^6 WBC (Scott, 1970), also blocked selective hydrolase release. When theophylline was added during preincubation with PGE(1) it acted further to reduce hydrolase release. The effect on enzyme release of PGA(2) is somewhat greater than the PGE(1) effect (Table I). Similar results are obtained following phagocytosis

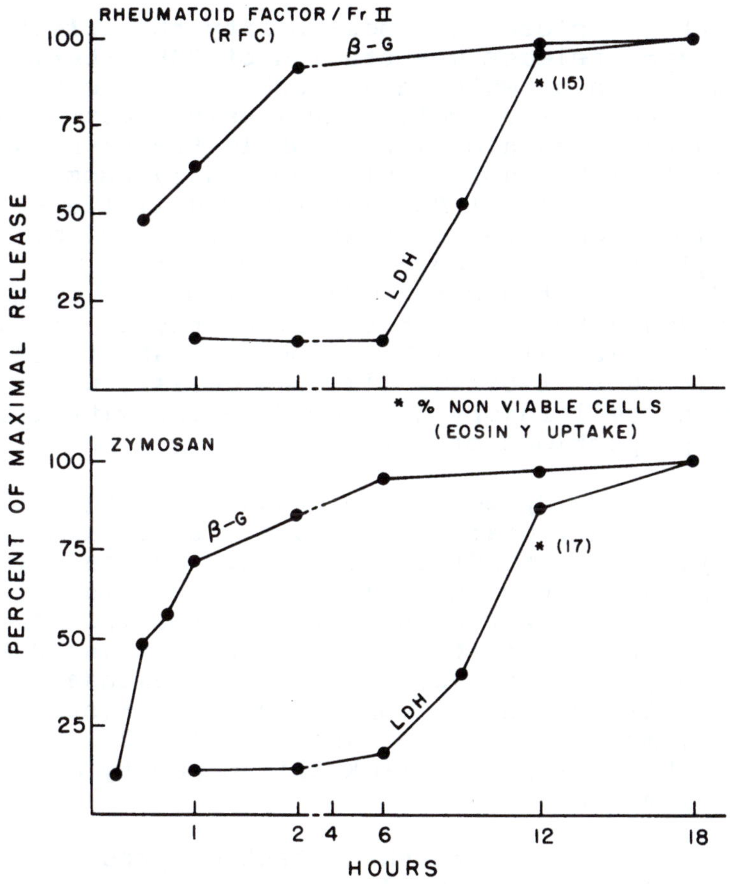

Figure 1. Enzyme release with respect to time. PMN (2 x 10^6) incubated at 37°C with RF-algG (RFC) or with zymosan. Maximal beta-glucuronidase (beta-G) release (100% at 18 hr) = 3.25 micrograms phenolphthalein/hr per 2 x 10^6 PMN from cells exposed to RFC and 5.36 micrograms phenolphthalein/hr per 2 x 10^6 PMN from cells exposed to zymosan. Maximal LDH release (100% at 18 hr) = 43.4 absorbancy units (A.U.) from cells exposed to RFC and 59.2 A.U. from cells exposed to zymosan.

of an immune complex (Table II). Dibutyryl-cyclic AMP also reduces enzyme release. PGE(1) and PGA(2) retarded hydrolase release from phagocytes at concentrations as low as 2.8 x 10^{-6} M (Table III). It therefore appeared that compounds which increase the intracellular concentration of cyclic AMP reduce selective extrusion of lysosomal acid hydrolase after particle ingestion.

To further explore this possibility the effect on lysosomal enzyme release was studied of PGF(1-beta) and PGF(2-alpha), both of which have much less effect than PGE(1) on intracellular cyclic AMP levels (Scott, 1970; Bourne, personal communication). While treatment with 100 microgram/ml PGE(1) consistently reduced lysosomal enzyme release, 100 microgram/ml PGF(1-beta) had little effect, and PGF(2-alpha) enhanced release of enzymes (Table IV). When theophylline was added together with the prostaglandins it acted with PGE(1) further to reduce hydrolase release but did not significantly potentiate the effects of PGF(1-beta) or PGF(2-alpha). PGE(1) did not cause loss of cell viability (1% dead cells), while PGF(2-alpha) induced cell injury (17% dead cells) with leak of substantial proportions of total cytoplasmic LDH.

PGE(2) and PGB(2) were also studied. PGE(2) has been reported to be "inflammatory" and affected by aspirin (Vane, 1971). Neither PGE(2) nor PGB(2) reduced hydrolase release from phagocytic cells (Table IV). Mild cell injury (5% dead cells) was inflicted by PGE(2). The data suggest that certain prostaglandins - PGE(1) and PGA(2) - can selectively inhibit lysosomal enzyme release from phagocytic cells and may be considered "anti-inflammatory" whereas "inflammatory" prostaglandins - PGE(2), PGB(2), PGF(1-beta), PGF(2-alpha) - lack this capacity and may injure cell membranes.

To further investigate the effect of prostaglandins on membranes we have done preliminary studies with isolated lysosomes. Rabbit liver lysosomes were obtained as previously described (Weissmann et al, 1963). The granules were incubated <u>in vitro</u> at 37°C for 90 min with prostaglandins, and the activity of beta-glucuronidase liberated from sedimentable granules into supernatant (20,000 x g for 10 min) was determined. This technique measures gross changes in the permeability of the lysosomal membranes to the contained enzymes. Agents that facilitate the release of enzymes from lysosomes have been termed "labilizers" and agents which retard release of enzymes are considered "stabilizers" of the lysosomal membrane (Weissmann, 1964). As seen in Table V, the majority of prostaglandins tested had little or no effect on enzyme release. PGE(2) had a significant, but apparently non-dose-dependent, stabilizing effect. It is clear that under these conditions (incubation in 0.34 M sucrose) the prostaglandins do not perturb the lysosomal membrane, but the significance of the PGE(2) effect is unknown at present. Further studies of the effect on lysosomes suspended in other media of prostaglandins over a wider dose range, are necessary.

The exact means are unclear whereby cyclic nucleotides, PGE(1) and PGA(2), and theophylline block the release of lysosomal hydrolases from PMN which have been exposed to zymosan or to immune complexes. We could not detect, by light microscopy, interference with particle uptake. In many cell types cyclic nucleotides modify processes thought to involve microtubules (Weissmann et al, 1971a). In fact, microtubule proteins have been found to be substrates for specific cyclic AMP activated protein kinases (Goodman et al, 1970). Since intact microtubules appear necessary for lysosomal movement (Freed and Lebowitz, 1970), it is possible that cyclic AMP acts to regulate the transit of lysosomes to heterophagic vacuoles or to the cell periphery. To more clearly demonstrate the relative effect on the cyclic AMP system of the various prostaglandins it will of course be necessary to quantitate the level within subpopulations of phagocytic cells of the nucleotides, adenylate cyclase, and protein kinase, since previous studies have not distinguished between neutrophils, eosinophils, monocytes, platelets or basophils (Bourne et al, 1971).

EFFECT OF PROSTAGLANDINS ON ADJUVANT ARTHRITIS

The documentation *in vitro* that PGE(1) and PGA(2) reduce release of lysosomal hydrolases from phagocytic cells, and evidence (Aspinall and Cammarata, 1969) that PGE(2) reduces swelling of the tibiotarsal joint in rats with adjuvant arthritis, suggested that prostaglandins might modify the inflammatory response. Adjuvant disease in the rat includes a severe and persistent polyarthritis which appears 10 to 14 days after a single intradermal injection of Freund's complete adjuvant (Pearson and Wood, 1959), and provides a convenient experimental model for *in vivo* evaluation of anti-inflammatory and/or immunosuppressive drugs. It has been observed (Zurier and Quagliata, in press) that a 21 day course of PGE(1) (500 micrograms b.i.d., s.c.) prevents and suppresses adjuvant arthritis while the same treatment with PGA(2) has no effect on the disease. Although PGE(1) treatment had no effect on delayed hypersensitivity to tuberculin purified protein derivative (PPD), it did effect a reduction in the antibody response to injected sheep red blood cells. Arthritis was prevented when rats were treated from day of adjuvant injection (day 0) or from day 7. When treatment began on day 14 the typical explosive course of the

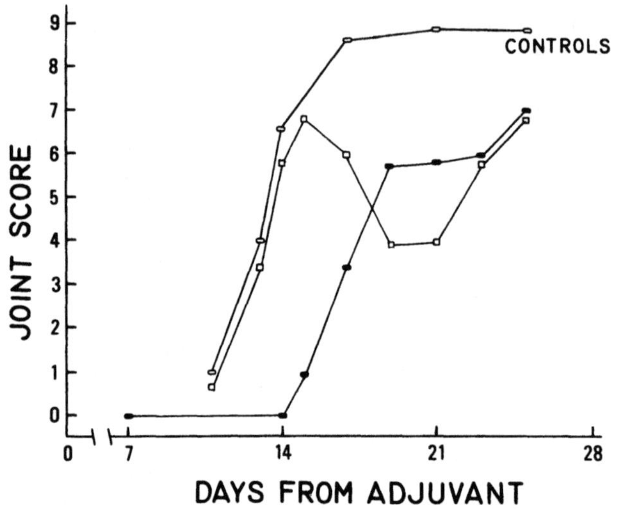

Figure 2. Effect of prostaglandin on adjuvant arthritis. Each point represents the mean arthritic score for the three uninjected paws (5 rats/group).
• PGE(1) 500 micrograms s.c. b.i.d. days 7-13.
□ PGE(1) 500 micrograms s.c. b.i.d. days 14-20.

arthritis, which had already started, was suppressed. Established inflammation was considerably reduced even when treatment began on day 21. Similar results have been obtained with PGE(2) treatment.

Seven days of treatment with PGE(1) is much less effective in suppressing arthritis than the 21 days of treatment (Fig. 2). If treatment is begun on day 0 (not shown) or day 7, onset of disease is delayed but severe arthritis subsequently develops. When PGE(1) is begun on day 14 there follows suppression of disease during the period of treatment and recrudescence of arthritis after PGE(1) is stopped. The pattern of response suggests an "anti-inflammatory" rather than an "immunosuppressive" action, though a partial effect on the humoral immune response cannot be excluded. We have compared this disease suppressive effect of PGE(1) (treatment from day 14 through day 20) with the effect of treatment with prostaglandins F(1-beta), F(2-alpha), and B(2). Adjuvant disease was induced in adult Holtzman male rats by injection in a hind paw of 600 micrograms killed Mycobacterium tuberculosis in mineral oil. All rats developed moderately severe polyarthritis by day 14 at which time they received either 500 micrograms PGE(1), PGF(1-beta), PGF(2-alpha), PGB(2), or buffer alone subcutaneously twice daily for 7 consecutive days. Whereas PGE(1) suppressed arthritis during the treatment period, PGF(1-beta), PGF(2-alpha), and PGB(2) had little or no effect (Fig. 3).

The mechanisms by which PGE(1) suppresses adjuvant arthritis are unknown but the biologic effects of prostaglandins suggest a number of possibilities. A stimulatory effect on the pituitary-adrenal axis is suggested by the finding that PGE(1) increases corticosteroidogenesis in perfused rat adrenal glands (Flack et al, 1969), and by the fact that PGE(1) stimulates release of pituitary ACTH in intact rats (deWied et al, 1969). In addition, it has been shown that PGE(1), but not PGA(1), stimulates beef adrenal steroidogenesis _in vitro_. However, although PGE(1) caused an elevation of serum corticosterone levels in normal rats, it had no effect on the corticosterone level in rats with adjuvant arthritis (Table VI). Delayed hypersensitivity to PPD was not suppressed in PGE(1) treated rats, but the reduction in anti-sheep red blood cell antibody titer suggests that an effect of PGE(1) on a restricted population of lymphocytes might contribute to suppression of adjuvant disease. Lymphocyte populations have cytolytic activity toward cells bearing antigenic determinants against which the lymphocytes are specifically sensitized (Rosenau, 1963). Lymphocytes - presumably sensitized - from rats with adjuvant arthritis, when injected intravenously to syngenic rats at the time of adjuvant injection, penetrate into the injected paw. Lymphocytes from normal rats do not enter the inflamed paw (Berney et al, 1971). It appears that flow of sensitized lymphocytes through the area of high antigen concentration may help mediate the persisting inflammation. Henney and Lichtenstein (1971) have recently shown that the lymphocyte-mediated cytolysis of target cells is impaired by drugs which increase intracellular cyclic AMP levels. Moreover, cyclic AMP, its diburyryl derivative, theophylline and PGE(1) prevent phytohemagglutinin induced transformation of lymphocytes (Hirschhorn et al, 1970; Smith et al, 1971). It has been shown (Weissmann et al, 1970) that PGE(1) reduces uptake and subsequent degradation of aggregated bovine serum albumin by mouse macrophages. It is possible that in the rat PGE(1) might interfere with uptake and/or processing of antigen by macrophages. The protection which PGE(1) affords against adjuvant arthritis might be due, in part, to its influence in reducing release of lysosomal enzyme from inflammatory and synovial lining cells. However, PGA(2) which also blocks lysosomal hydrolase release _in vitro_, does not suppress adjuvant

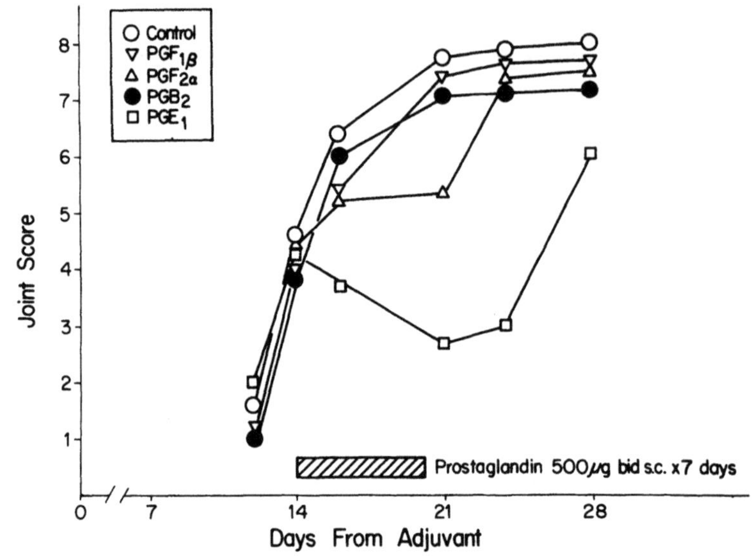

Figure 3. Effect of prostaglandin on adjuvant arthritis. Each point represents the mean arthritic score for the three uninjected paws (5 rats/group).

arthritis, and PGE(2), which does not affect enzyme release from phagocytic cells does ameliorate adjuvant disease. It is clear, then, that the slight molecular differences between prostaglandin classes have profound effects on their biologic activities. Since, of the prostaglandins studied, only PGE(1) and PGE(2) suppress adjuvant arthritis, and since changes in the parent structure generally result in loss of biologic activity, whatever prostaglandin effect is responsible for suppression of adjuvant disease appears dependent - structurally - on the combination of a ketone function at C9 and absence of double bonds in the cyclopentane ring. Koopman et al (1971) have demonstrated that PGE(1) and PGE(2) were ten times as potent as PGA(1) and PGA(2) in their ability to inhibit the IgE mediated release of SRS-A in rats, and that PGF(1-alpha), PGF(2-alpha), PGF(2-beta), and PGB(1) were inactive.

EFFECT OF BEE VENOM ON ADJUVANT ARTHRITIS

Stimulation of cells and tissues, whether by mechanical, hormonal, or neurologic means, results in increased biosynthesis of the prostaglandins (Ramwell and Shaw, 1970). It has been suggested that during endocytosis phospholipases are freed from the lysosomes of

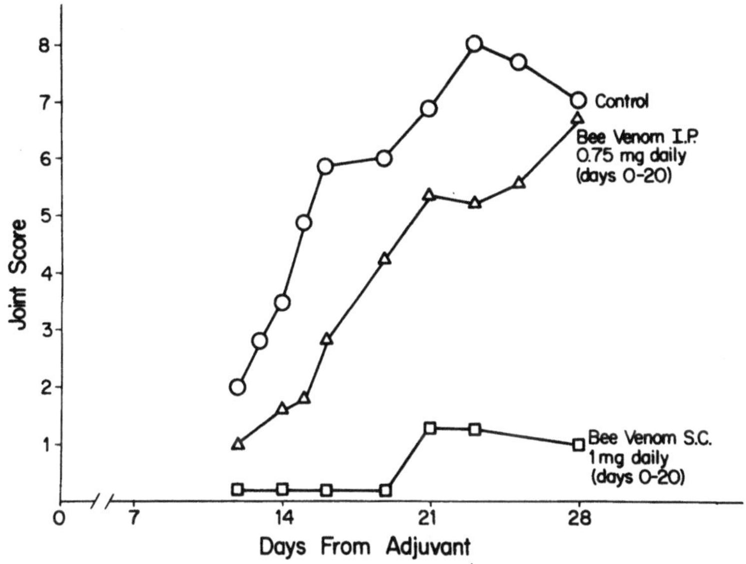

Figure 4. Effect of bee venom on adjuvant arthritis. Each point represents the mean arthritic score for the three uninjected paws (10 rats/group).

phagocytes, and that these attack cell membrane phospholipid to yield arachidonic acid which is converted to prostaglandin by freely available tissue enzymes (Anderson et al, 1971). Phospholipase is one of the most active components of a number of animal venoms. Bee venom (Eliasson, 1959), and snake venom (Vogt et al, 1969), both containing phospholipase A, release prostaglandins from tissues. It was therefore considered that injections of venom to rats with adjuvant arthritis might have an effect similar to prostaglandin treatment. Because of its venerable position in the medical and folk lore of Europe and Asia we chose to use the venom of the honey bee (*Apis mellifera*). It has been used by no less eminent physicians than Hippocrates and Galen and is said to be most effective in treatment of the rheumatic diseases. It is widely used even at the present time in many regions of the U.S.S.R. (Zaitzev and Poriadin, 1964).

The effect on adjuvant arthritis of a 21-day course (day 0 through day 20) of whole bee venom (4 mg/Kg s.c.) is documented in Fig. 4. Suppression of disease is similar to that obtained during PGE(1) treatment. It has been demonstrated (Couch and Benton, 1971) that comparable s.c. doses of bee venom cause a significant increase in the serum corticosterone level within the first hr after injection with return to normal by the fourth hr, whereas

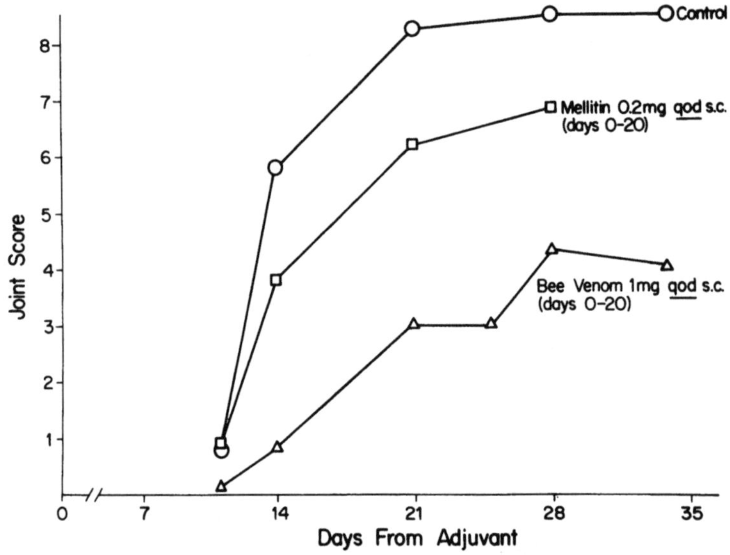

Figure 5. Effect of bee venom and melittin on adjuvant arthritis. Each point represents the mean arthritic score for the three uninjected paws (10 rats/group).

a similar dose injected i.p. caused a sustained elevation of the corticosterone level over a greater than 8 hr period. Administration of bee venom i.p. was not effective in adjuvant arthritis (Fig. 4) suggesting that disease suppression was not mediated through stimulation of the adrenal cortex. In addition, we were not able to demonstrate an elevated corticosterone level following bee venom injection to adjuvant arthritic rats (Table VI). However, we may have missed an elevation during the first hour. The same is true of our findings following prostaglandin injection in adjuvant arthritis and this point wants clarification.

Now, this validation of folk wisdom in an experimental system may or may not be related tc our observations with prostaglandins. Bee venom contains such other biologically active substances as melittin, apamin, and hyaluronidase, each of which needs to be studied in isolation. Our earliest studies in this regard indicate that melittin, the amphipathic polypeptide which disrupts lipid membranes (Sessa et al, 1969), has little effect upon adjuvant arthritis (Fig. 5) and we are consequently determining the action of purified apamine and bee venom phospholipase A. Whatever the exact mechanism proves to be whereby injections of bee venom affect experimental arthritis in a fashion similar to prostaglandins, the latter agents may prove to be the mediators of a wide variety of noxious and beneficial effects of naturally occurring venoms.

SUMMARY

Selective release of lysosomal materials from living cells has been documented when polymorphonuclear leukocytes (PMN) ingest particles. Agents which elevate intracellular levels of cyclic AMP in leukocytes reduce lysosomal hydrolase release from phagocytic cells. The effect on enzyme release of prostaglandins was studied. Human leukocytes were incubated with several prostaglandins prior to exposure to particles (zymosan or immune complexes). PGE(1) and PGA(2) reduce lysosomal enzyme release, but PGE(2), PGF(1-beta), and PGB(2) do not affect enzyme release. PGF(2-alpha) caused increased cytoplasmic enzyme release and cell death.

The effect on adjuvant arthritis of several prostaglandins was determined. PG was administered for 7 days (days 14-20 after adjuvant injection). PGE(1) suppressed arthritis during the treatment period but there was recrudescense of arthritis when treatment was stopped. PGF(2-alpha), PGF(1-beta), and PGB(2) had little or no effect on the arthritis.

The effect on adjuvant arthritis of bee venom, a phospholipase-A-containing venom which releases prostaglandins from tissues, was studied. Bee venom, given to rats for 21 days from day of adjuvant injection, effectively suppressed arthritis.

Possible mechanisms to account for the observed results are discussed. The data suggest that certain prostaglandins may be useful agents in controlling the inflammatory response.

TABLE I

INHIBITION OF ENZYME RELEASE FROM HUMAN LEUKOCYTES EXPOSED TO ZYMOSAN*

Compound	Concentration (M)	N	Mean % of Control Beta-Glucuronidase Release
None	(Control)	10	100.0
cyclic AMP	10^{-3}	5	95.5
Theophylline	10^{-3}	5	69.2
cyclic AMP +	10^{-3}		
Theophylline	10^{-3}	5	36.2
PGE(1)	2.8×10^{-4}	5	67.4
PGE(1) +	2.8×10^{-4}		
Theophylline	10^{-3}	5	46.1
PGA(2)	2.8×10^{-4}	4	39.0
PGA(2) +	2.8×10^{-4}		
Theophylline	10^{-3}	2	43.2

*1 hr incubation with drug alone followed by 1 hr incubation with drug and 10^7 zymosan particles.

TABLE II

INHIBITION OF ENZYME RELEASE FROM HUMAN LEUKOCYTES EXPOSED TO IMMUNE COMPLEX*

Compound	Concentration (M)	N	Mean % of Control Beta-Glucuronidase Release
None	(Control)	7	100.0
PGE(1)	2.8×10^{-4}	7	56.3
Theophylline	10^{-3}	7	54.9
PGE(1) +	2.8×10^{-4}		
Theophylline	10^{-3}	5	36.2
PGA(2)	2.8×10^{-4}	2	40.0
PGA(2) +	2.8×10^{-4}		
Theophylline	10^{-3}	2	33.0
Dibutyryl-cyclic AMP	10^{-3}	7	51.4

*1 hr incubation with drug alone followed by 1 hr incubation with 0.5 ml aggregated IgG-Rheumatoid Factor complex.

TABLE III

INHIBITION OF ENZYME RELEASE FROM HUMAN LEUKOCYTES*

Compound	Concentration (M)	Mean % of Control Beta-Glucuronidase Release
None	(Control)	100.0
PGE(1)	2.8×10^{-4}	65.9
PGE(1)	2.8×10^{-5}	77.5
PGE(1)	2.8×10^{-6}	83.5
PGA(2)	2.8×10^{-4}	41.0
PGA(2)	2.8×10^{-5}	73.2
PGA(2)	2.8×10^{-6}	80.8

*1 hr incubation with drug followed by 1 hr incubation with 10^7 zymosan particles.

TABLE IV

INHIBITION OF ENZYME RELEASE FROM HUMAN LEUKOCYTES*

Compound	Concentration (M)	N	Mean % of Control Beta-Glucoronidase Release
None	(Control)	10	100.0
PGE(1)	2.8×10^{-4}	5	68.2
Theophylline	10^{-3}	5	67.9
PGE(1) +	2.8×10^{-4}		
Theophylline	10^{-3}	5	49.1
PGF(1-beta)	2.8×10^{-4}	6	88.6
PGF(1-beta) +	2.8×10^{-4}		
Theophylline	10^{-3}	6	64.7
PGF(2-alpha)	2.8×10^{-4}	5	135.8
PGF(2-alpha) +	2.8×10^{-4}		
Theophylline	10^{-3}	5	125.1
PGE(2)	2.8×10^{-4}	6	89.8
PGE(2) +	2.8×10^{-4}		
Theophylline	10^{-3}	6	65.6
PGB(2)	2.8×10^{-4}	6	88.2
PGB(2) +	2.8×10^{-4}		
Theophylline	10^{-3}	6	66.2

*1 hr incubation with drug alone followed by 1 hr incubation with 10^7 zymosan particles.

TABLE V

EFFECT OF PROSTAGLANDINS ON ISOLATED LYSOSOMES*

Concentration (M)	Beta-Glucuronidase Release (% of Controls) PGE(1)	PGE(2)	PGF(2-alpha)	PGA(2)	PGB(2)
None (Control)	100	100	100	100	100
2.8×10^{-4}	108	71	117	108	118
2.8×10^{-5}	102	69	100	102	114
2.8×10^{-6}	105	96	108	111	112
2.8×10^{-7}	100	77	103	103	98

*Rabbit liver lysosomes in 0.34 M sucrose incubated with prostaglandins at 37°C for 90 min.

TABLE VI

EFFECT OF PROSTAGLANDIN E(1) AND BEE VENOM ON SERUM CORTICOSTERONE LEVELS IN NORMAL AND ADJUVANT ARTHRITIC RATS

Condition of Rat	Drug & Route of Injection	Mean & Range Serum Corticosterone* (microgram/100 ml)
Normal	Saline s.c.	13.3 (10-18)
Adjuvant Arthritis	Saline s.c.	34.3 (31-39)
Normal	PGE(1) 500 micrograms s.c.	96.0 (93-98)
Adjuvant Arthritis	PGE(1) 500 micrograms s.c.	39.5 (36-40)
Normal**	Whole Bee Venom 4 mg/Kg s.c.	1 hr 50.7±10.5
	Whole Bee Venom 4 mg/Kg s.c.	4 hr 9.9±1.8
	Whole Bee Venom 4 mg/Kg i.p.	1 hr 49.4±6.5
	Whole Bee Venom 4 mg/Kg i.p.	4 hr 22.1±4.0
Adjuvant Arthritis	Whole Bee Venom 4 mg/Kg s.c.	23.4 (16-43)

*3 hr following s.c. injection of drug.

**Not done by us. Levels (±S.E.) 1 and 4 hr after injection (Couch, T. L., and Benton, A. W., Toxicon, 9:000, 1971).

ACKNOWLEDGMENTS

We wish to thank Dr. John Pike of the Upjohn Company, Kalamazoo, Michigan, for his generous contribution of the prostaglandins used in these studies and Dr. Ralph Peterson of The New York Hospital in whose laboratory corticosterone levels were determined.

The work described in this paper was aided by Grant No. AM-11949 from the National Institutes of Health and the Glenn B. and Gertrude P. Warren Foundation. R. B. Zurier is a Special Fellow of the National Institutes of Health Grant No. AM-50489. G. Weissmann is a Career Scientist of the Health Research Council of the City of New York (I-467).

REFERENCES

Anderson, A. J., Brocklehurst, W. E., and Willis, A. L., 1971, Evidence for the role of lysosomes in the formation of prostaglandins during carrageenin induced inflammation in the rat, Pharmacol. Res. Commun. 3:13.

Aspinall, R. L., and Cammarata, P. S., 1969, Effect of Prostaglandin E(2) on adjuvant arthritis, Nature, 224:1320.

Berney, S., Bishko, F., and Quagliata, F., 1971, Distribution of normal and sensitized lymphoid cells with adjuvant induced arthritis, Arth. & Rheum., 14:370.

Bourne, H. R., Lehrer, R. I., Cline, M. J., Melmon, R. J., 1971, Cyclic 3',5'-adenosine monophosphate in the human leukocyte: Synthesis, degradation and effects on neutrophil candidacidal activity, J. Clin. Invest., 50:920.

Couch, T. L., and Benton, A. W., 1971, The effect of the venom of the honey bee Apis Mellifera L., on the adrenocortical response of the adult male rat, Toxicon, 9:000.

deDuve, C., and Wattiaux, R., 1966, Functions of lysosomes, Ann. Rev. Physiol., 28:435.

deWied, D., Witter, A., Versteeg, D. H. G., and Mulder, A. H., 1969; Release of ACTH by substances of central nervous system origin, Endocrinol., 85:561.

Dingle, J. T., 1968, Vacuoles, vesicles and lysosomes, Brit. Med. Bull., 24:141.

Eliasson, R., 1959, Acta Physiol. Scand., 46, Suppl. 158:1.

Flack, J. D., Jessup, R., and Ramwell, P. W., 1969, Prostaglandin stimulation of rat corticosteroidogenesis, Science, 163:691.

Freed, J. J., and Lebowitz, M. M., 1970, The association of a class of saltatory movements with microtubules in cultured cells, J. Cell Biol. 45:334.

Goodman, D. B. P., Rasmussen, H., DiBella, F., and Guthrow, C., 1970, Cyclic adenosine 3',5'-monophosphate-stimulated phosphorylation of isolated neurotubule subunits, Proc. Nat. Acad. Sci., 67:652.

Henney, C. S., and Lichtenstein, L. M., 1971, The role of cyclic AMP in the cytolytic activity of lymphocytes, J. Immunol., 107:610.

Hirschhorn, R., Grossman, J., and Weissmann, G., 1970, Effect of cyclic 3',5'-adenosine monophosphate and theophylline on lymphocyte transformation, Proc. Soc. Exp. Biol. & Med., 133:1361.

Karnovsky, M. L., Baehner, R. L., Githens, S., Simmons, S., and Glass, E. A., June 1970, Correlations of metabolism and function in various phagocytes, Excerpta Medica International Congress, Series, No. 229, In Immunopathology of Inflammation, p. 121.

Koopman, W. J., Orange, R. P., and Austen, K. F., 1971, Prostaglandin inhibition of the immunologic release of slow reacting substance of anaphylaxis in the rat, Proc. Soc. Exp. Biol. & Med. 137:64.

Metchnikoff, E., 1905 (1968 reprint), Immunity in infective diseases, Johnson Reprint Corp., New York.

Pearson, C. M., and Wood, F. D., 1959, Studies of polyarthritis and other lesions induced in rats by injection of mycobacterial adjuvant. I. General clinical and pathological characteristics and some modifying factors, Arth. & Rheum. 2:440.

Ramwell, P. W., and Shaw, J., 1970, Biological significance of the prostaglandins, Recent Progress Horm. Res., Vol. 26, Academic Press, N. Y.

Rosenau, W., 1963, In Cell-Bound Antibodies, Ed. by B. Amos and H. Koprowski, p. 75, Wistar Inst. Press, Philadelphia.

Scott, R. E., 1970, Effects of prostaglandins, epinephrine and NaF on human leukocyte, platelet and liver adenylate cyclase, Blood, 35:514.

Sessa, G., Freer, J. H., Colacicco, G., and Weissmann, G., 1969, Interaction of a lytic polypeptide, melittin, with lipid membrane systems, J. Biol. Chem. 244:3575.

Smith, J. W., Steiner, A. L., Parker, C. W., 1971, Human lymphocyte metabolism: Effects of cyclic and noncyclic nucleotides on stimulation by phytohemagglutinin, J. Clin. Invest. 50:442.

Vane, J. R., 1971, Inhibition of prostaglandin synthesis as a mechanism of action for aspirin-like drugs, Nature New Biol., 231:232.

Vogt, W., Meyer, J., Kunze, H., Lufft, E., and Babilli, S., 1969, Enststehung von SRS-C in der durchstromten Meerschweinchenlunge durch phospholipase A. Identifizierung mit prostaglandin., Naunyn-Schmiedebergs Arch. Exp. Path. Pharmak. 262:124.

Weissmann, G., 1967, The role of lysosomes in inflammation and disease, Ann. Rev. Med. 18:97.

Weissmann, G., Dukor, P., and Zurier, R. B., 1971a, Effect of cyclic AMP on release of lysosomal enzymes from phagocytes, Nature New Biol., 231:131.

Weissmann, G., and Dukor, P., 1970, The role of lysosomes in the immune response, Adv. Immunol. 12:283.

Weissmann, G., Keiser, H., and Bernheimer, A. W., 1963, Studies on lysosomes. III. The effects of streptolysins O and S on the release of acid hydrolases from a granular fraction of rabbit liver, J. Exp. Med. 118:205.

Weissmann, G., 1964, Labilization and stabilization of lysosomes, Fed. Proc. 23:1038.

Weissmann, G., Dukor, P., and Sessa, G., June 1970, Studies on lysosomes: Mechanisms of enzyme release from endocytic cells and a model for latency *in vitro*. Excerpta Medica International Congress Series No. 229, Immunopathology of Inflammation, p. 107.

Weissmann, G., Spilberg, I., and Krakauer, K., 1969, Arthritis induced in rabbits by lysates of granulocyte lysosomes, Arth. & Rheum., 12:103.

Weissmann, G., Zurier, R. B., Spieler, P., and Goldstein, I., 1971b, Mechanisms of lysosomal enzyme release from leukocytes exposed to immune complexes and other particles, J. Exp. Med., 134:149s.

Zaitzev, G. P., and Poriadin, V. T., 1964, Use of the agriculture products in man, in Honeybee and the Health of Man, Ministry of Agriculture USSR, p. 52, Moscow.

Zucker-Franklin, D., and Hirsch, J. G., 1964, Electron microscope studies on the degranulation of rabbit peritoneal leukocytes during phagocytosis, J. Exp. Med. 120:569.

Zurier, R. B., and Quagliata, F., (in press), Effect of prostaglandin E(1) on adjuvant arthritis, Nature.

DISCUSSION

The following discussion covers the paper by Zurier and Weissmann, which was presented by Weissmann.

KALEY: Did you test PGE(2) in the adjuvant arthritis model? If so, how effective was it when compared with PGE(1)?

WEISSMANN: Yes. The membrane stabilization effect of PGE(2) is something we have found only with liver lysosomes. Since it worked in the direction of cortisol, there is a marked discrepancy there. I have no explanations for the discrepancy at this moment, and I do not know what happens to PGE(2) when injected.

GLENN: There are a few things that worry me specifically about the adjuvant arthritis. That is, you can swim these animals for 15 min a day, twice a day, and inhibit their disease. You can store them in the refrigerator for a couple of hours a day, for a week, and inhibit the disease. You can inject them with histamine or serotonin and inhibit the disease. We believe, in contrast to your data, that PGE(1) and PGE(2) inhibition of adjuvant arthritis is not specific. It is probably related to stimulation of the adrenals.

WEISSMANN: Myles, what you are asking me to do is to suggest that we are studying artifacts. I would suggest that what you call "non-specific" is only non-specific in the mind of the beholder. We have indeed studied corticosterone levels in the adjuvant-treated animals with PGE(1), and their corticosterone levels taken during the day really are not very elevated compared to the controls of adjuvant arthritic rats. Adjuvant arthritic rats have very high levels of corticosterone throughout the development of their lesion, far above those of normal rats. I don't know how much more we could stimulate them. There are, of course, a number agents that don't influence adjuvant arthritis, PGA compounds, for example. These, I think, constitute reasonable controls, so when you say PGE(1) is "not specific" I would say perhaps, but in respect to other prostaglandins the structure of PGE(2) renders it specific. But if you suggest to me that dunking rats in cold water or putting them in the refrigerator can suppress adjuvant arthritis, I could suggest that these procedures are doing terrible things to the animals' microtubules, via the cyclase system. I believe, as I am sure that Dr. Rasmussen does, that these structural proteins act as substrates for protein kinases. Whatever the cellular mechanism be whereby these "non-specific" agents suppress adjuvant arthritis, the fact that there is structural specificity to prostaglandin in suppressing arthritis leads me to think that further trials are indicated.

FRIEDBERG: Would you tell us what the side effects of the prostaglandin injections were? What did the rats do after the injection?

WEISSMANN: Our rats were slightly somnolent. They weren't anesthetized. They did not develop much diarrhea, which I would have expected.

FRIEDBERG: So there were no adverse side effects. Did they lose weight?

WEISSMANN: With PGE(1) they lost hair at the site of the injection, which I think everyone has observed.

FRIEDBERG: Secondly, could you describe the blood picture to us, and thirdly, would you give some indication of what the changes in circulating antibody were, in view of the fact that the condition may have an immunologic basis?

WEISSMANN: Well, two parameters have been tested with PGE(1) (Zurier, R. B., and Quagliata, F., Effect of prostaglandin E(1) on adjuvant arthritis, Nature, in press). They've tested two parameters of immune response in the adjuvant arthritic rat: the response to PPD by ear volume and the response to heterologous sheep red blood cells. The PPD responses were not significantly diminished in the prostaglandin-treated animals. However, there was, I think, a modest, twofold diminution in humoral response to heterologous red blood cells in the PGE(1) treated animals. What immune mechanism do you think is responsible for adjuvant arthritis?

FRIEDBERG: Certainly it is possible that it has an immune basis.

WEISSMANN: If one has been implicated, it might be the one that Dr. Lichtenstein is interested in, namely, the effect of sensitized lymphocytes meeting PPD in the tissues. But I don't think that is at all clear cut. We are studying a direct action on an immune system. I think what one is really studying is an anti-inflammatory mechanism.

FRIEDBERG: What about the blood picture?

WEISSMANN: Very little seems to happen.

ANGGARD: Like the previous speaker I am concerned about the kinetics of the situation. It seems that the injected prostaglandin would in your case be a "hit and run" drug because the elimination of prostaglandins subcutaneously injected in rats is fast. Also 500 micrograms is a very large dose for a rat and would have very profound pharmacological effects such as lowering of blood pressure.

WEISSMANN: I don't disagree with you one bit.

GLENN: One of the questions was concerned with what happens in the blood. We determined the ESRs and the fibrinogen concentrations were elevated. It turns out that when one gives non-steroidal anti-inflammatory drugs and/or steroidal anti-inflammatory drugs, these parameters are lowered, but not with prostaglandins. And I can also attest that the prostaglandins cause very bizarre reactions in rats. 500 micrograms twice a day in rats is an extremely large dose of PGE(1). We attempted to use the vasodilator alcohol as a control. At a dosage which did not approach systemic effects, one obtains acute inflammation in addition to inhibition of adjuvant induced arthritis.

WEISSMANN: Well, may I suggest that if I were particularly interested in the neurogenic basis of inflammation, or indeed an effect on microtubules or structural proteins within cells, ethanol is just the agent I'd pick to act upon these elements.

GLENN: I think you are doing a similar thing with PGE(1).

WEISSMANN: I would suggest indeed that the two may share a common final pathway, but one can't realistically discuss in terms of modern cell biology what happens when one gives 500 micrograms of PGE(1) to a rat. I agree with the other speakers that this is only one of the models for a chronic inflammatory disease of the joint.

GLENN: But the implication of what you have said was quite different. What I want to know is, according to your data, there should be a concentration of PGE(1) that would enhance the arthritic response. Wasn't there a concentration of cyclic AMP that stimulated this whole process?

WEISSMANN: I think that at the moment the answer is yes. If inflammation in the whole animal is related to discharge of hydrolases from phagocytic cells, we ought to have the same kind of biphasic effect that one sees *in vitro*. I just don't know what happens when one exhibits these agents at lower levels to animals. *In vitro*, there is the biphasic effect of cyclic AMP which is striking and really challenging.

THE ROLE OF PROSTAGLANDINS IN THE IMMUNE RESPONSE

C. W. Parker

Washington University Schol of Medicine
Department of Internal Medicine
St. Louis, Mo.

Prostaglandins could exert an important influence on the immune response at at least six levels: (1) As a part of the control mechanism which is altered when lymphocytes dedifferentiate and replicate in response to antigen. (2) As short or long range messengers in cell-cell interactions between lymphoid cells. Thus prostaglandins might be involved in the interactions between thymic (T) and bone marrow (B) cells that are required for the full development of cellular and humoral immunity (Hartman, 1971). Or they might participate in the phenomenon of antigenic competition where an ongoing immune response interferes nonspecifically with the effectiveness of an unrelated antigen (Moller, 1971). (3) As regional hormones which might influence the level of differentiation of lymphoid cells in their local environment. Thus, local differences in prostaglandin levels in the thymus and peripheral lymphoid tissues might be involved in thymic cell differentiation in the periphery (Owen & Raff, 1970). (4) As agents which modulate the response of sensitized cells to antigen and alter the release of small molecular weight and macromolecular weight and macromolecular mediators of immediate and delayed hypersensitivity. (5) As agents which alter the response of migrating leukocytes to chemotactic stimuli and thereby influence immunologically mediated inflammation. (6) As agonists which directly mediate smooth muscle and vascular end organ responses or as modulators of the response of those tissues to other pharmacological mediators of immediate hypersensitivity.

While present information does not clearly establish precisely what (if any) the physiological role for prostaglandins in the above processes is, there is increasing evidence that exogenous prostaglandins are capable of influencing the immune response at each of these levels. Information bearing on the possible action of prostaglandins in mediator release and inflammation will be considered elsewhere in this symposium. In this Section categories (1) and (6) (the role of prostaglandins in lymphocyte response to antigen and prostaglandin effects on smooth muscle and vascular end organ responsiveness) will be discussed in some detail. In the discussion of end organ responsiveness particular emphasis will be given to the problem of human bronchial asthma and the evidence for altered tissue responsiveness to catecholamines in this disease.

LYMPHOCYTE RESPONSES TO ANTIGEN AND OTHER MITOGENS

Previous studies from this laboratory have established that human peripheral blood lymphocytes contain adenylate cyclase and that the enzyme can be stimulated by catecholamines and fluoride in broken cell preparations (Table I) (Smith et al, 1969a, 1969b, 1970). The cyclic AMP level in resting, intact lymphocytes averages about 6.0 picoM/10^7 cells with a 2-20 fold increase in the cyclic AMP concentration in response to maximal stimulatory concentrations of isoproterenol.

A 3-30 fold increase in the cyclic AMP concentration in intact lymphocytes can be obtained with a variety of prostaglandins (Table II). Members of the E, A, and F series are all stimulatory with PGF(1-alpha) being the least effective on a molar basis (Smith et al, 1971a). PGE(1) and PGA(1) are slightly more effective stimulators than PGE(2) and PGA(2) indicating the importance of a single olefinic bond for maximal stimulatory activity. Studies with broken cell preparations and PGE(1) indicate that the increase in cyclic AMP is due to a direct stimulation of lymphocyte adenylate cyclase (Table I). The threshold for PGE(1) stimulation of cyclic AMP formation in intact lymphocytes is about 30 picoM, which is well above the estimated PGE(1) concentration range in human serum under resting conditions. PGE(1) stimulation of adenylate cyclase is blocked poorly or not at all by D,L-propranolol (a beta-blocking agent) except at very low prostaglandin concentrations (Table III), suggesting that the prostaglandin receptor differs from the beta-adrenergic receptor(s). By contrast, D,L-propranolol markedly inhibits the cyclic AMP response to isoproterenol, even at high isoproterenol concentrations.

In view of the marked activation of lymphocyte adenylate cyclase activity by members of the prostaglandin family and the far reaching influence of cyclic AMP on intracellular metabolism (Robison et al, 1968), an effect of prostaglandins on lymphocyte behavior in tissue culture might be anticipated. Plant mitogens such as Phaseolus vulgaris phytohemagglutinin (PHA) stimulate lymphocytes to undergo differentiation and proliferation. Similar effects are induced by antigen in the presence of lymphocytes that have been sensitized by prior exposure with antigen in vivo. The response in stimulated lymphocytes can be followed by measuring the increase in rate of incorporation of radiolabeled DNA, RNA and protein precursors into cellular macromolecules. Prostaglandins, at concentrations of 1-20 microM and higher, inhibit DNA (Fig. 1), protein and RNA synthesis in PHA and antigen stimulated cells (Smith et al, 1971b). The inhibition of lymphocyte stimulation in the presence of prostaglandins is presumably due to elevated intracellular cyclic AMP levels. Theophylline and isoproterenol, which also increase intracellular cyclic AMP concentrations, are also inhibitory. Moreover, at high prostaglandin concentrations PGF(1-alpha) is the least effective inhibitor, correlating its lesser activity (at maximal stimulatory concentrations) in elevating cyclic AMP levels. In further support of cyclic AMP as the probable mediator of prostaglandin inhibition of lymphocyte transformation, dibutyryl-cyclic AMP at concentrations in the 10-100 microM range also inhibits the lymphocyte response. Dibutyryl-cyclic AMP is known to enter cells and simulate many of the effects of intracellular cyclic AMP (Robison et al, 1968).

Cyclic AMP can also reduce the mitogenic response in nonlymphocytic cells. It is of considerable interest that malignant cell lines seem to be particularly susceptible to cyclic AMP inhibition of cell growth (Heidrick and Ryan, 1970; Shepard, 1971; Yang and Vas, 1971). The effect is reversible and not usually associated with cell death. As long as cyclic AMP is present malignant cells assume an appearance more like that of normal cells (Johnson et al, 1971, section by Johnson, this Symposium). But as soon as the cyclic nucleotide is removed, the cells resume full mitotic activity and revert to their previous abnormal morphology.

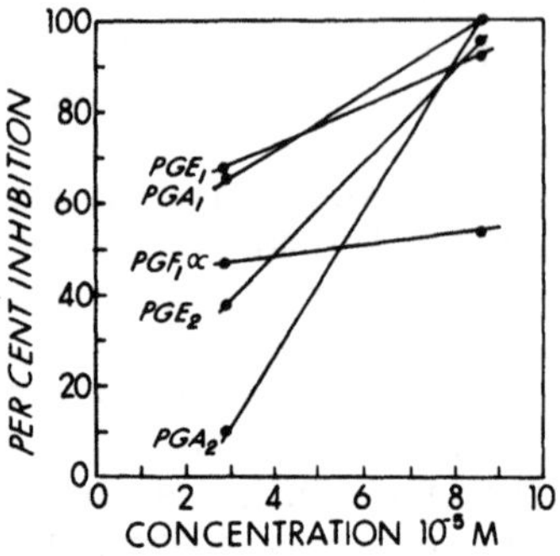

Figure 1. Inhibition of DNA synthesis by prostaglandins. Human peripheral blood leukocytes were incubated for 52 hr with PHA with thymidine-^{3}H present during the final 4 hr of culture. The prostaglandins were added at 0 time, just before PHA. (Taken from Smith et al, 1971b).

The basis for the inhibition of lymphocyte transformation when intracellular cyclic AMP concentrations are increased for sustained periods is not clear. PHA itself produces an early increase in lymphocyte cyclic AMP concentration beginning within one min and persisting for about two hr followed by prolonged fall (6-24 hr) (Fig. 2) to below the level in control cells (Smith et al, 1971b). In human and pig lymphocytes low concentrations of cyclic AMP (and dibutyryl-cyclic AMP, 10-100 picoM) stimulate rather than inhibit protein, DNA and RNA synthesis (Smith et al, 1969b, 1970, 1971b; Hirschhorn et al, 1970; Cross and Ord, 1971). The timing of low dose cyclic nucleotide mediated stimulation of lymphocyte metabolism closely conforms to that of the alterations in cell metabolism induced by PHA, suggesting that the two stimuli produce a qualitatively similar or identical response (Cross and Ord, 1971; Smith et al, 1971b). Dibutyryl-cyclic AMP also stimulates DNA synthesis in rat thymic lymphocytes in vitro (Whitfield et al, 1970), and causes an increase (or decrease depending on the concentratation) in the number of lymphoid cells producing antibody to foreign erythrocytes in vivo and in vitro (Braun and Ishizuka, 1971). Evidently, then, cyclic AMP can either stimulate or inhibit lymphoid cell proliferation depending on the intracellular level and how long it is maintained.

The increase in lymphocyte cyclic AMP concentration in response to PHA is observable within 30-60 sec and precedes other previously demonstrated metabolic alterations in PHA stimulated cells (Smith et al, 1971a).

The rapidity of the response and the fact that cyclic AMP and dibutyryl-cyclic AMP can initiate alterations in lymphocyte metabolism which are indistinguishable in tempo from the PHA induced alterations could be taken to indicate that cyclic AMP is an early obligatory intracellular messenger for PHA. Gene activation in this situation could be due to cyclic AMP modulated alterations in phosphokinase activity with secondary increases in histone phosphorylation (Pogo et al, 1966). Cyclic AMP is known to promote histone phosphorylation in rat liver (Langan 1969) and recent evidence would indicate that it may also do so in pig and human lymphocytes (Cross and Ord, 1971; Smith and Parker, 1971, unpublished). The phosphorylation of histones complexed to DNA could reduce the stability of the complex and result in unmasking of latent genetic information. An effect of cyclic AMP at the transcriptional level is already indicated in E.coli (Pastan, this Symposium) (Zubay et al, 1970) where cyclic AMP is directly implicated in the derepression of beta-galactosidase synthesis. Cyclic AMP also appears to play a central role in isoproterenol induced DNA synthesis in salivary tissues (Stein and Baserga, 1970). Despite the growing evidence that cyclic AMP can initiate gene activation it would be premature to conclude that it is the sole or even the most important intracellular messenger in lymphocyte activation by PHA. The fact that high concentrations of cyclic AMP inhibit transformation must be kept in mind. (However, it could be argued that this inhibition occurs at a later stage of the cell cycle or that the optimal concentration of cyclic AMP for activation of phosphokinase activity has been exceeded.) More puzzling is the observation that isoproterenol, which alters intracellular cyclic AMP levels in lymphocytes to the same degree as PHA, fails to induce lymphocyte transformation (Smith et al, 1971b). This is true even when cells pulsed with isoproterenol so as to simulate the time course of PHA induced changes in cyclic AMP concentrations. The possibility remains that isoproterenol and PHA stimulate different intracellular cyclic AMP pools and that only the PHA alterable pool is involved in cell activation. PHA and isoproterenol effects on lymphocyte cyclic AMP levels at maximal stimulatory concentrations of each are additive, a finding which is compatible with the separate pool theory (Table IV). Clearly additional studies are needed at the whole cell and subcellular level to clarify the role of cyclic AMP in lymphocyte differentiation and replication. But new technology (such as the development of better histochemical techniques for the localization and quantitation of adenylate cyclase and cyclic AMP) may be required to resolve the dilemma.

On the basis of the above discussion it seems clear that the prostaglandins are capable of exerting major effects on cell differentiation and replication. The eventual use of the prostaglandins (and other agents which produce persistent elevations in intracellular cyclic AMP) as well as cyclic AMP itself for immunosuppression and the treatment of malignancy deserves serious consideration. Unfortunately, the relatively high concentrations of these agents needed for inhibition of cell proliferation *in vitro* would increase the possibility of marked side effects *in vivo*, limiting the success of this approach. However, as more is learned about the individual prostaglandins and the way in which they interact with normal tissues, a greater measure of selectivity may be achieved. Apart from possible pragmatic considerations further information in regard to prostaglandin metabolism and the possible role of prostaglandins in short range interactions between individual components in complex tissue mixtures seems particularly important. Participation of the prostaglandins as regional hormones in the control of tissue differentiation and organization would be consistent with the known chemical and biological properties of these substances.

END ORGAN RESPONSIVENESS AND HUMAN BRONCHIAL ASTHMA

The bronchial musculature is under the control of the autonomic nervous system and circulating catecholamines. Beta-adrenergic stimulating agents ordinarily produce bronchodilatation whereas beta-blocking agents and cholinergic stimuli cause bronchoconstriction. These effects are almost certainly exerted through the adenylate cyclase system, bronchodilation being associated with an increase in intracellular cyclic AMP concentrations and bronchoconstriction with a fall in cyclic AMP concentration. In view of the potent activity of the prostaglandins in other adenylate cyclase systems, not unexpectedly the prostaglandins have marked inhibitory or stimulatory effects on bronchial smooth muscle tone. *In vitro* human bronchial muscle is relaxed by PGE(1) and PGE(2) but contracted by PGF(2-alpha) (Sweatman and Collier, 1968; Sheard, 1968). PGF(2-alpha) also inhibits the bronchodilator response to isoproterenol (Adolphson et al, 1969). *In vivo* PGE(1) has little effect on forced expiratory velocity in normal human subjects but produces bronchodilatation in patients with asthma (Cuthbert,

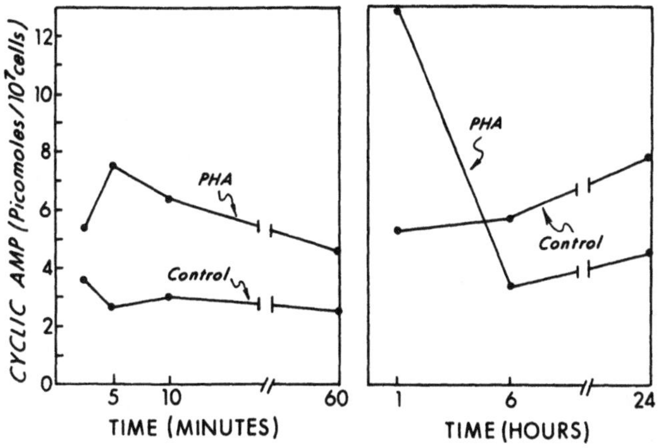

Figure 2. Cyclic AMP concentration vs time in PHA-stimulated lymphocytes. 10^7 lymphocytes in a final volume of 0.7 ml were incubated in Eagle's medium with either 0.05 ml of PHA-P or 0.05 ml of Gey's balanced salt solution for various times at 37°C. The cells were then centrifuged for 2 min at 250 g, the supernatants removed, and the cell pellets immediately frozen. Two separate experiments are shown. (Taken from Smith et al, 1971a).

1969). PGE(1) and PGE(2) also cause bronchial and tracheal muscle relaxation in several animal species (Main, 1964; Large et al, 1969; Adolphson and Townley, 1970; Rosenthale et al, 1970). The role of the prostaglandins in normal respiratory physiology is unknown. PGE(2) and PGF(2-alpha) have been demonstrated in human lung (Cuthbert, 1969), and Horton (1969) has suggested that an abnormal PGF to PGE ratio or an increase in PGF concentration might be involved in the pathogenesis of human asthma. Unfortunately, information in regard to possible alterations in prostaglandin metabolism in patients with asthma is not yet available.

Even if prostaglandin metabolism does not prove to be altered in human asthma, there is the possibility that members of the PGE series might be useful therapeutically as bronchodilator agents in this disease. Interest in the possible use of prostaglandins for this purpose has been stimulated by the relative ineffectiveness of catecholamines in certain patients with asthma and the increasing evidence that many patients with active asthma have partial beta-adrenergic blockade:

1. Patients with active bronchial asthma may have a reduced or even paradoxical bronchial response to

adrenaline and isoproterenol (given either by parenteral injection or aerosol) (Keighley, 1969; Reisman, 1969). This is especially true following excessive use of aerosol preparations containing beta-adrenergic agonists. Patients with asthma also exhibit exquisite sensitivity to the bronchoconstrictor effects of histamine and mecholyl (Itkin, 1967).

2. Following the systemic administration of adrenaline, patients with recent asthma have a less marked metabolic response (less striking changes in blood glucose, lactate, pyruvate and - in one study - nonesterified fatty acids) than normal controls (Lockey et al, 1967; Middleton and Finke, 1968).

3. As already discussed, normally, beta-adrenergic agents inhibit lymphocyte activation by PHA and antigen. The peripheral blood lymphocytes of patients with bronchial asthma are less subject to catecholamine inhibition than the lymphocytes of normal controls (Smith and Parker, 1971, unpublished).

4. In recent studies in this laboratory a convenient _in vitro_ system for the study of cyclic AMP metabolism in patients with asthma has been identified (Smith and Parker, 1970). We found that the peripheral blood leukocytes of patients with asthma exhibit a relatively small increase in cyclic AMP in response to high concentrations of isoproterenol (added _in vitro_) by comparison with leukocytes from normal individuals (Table V) Smith and Parker, 1970). Thus the mean increase in cyclic AMP in the leukocytes of patients with asthma in the presence of 10 mM isoproterenol was less than twofold, whereas a fivefold increase occurred with the leukocytes of control subjects. In view of the reduced cyclic AMP response _in vitro_ it seems reasonable to assume that the decreased metabolic response to the infusion of adrenaline in patients with asthma is due to a diminished ability of adrenaline to increase the intracellular cyclic AMP concentration in relevant tissues. Certainly there is no need to postulate that reduced cellular responsiveness to cyclic AMP (as might occur secondary to cyclic AMP inhibitors or reduced phosphokinase sensitivity to cyclic AMP) or an alteration in adrenaline metabolism or distribution are major factors in the altered _in vivo_ response. Preliminary results with broken cell preparations indicate that the diminution in the cyclic AMP response to isoproterenol is at least in part at the level of adenylate cyclase itself with a decrease in the rate of conversion of known amounts of ATP to cyclic AMP.

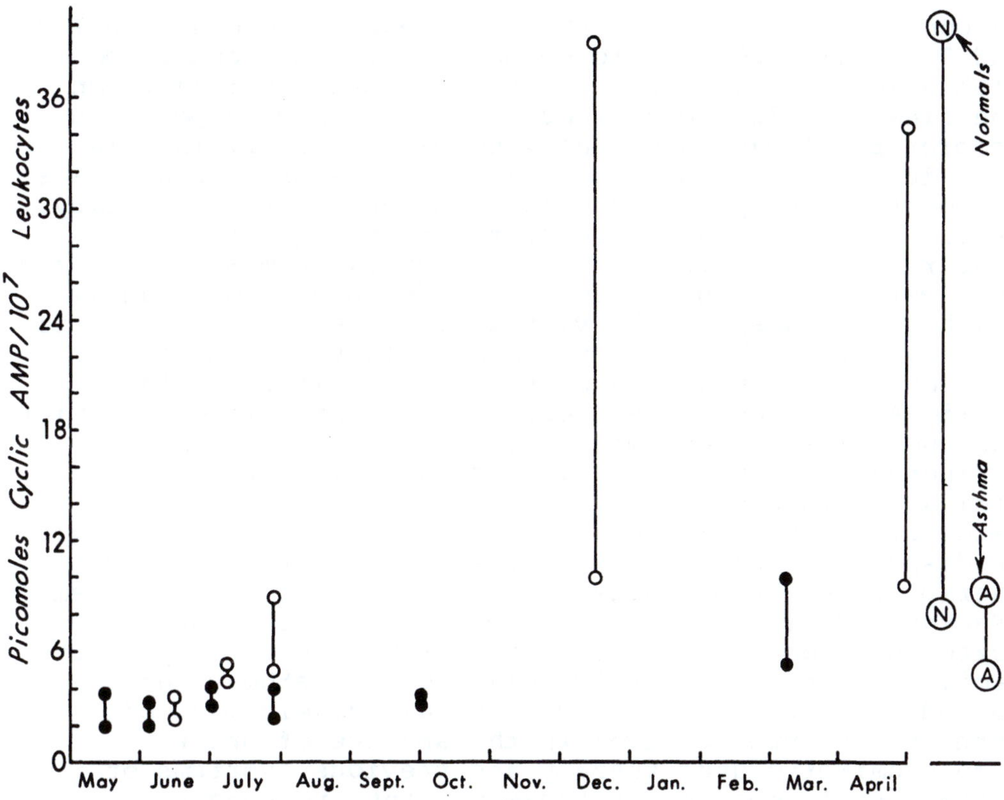

Figure 3. Serial changes in the leukocyte cyclic AMP response (37°C, 30 min) to isoproterenol (10 mM) in two patients with asthma. The lower dot is the unstimulated control, the upper dot is the isoproterenol response; open circles are for M. S. leukocytes (symptoms of asthma from May-September 15; largely symptom free thereafter); closed circles are for M. H. leukocytes (symptoms of asthma from June-July) essentially symptom free thereafter. Mean responses with leukocytes from 29 asthmatic patients and 23 normal subjects are shown on the right.

However, additional factors in the reduced cyclic AMP response such as reduced substrate concentration, increased cyclic AMP hydrolysis or increased release of cyclic AMP from cells are not yet excluded.

An altered leukocyte cyclic AMP response to isoproterenol in association with acute asthma may be succeeded by considerably improved or even normal responsiveness in symptom free periods. The results of serial studies in two patients with asthma are shown in Fig. 3. With leukocytes from M. H. (open circles) responsiveness to cyclic AMP was clearly increased on 7-9 and 7-23 (although not to normal levels) by comparison with 6-14, the time of the acute asthmatic episode. On the final two determinations the cyclic AMP response to isoproterenol was essentially normal. Clearly in this individual the diminished cyclic AMP response on 6-14 was temporary, perhaps associated with catecholamine therapy, or the precipitating cause of the asthma (acute respiratory infection). The leukocytes from M. S. (closed circles) on the other hand, responded poorly over an 11 month period despite the virtual absence of antiasthmatic therapy during the final 5 months. In this and similar patients there is a more sustained alteration in the cyclic AMP response, well beyond the period in which prior catecholamine therapy might be expected to continue to influence the response. Further support for a chronic diminution in the leukocyte cyclic AMP response to catecholamines in association with asthma (or an allergic predisposition to asthma) in the absence of antiasthmatic therapy comes from studies of patients with atopic eczema (Parker and Eisen, 1971). Atopic eczema is a chronic eczematous dermatitis associated with a strongly positive family history of allergic rhinitis and asthma, and a marked predisposition to the eventual development of asthmatic symptoms. Even in the absence of prior sympathomimetic drug therapy and previous symptoms of asthma the leukocytes of these patients respond very poorly to isoproterenol (Table V).

The question arises as to whether the reduced leukocyte cyclic AMP response to catecholamines in association with bronchial asthma extends to other hormonal agents which act directly on the adenylate cyclase system. Work in progress with leukocytes from patients with asthma indicates that the cyclic AMP response to PGE(1) (30 picoM) is frequently normal or only moderately reduced despite a marked reduction in the isoproterenol response. The fact that PGE(1) is a more effective stimulus to adenylate cyclase than isoproterenol suggests that there is partial beta-adrenergic blockade in human asthma and that PGE(1) can partially or completely bypass the block. The alternative explanation, that PGE(1) is simply a much more effective stimulus to leukocyte adenylate cyclase than isoproterenol at the concentrations used, is not supported by studies with leukocytes from normal controls. The effectiveness of PGE(1) under conditions in which there is a poor response to isoproterenol is consistant with the view that in human leukocytes the prostaglandin receptor is independent of the beta-adrenergic receptor, as already suggested by the relative ineffectiveness of propranolol as an inhibitor of the human lymphocyte response to PGE(1) (Smith et al, 1971a). Similar conclusions can be drawn in regard to the receptor in guinea pig tracheal muscle, where D,L-propranolol blockade prevents the expected response to isoproterenol but not the response to PGE(1) (Adolphson and Townley, 1970). Taken together the results of PGE(1) stimulation of leukocyte cyclic AMP levels and the relative effectiveness (usually) of PGE(1) as a stimulator of adenylate cyclase in beta-blocked tissues, the possible therapeutic usefulness of PGE(1) and PGE(2) in asthma becomes a matter of considerable interest. PGE(1) has been demonstrated to produce bronchodilatation in patients with asthma (Cuthbert 1969), but information on the effectiveness of PGE(1) in patients with status asthmaticus who are refractory to epinephrine is not yet available.

TABLE I

ADENYLATE CYCLASE IN LYMPHOCYTES

Experimental conditions	Adenylate cyclase activity, picoM cyclic AMP/mg protein per 10 min
Control	13.6±1.6
Na Fluoride (10 mM)	202 ±3.6
PGE(1) (0.2 mM)	165 ±11.6
PHA-P (1:16)*	39.0± 2.9
Isoproterenol (0.35 mM)	36.2+14.9

*5 microliters of stock Difco PHA-P (1 vial reconstituted in 5 ml) in a final volume of 30 microliters.

Adenylate cyclase determinations were based on the conversion of uniformly labeled ^{14}C labeled ATP to cyclic AMP in broken cell preparations containing 0.375 mM ATP and an ATP regenerating system. Each value is the mean of three separate determinations, ±SEM. The cell preparation contained 98% lymphocytes. The above values are corrected for the boiled tissue blank (12.6±2.3). Taken from Smith et al, 1971a).

TABLE II

EFFECT OF PROSTAGLANDINS ON LYMPHOCYTE CYCLIC AMP

Prostaglandin	Cyclic AMP picoM/10^7 cells
PGE(1)	30
PGE(2)	28
PGA(1)	24
PGA(2)	21
PGF(1-alpha)	14
EtOH-NaCl control	2.5
Gey's control	2.2

Prostaglandins at 2 x 10^{-4} M/l and lymphocytes (10^7/ml) were incubated at 37°C for 1 hr. The cells were isolated by centrifugation and snap frozen in liquid nitrogen. Cyclic AMP was determined by radioimmunoassay. Each value represents the average of four determinations. Results are expressed in picoM/10^7 cells (uncorrected for recovery). (Taken from Smith et al, 1971a).

TABLE III

HUMAN LYMPHOCYTES. EFFECT OF PHENTOLAMINE, AND D,L-PROPRANOLOL ON CYCLIC AMP STIMULATION BY ISOPROTERENOL, AND PGE(1)

Condition	Cyclic AMP picoM/10^7 lymphocytes
Buffer control	1.1
Isoproterenol, 10 mM/l	12.0
Isoproterenol, 10 mM/l + phentolamine, 20 microM/l	11.0
Isoproterenol, 10 mM/l + propranolol, 50 microM/l	2.0
Isoproterenol, 10 mM/l + propranolol, 20 microM/l	5.0
PGE(1), 1 x 10^{-7} M/l	18.0
PGE(1), 1 x 10^{-7} M/l + phentolamine, 20 microM/l	20.0
PGE(1), 1 x 10^{-7} M/l + propranolol, 50 microM/l	10.0
PGE(1), 1 x 10^{-7} M/l + propranolol, 20 microM/l	13.0
PGE(1), 1 x 10^{-5} M/l	32.0
PGE(1), 1 x 10^{-5} M/l + propranolol, 50 microM/l	31.0

Each tube contained 5 x 10^6 lymphocytes in 0.6 ml of Gey's solution. The lymphocytes were incubated with blocking agents (or control buffer solution) for 10 min at 37°C before the addition of the adenylate cyclase stimulators. The cells were harvested after an additional 5 min at 37°C. (Cyclic AMP values are uncorrected for recovery.) (Taken from Smith et al, 1971a).

TABLE IV

THE EFFECT OF PHA ALONE AND IN COMBINATION WITH ISOPROTERENOL ON LYMPHOCYTE CYCLIC AMP CONCENTRATIONS

Condition	Cyclic AMP picoM/10^7 lymphocytes
Buffer control	1.3
PHA 1:3	16.0
PHA 1:7	15.5
PHA 1:14	10.0
PHA 1:3 + Isoproterenol, 40 mM/l	28.0
PHA 1:7 + Isoproterenol, 20 mM/l	31.0
PHA 1:14 + Isoproterenol, 10 mM/l	30.0
Isoproterenol, 40 mM/l	13.0
Isporoterenol, 20 mM/l	12.0
Isoproterenol, 10 mM/l	9.0

1 x 10^7 lymphocytes were incubated at 37°C for 5 min in Gey's solution. (Cyclic AMP values are uncorrected for recovery.) (Modified from Smith et al, 1971a).

TABLE V

LEUKOCYTE CYCLIC AMP RESPONSE TO 10 mM ISOPROTERENOL

Diagnosis	Number of Determin-ations	Mean Cyclic AMP Concentration picoM/10^7 leukocytes	
		Control	Isoproterenol
Asthma	29	5.6	9.2
Control	32	8.0	40.2
Hay Fever	18	7.4	33.0
Atopic Eczema	23	6.3	13.1

Mixed leukocytes from heparinized human peripheral blood were incubated for 30 min at 37°C in the presence and absence of 10 mM isoproterenol.

ACKNOWLEDGMENTS

The author of this paper is a Career Development Awardee of the National Institute of Allergy and Infectious Diseases. His work was supported by grants of the U. S. Public Health Service.

REFERENCES

Adolphson, R. L., and Townley, R. G., 1970, A comparison of the bronchodilator activities of isoproterenol and prostaglandin E(1) aerosols, J. Allergy 45:119.

Adolphson, R. L., Kennedy, T. J., Reeb, R., and Townley, R. G., 1969, Effect of beta-adrenergic blockade and prostaglandin F(2-alpha) on isoproterenol and theophylline bronchodilation, J. Allergy 43:176.

Braun, W., and Ishizuka, M., 1971, Antibody formation: Reduced responses after administration of excessive amounts of nonspecific stimulators, Proc. Nat. Acad. Sci. 68:1114.

Cross, M. E., and Ord, M. G., 1971, Changes in histone phosphoyrylation and associated early metabolic events in pig lymphocyte cultures transformed by phytohemagglutinin or 6-N, 2-O Dibutyryladenosine-3',5-cyclic monophosphate, Biochem. J. 124:241.

Cuthbert, M. F., 1969, Effect on airways resistance of prostaglandin E(1) given by aerosol to healthy and asthmatic volunteers, Brit. Med. J. 4:723.

Hartmann, K., 1971, Induction of a hemolysin response *in vitro*. II. Influence of the thymus-derived cells during the development of the antibody producing cells, J. Exp. Med. 133:1325.

Heidreck, M. L., and Ryan, W. L., 1970, Metabolism of 3',5'-cyclic AMP by strain L cells, Biochim. Biophys. Acta 237:301.

Hirschhorn, R., Grossman, J., and Weissman, G., 1970, Effect of cyclic AMP and theophylline on lymphocyte transformation, Proc. Soc. Exp. Biol. Med. 133:1361.

Horton, E. W., 1969, Hypotheses on physiological roles of prostaglandins, Physiol. Rev. 49:122.

Itkin, I. H., 1967, Bronchial hypersensitivity to mecholyl and histamine in asthma subjects, J. Allergy 40:245.

Johnson, G. S., Friedman, R. M., and Pastan, I., 1971, Restoration of several morphological characteristics of normal fibroblasts in sarcoma cells treated with adenosine-3',5'-cyclic monophosphate and its derivatives, Proc. Nat. Acad. Sci. 68:425.

Keighley, J. F., 1966, Iatrogenic asthma associated with adrenergic aerosols, Ann. Int. Med. 65:985.

Kischer, C. W., 1969, Accelerated maturation of chick embryo skin treated with a prostaglandin, PGB(1): An electron microscopic study, Am. J. Anat. 124:491.

Langan, T., 1969, Phosphorylation of liver histone following the administration of glucagon and insulin, Proc. Nat. Acad. Sci. 64:1276.

Large, B. J., Leswell, P. F., and Maxwell, D. R., 1969, Bronchodilator activity of an aerosol of prostaglandin E(1) in experimental animals, Nature 224:78.

Lockey, S. D., Glennon, J. A., and Reed, C. W., 1967, Comparison of the metabolic responses of normal and asthmatic subjects to epinephrine, J. Allergy 30:102.

Main, I. H. M., 1964, The inhibitory actions of prostaglandins on respiratory smooth muscle, Brit. J. Pharm. & Chemotherapy 22:511.

Middleton, E., and Finke, S. R., 1968, Metabolic response to epinephrine in bronchial asthma, J. Allergy 42:288.

Moller, G., 1971, Suppressive effect of graft versus host reactions on the immune response to heterologous red cells, Immunol. 20:597.

Owen, J. J. T., and Raff, M. C., 1970, Studies on the differentiation of thymus derived lymphocytes, J. Exp. Med. 132:1216.

Parker, C. W., and Eisen, A. Z., 1971, in preparation.

Pogo, B. G. T., Alfrey, V. G., and Mirsky, A. E., 1966, RNA synthesis and histone acetylation during the course of gene activation in lymphocytes, Proc. Nat. Acad. Sci. 55:805.

Reisman, R. E., 1969, Asthma induced by adrenergic aerosols, J. Allergy 46:162.

Robison, G. A., Butcher, R. W., and Sutherland, E. W., 1968, Cyclic AMP, Ann. Rev. Biochem. 37:149.

Rosenthale, M. E., Dervinis, A., Begany, A. J., Lapidus, M., and Gluckman, M. I., 1970, Bronchodilator activity of prostaglandin E(2) when administered by aeroscl to three species, Experientia 26:1119.

Sheard, P., 1968, The effect of prostaglandin E(1) on isolated bronchial muscle from man, J. Pharm. Pharmac. 20:232.

Shepard, J. R., 1971, Restoration of contact-inhibited growth to transformed cells by diburyryl-adenosine 3',5'-cyclic monophosphate, Proc. Nat. Acad. Sci. 68:1316.

Smith, J. W., Steiner, A. L., Newberry, W. M., and Parker, C. W., 1969a, The effect of dibutyryl-cyclic adenosine-3',5'-monophosphate (DC-AMP) on human lymphocyte stimulation by phytohemagglutinin (PHA), Fed. Proc. 28:566.

Smith, J. W., Steiner, A. L., Newberry, W. M., and Parker, C. W., 1969b, Cyclic nucleotide inhibition of lymphocyte transformation, Clin. Res. 17:549.

Smith, J. W., and Parker, C. W., 1970, The responsiveness of leukocyte cyclic AMP to adrenergic agents in patients with asthma, Proc. Cent. Soc. for Clin. Res. 43:76.

Smith, J. W., Steiner, A. L., and Parker, C. W., Early effects of phytohemagglutinin (PHA) on lymphocyte cyclic AMP levels, Fed. Proc. 29:369.

Smith, J. W., Steiner, A. L., Newberry, W. M., and Parker, C. W., 1971a, Cyclic adenosine 3',5'-monophosphate in human lymphocytes. Alterations after phytohemagglutinin stimulation, J. Clin. Invest. 50:432.

Smith, J. W., Steiner, A. L., and Parker, C. W., 1971b, Human lymphocyte metabolism. Effects of cyclic and noncyclic nucleotides on stimulation by phytohemagglutinin, J. Clin. Invest. 50:442.

Smith, J. W., and Parker, C. W., 1971, Unpublished data.

Stein, G., and Baserga, R., 1970, The synthesis of acidic nuclear proteins in the prereplicative phase of the isoproterenol-stimulated salivary gland, J. Biol. Chem. 245:6097.

Sweatman, W. J. F., and Collier, H. O. J., 1968, Effects of prostaglandins on human bronchial muscle, Nature 217:69.

Whitfield, J. F., McManus, J. P., and Gillan, D. J., 1970, Cyclic AMP mediation of bradykinin induced stimulation of mitotic activity and DNA synthesis in thymocytes, Proc. Soc. Exp. Biol. Med. 133:1270.

Yang, T. J., and Vas, S. I., 1971, Growth inhibitory effects of adenosine 3',5'-monophosphate on mouse leukemia L-5178-Y-R cells in culture, Experientia 27:442.

Zubay, G., Schwartz, D., and Beckwith, J., 1970, Mechanism of activation of catabolite-sensitive genes. A positive control system, Proc. Nat. Acad. Sci. 66:104.

DISCUSSION

BRAUN: If I understand you correctly, you did get a modest amount of lymphocyte transformation with prostaglandins alone as measured by increased incorporation of tritiated thymidine.

PARKER: No, as I recall, we have not used prostaglandin alone. We have used dibutyryl-cyclic AMP, isoproterenol and theophylline alone and obtained stimulation only with dibutyryl-cyclic AMP.

BRAUN: In the studies which we have done with mouse spleen lymphocytes we found a situation that indicates that adenylate cyclase stimulation alone, which for example can be obtained by exposure to double-stranded synthetic polynucleotides such as poly A:U, may not suffice to yield lymphocyte transformation in terms of increased DNA synthesis. Thus, a mere stimulation of adenylate cyclase activity did not mimic the effects of a mitogen or an antigen, which also produce adenylate cyclase stimulation but in addition stimulate DNA synthesis. Therefore, we have wondered whether the adenylate cyclase system is only an amplification system for a specific signal (antigen, mitogen) which also affects adenylate cyclase activity but may require support from a general non-specific stimulator of adenylate cyclase. In other words, do we have to distinguish between specific activation signals and supporting non-specific effects, both capable of modifying adenylate cyclase activity?

PARKER: I don't believe you can use our data to make a strong argument for this, but it would be one way of interpreting our data.

BRAUN: The distinction that I am trying to make between specific and non-specific effects certainly seems to apply in the case of modifying antibody responses where adenylate cyclase activation, per se, cannot activate antibody formation by so-called B cells, but can dramatically alter the behavior of antigen-activated cells. One would suspect that the antigenic signal is akin to the first messenger and that adenylate cyclase can both mediate intracellular responses and amplify the magnitude of the response.

PARKER: There is significant (but limited) overlap between the normal asthmatic groups, but the differences are highly significant ($p < 0.001$) statistically. Variation in serial determination made in the same normal subject is substantial (up to a factor of 2) but total or nearly total unresponsiveness as seen in some asthmatic patients is quite reproducible and in one patient was demonstrated on five separate occasions over a 2 wk period. Our total data include more than 50 asthmatic and normal patients, many of whom have been examined on more than one occasion and at a number of isoproterenol concentrations and incubation times.

BOURNE: I have two questions. One is relative to the lymphocytes and has to do with the concentration of phytohemagglutinin (PHA). We have been able to reproduce all of your data, with the prostaglandins and the catecholamines. We could not get any of those drugs to increase thymidine incorporation. We could not get PHA at any concentration to have an effect on cyclic AMP. This suggests to me that we are not really dealing with different pools in the same cell, but it is quite likely we are dealing with different cells. The ones that respond to PHA by stimulation of adenylate cyclase may not be the same cells that respond to stimulation to PHA by increasing thymidine incorporation. My second question has to do with asthmatics and the concentration of isoproterenol with time. We have attempted to look at time courses of cyclic AMP stimulation in response to isoproterenol but we have never had the courage to go up to 10 mM. At 10^{-4} concentration the maximum cyclic AMP concentration is at about 10 min and decays at a variable rate thereafter. I wonder if the metabolism of isoproterenol could be the difference between asthmatic and normal patients.

PARKER: As far as the second question is concerned, we have looked at a wide range of isoproterenol concentrations (0.001-2000 microM) and incubation times (1-240 min) and wherever some degree of leukocyte stimulation is obtained we see essentially the same differences. We have used 10 mM isoproterenol and 30 min in a sizable number of experiments because they give us reproducible rises in cyclic AMP with leukocytes from normal subjects, but we have a good deal of other data under other conditions that support the general conclusions. As far as the metabolism of isoproterenol is concerned, it is conceivable that their effects could be exerted through an alteration in the metabolism of isoproterenol. However, we see differences in leukocyte stimulation between asthmatic and nonasthmatic patients as early as 1 min after isoproterenol stimulation, so that seems unlikely. Coming back to your first question, you asked about phytohemagglutinin concentrations, and we have indeed used relatively high concentrations of this material. However, we also get significant cyclic AMP stimulation at levels which are maximally mitogenic. The problem here may be the rate at which the cells are stimulated. If you recall the earlier slide on the time course of PHA effects on lymphocyte cyclic AMP, the level of cyclic AMP was high early, but later the level of cyclic AMP was below that in control cells. Perhaps the effect of a high phytohemagglutinin concentration is to activate all of the cells rapidly. Quite possibly at lower PHA concentrations, activation of cells may take place over a period of some hours. Therefore, some of the cells might already have been activated and fallen below the level in control cells at a time when other cells were being activated for the first time. Another possibility is that at high phytohemagglutinin concentrations, PHA may continue to stimulate adenylate cyclase at a time in the cell cycle when cyclic AMP ought to be falling. And in fact we know that persistently high concentrations of cyclic AMP inhibit transformation. In this circumstance there would not be an exact parallel between cyclic AMP stimulation and mitogenesis.

KORNBERG: I have been curious to know what sodium fluoride does.

PARKER: I think many people, including Dr. Sutherland, would like to have the answer to that. There is an increase in maximal velocity of the enzymatic reaction without any change in the affinity of the adenylate cyclase for ATP.

KORNBERG: You are saying that sodium fluoride acts directly on that protein.

PARKER: Yes, that certainly appears to be the case in essentially all adenylate cyclase systems that have been well studied, and our own results in lymphocytes are consistent with this interpretation. Of course, adenylate cyclase has been difficult to purify and we don't know what the effect of fluoride would be on the pure enzyme.

EDELMAN: In almost all systems sodium fluoride gives maximum activation of adenylate cyclase, and the levels are usually higher than can be obtained with a single hormone or a combination of hormones. In most systems pre-stimulation with fluoride blocks an additional response with any other stimulus. It appears, therefore, that the maximum capability of the enzyme is achieved with fluoride. This suggests a direct effect on the enzyme.

RAMWELL: In leukocyte homogenates Butcher and Baird find that PGE(1) is more effective than fluoride in stimulating cyclic AMP accumulation. One could argue that if the experiments were done at the pH optima of the two agents on cyclase, the effect of fluoride may be greater. However, this type of data is clearly necessary before one can construct good arguments.

KORNBERG: How about the soluble cyclic AMP systems that have been isolated from bacteria?

EDELMAN: I do not know the answer to that question.

THE ROLE OF PROSTAGLANDINS IN THE REGULATION OF GROWTH AND MORPHOLOGY OF TRANSFORMED FIBROBLASTS

G. S. Johnson, I. Pastan, C. V. Peery,
J. Otten, and M. Willingham

Lab. of Molecular Biology, National Cancer Inst.
National Institutes of Health, Bethesda, Md.

The use of $N^6,O^{2'}$-dibutyryl 3',5'-adenosine monophosphate (dbc-AMP) has demonstrated that cyclic AMP regulates several functions of fibroblasts (Table I) (Johnson et al, 1971a and 1971b; Sheppard, 1971; Hsie and Puck, 1971). Since these functions are frequently altered by transformation, the regulation of cyclic AMP metabolism is of obvious inportance to cancer research.

We have found that following addition of dibutyryl-cyclic AMP, several characteristics of transformed fibroblasts are restored towards normal (Johnson et al, 1971a and 1971b). The cellular processes are elongated, the cells frequently align in parallel, the cells tend to spread out and occupy a greater area in contact with the substratum, and the cells are more difficult to remove from the substratum. Also, within minutes after addition of dibutyryl-cyclic AMP, the motion of the cells is markedly decreased (Johnson et al, 1971c).

It is not clear what regulates cyclic AMP levels in fibroblasts. The prostaglandins, which are of widespread occurrence and may function to regulate cyclic AMP levels in several systems, may play a role in this regulation.

We have shown that prostaglandins activate adenylate cyclase in cell extracts of some but not all cells (Peery et al, 1971). Prostaglandins activate adenylate cyclase in L-929 cell extracts (Table II). PGE(1) is by far the most effective. PGE(2) is less active, but PGB(1), PGA(2), and PGF(2-alpha) have very little effect. Similar results are obtained in extracts of mouse embryo fibroblasts and 3T3 cells (Table III), but the adenylate cyclase of the SV40 transformed 3T3 cells is unresponsive to all of the prostaglandins. It should be noted that this loss of responsiveness is not a property of SV40 transformation since other SV40 transformed lines have retained the response to PGE(1) (Peery et al, 1971).

Since PGE(1) activates adenylate cyclase in extracts of some but not all cells, addition of PGE(1) to intact cells which have a responsive adenylate cyclase should elevate intracellular cyclic AMP levels. The increased cyclic AMP should then alter several aspects of the behavior of fibroblasts (Table I). However, in cells such as the 3T3-SV40 line, the adenylate cyclase should not be stimulated and no response should occur.

In L-929 cells cyclic AMP levels increase ten-fold within 5 min after addition of PGE(1) and after 1 hr the levels slowly decrease (Otten, Johnson, Pastan, unpublished data). This elevation of cyclic AMP levels is followed by the expected changes in morphology and growth (Johnson and Pastan, 1971). At low cell density, the cellular processes are considerably extended and the cells acquire a marked spindly appearance. If the cells are treated with PGE(1) for extended periods of time (with frequent medium changes), the cells continue to divide, but at a much slower rate, and at confluency the cells lose the spindly appearance and become more round in shape. This morphological change is not observed in treatment of the L-929 cells with dibutyryl-cyclic AMP where the cells retain the spindly appearance at confluency and occasionally a parallel alignment of the cells is observed. This suggests that the effect of PGE(1) on confluent cells is independent of cyclic AMP. Also, the motion of the L-929 cells across the substratum is considerably decreased and the cells are more difficult to remove from the substratum.

Within a few hours after the addition of PGA(2) to L-929 cells, and also mouse embryo fibroblasts or BHK-21 fibroblasts, the cells withdraw the cellular processes and detach from the substratum. Since PGA(2) is essentially ineffective in activating adenylate cyclase, this effect may be independent of cyclic AMP.

PGE(2) is moderately active in stimulating adenylate cyclase activity. Accordingly, within a few hours after addition of PGE(2) to a culture of L-929 cells, the cellular processes are extended, but less so than after addition of PGE(1). However, after about 2 days incubation, the cells begin to withdraw the cellular processes and detach from the substratum. This effect is possibly due to metabolism of the PGE(2) into PGA(2).

No increase in cyclic AMP levels or alterations in morphology could be detected during treatment of 3T3-SV40 cells with PGE(1). Also, Table IV shows that PGE(1) at 50 micrograms/ml has no effect on the growth of these cells.

These results suggest that prostaglandins may play a role in the regulation of cyclic AMP levels in the intact cells. Thus, prostaglandins may be important in the regulation of growth and morphology of normal cells and defects in prostaglandin metabolism may explain in part some of the abnormal properties of cancer cells. A detailed analysis of prostaglandin levels in normal and transformed cells would be beneficial to understanding this possibility.

SUMMARY

Many lines of mouse and rat fibroblasts including mouse L-929 cells contain an adenylate cyclase that is activated by prostaglandins. Other lines of fibroblasts including one line transformed by SV40 virus have lost the prostaglandin response. The adenylate cyclase from L-929 cells is stimulated by prostaglandins in the following order of effectiveness: PGE(1) > PGE(2) > PGF(2-alpha). In L-929 cells, PGE(1) elevates cyclic AMP levels. Associated with the increase in cyclic AMP levels, the following responses are observed: (1) decreased cell motility, (2) decreased growth rate, (3) increased adherence of the cells to the substratum, and (4) restoration of some morphologic characteristics. In the SV40 line containing an adenylate cyclase unresponsive to prostaglandins, none of these responses are observed. Thus, prostaglandins acting through the adenylate cyclase system seem to control many important aspects of the behavior of fibroblasts.

However, different results are observed with PGA(2): within a few hours cells withdraw cellular processes and detach from the substratum. Since PGA(2) is ineffective in stimulating adenylate cyclase activity, this effect may be independent of cyclic AMP.

TABLE I

PROPERTIES OF FIBROBLASTS UNDER CONTROL OF CYCLIC AMP

1) morphology
2) growth
3) motion
4) adhesion to substratum

TABLE IV

EFFECT OF PGE(1)

ON THE GROWTH OF SV40 TRANSFORMED 3T3 CELLS

	Cells/cm²	
Time (hr)	Control	PGE(1) (50 microgram/ml)
0*	54,000	57,000
6	79,200	83,600
19	154,000	157,000
25	173,000	180,000
43	250,000	250,000

*Prostaglandin was added 24 hr after planting.

TABLE II

ACTIVATION OF L-929 ADENYLATE CYCLASE BY PROSTAGLANDINS*

Activity (picoM cyclic AMP formed/mg protein/10 min)						
basal	F^- (8.5 mM)	PGE(1)	PGE(2)	PGB(1)	PGA(2)	PGF(2-alpha)
30	1050	1220	275	58	43	62

TABLE III

ADENYLATE CYCLASE ACTIVITY OF 3T3 AND SV40-3T3 CELLS*

	Activity (picoM cyclic AMP formed/mg protein/10 min)				
	basal	PGE(1)	PGE(2)	PGB(1)	PGF(2-alpha)
mouse embryo fibroblasts	223	920	386	395	280
3T3	29	590	53	103	47
SV40-3T3	32	40	25	47	36

*Concentration of prostaglandins is 50 microgram/ml.

REFERENCES

Hsie, A., and Puck, T. T., 1971, Morphological transformation of Chinese hamster cells by dibutyryl adenosine cyclic 3',5'-monophosphate and testosterone, Proc. Nat. Acad. Sci. U.S.A. 68:358.

Johnson, G. S., and Pastan, I., 1971, Prostaglandins alter the growth and morphology of fibroblasts, J. Nat. Cancer Inst. (in press).

Johnson, G. S., Friedman, R. M., and Pastan, I., 1971a, Restoration of several morphological characteristics of normal fibroblasts in sarcoma cells treated with adenosine 3',5'-cyclic monophosphate and its derivatives, Proc. Nat. Acad. Sci. U.S.A. 68:425.

Johnson, G. S., Friedman, R. M., and Pastan, I., 1971b, Cyclic AMP treated sarcoma cells acquire several morphological characteristics of normal fibroblasts, Ann. N. Y. Acad. Sci. 185:413.

Johnson, G. S., Morgan, W. D., and Pastan, I., 1971c, Regulation of cell motility by cyclic AMP, Nature New Biol. (in press).

Peery, C. V., Johnson, G. S., and Pastan, I., 1971, Adenylate cyclase in normal and tranformed fibroblasts in tissue culture: activation by prostaglandins, J. Biol. Chem. 246:5785.

Sheppard, J. R., 1971, Restoration of contact-inhibited growth in transformed cells by dibutyryl adenosine 3',5'-cyclic monophosphate, Proc. Nat. Acad. Sci. U.S.A. 68:1316.

DISCUSSION

The following discussion covers the paper by Johnson et al, which was presented by Dr. Pastan.

RASMUSSEN: I have noticed from your data that you have a maximal effect of PGE(1) with 5 microgram/ml on adenylate cyclase activity.

JOHNSON: Concentrations as high as 10 microgram/ml PGE(1) were necessary to see any morphological changes. We routinely did experiments at 50 microgram/ml, where we observed large effects. These higher concentrations were required presumably due to metabolism of the prostaglandin or to binding to the fetal calf serum (10%) in the growth medium. The 10-fold increase in intracellular levels of cyclic AMP was observed with 50 microgram/ml PGE(1). We have not yet tested this at lower concentrations. Does anyone know whether serum binds prostaglandins?

KESSLER: In studies on the isolation of prostaglandin from plasma or animals in shock we have found that normal plasma, after extraction with ethyl acetate at low pH, showed evidence of prostaglandin-like activity in the supernatant but none in the residue. By contrast, in the shock plasma we get exactly the reverse effect. These opposing observations in normal and shock plasma have led us to raise the same question about shock. Is prostaglandin bound to plasma protein in shock?

WEISSMANN: In your transformed and untransformed cells, have you also done ultrastructural studies to determine the effects of prostaglandins on structural proteins, particularly microtubules or microfilaments?

JOHNSON: We are doing that; colchicine will prevent cell stretching.

WEISSMANN: What about cytochalasin B?

JOHNSON: We have not tested that in our system.

WEISSMANN: What elements, do you think, are mediating the structural changes which you have shown?

JOHNSON: I assume it's through microtubules and microfilaments.

WEISSMANN: Similar work has suggested that following exhibition of dibutyryl-cyclic AMP, microtubules may be affected (experiments, I believe, from the laboratories of Puck and Porter).

JOHNSON: We have confirmed their results.

BRAUN: I have some data that do not deal with prostaglandins, but rather with the effects of endogenous cyclic AMP stimulation on the rate of growth of tumor cells in vivo. Our studies have been based on our prior demonstration (1971, J. Immunol., 106:1399) that synthetic double-stranded polynucleotides, such as poly A:U or poly I:C, are potent stimulators of adenylate cyclase activity. Subsequently, in studies on immune responses (1971, J. Immunol. 107:1027, 1036) we found that poly A:U, particularly when combined with theophylline, produced a very dramatic stimulation of antibody formation at low levels of theophylline and inhibition of the response at high levels. We found that with high levels of theophylline, in the presence of poly A:U, it is easier to turn off B cells, i.e., the antibody-forming cells, than T cells, i.e., the thymus-dependent cells that are responsible for cell-mediated immunity. Since immune responses to syngenic tumor cells depress tumor growth if they are due to T cell function but enhance tumor growth when they are due to B cell function, which leads to so-called blocking antibody, we wondered whether we might obtain favorable immune response in tumor-bearing hosts by using poly A:U and theophylline as stimulators of endogenous cyclic AMP levels. We used two test systems; one is a Rauscher leukemia virus-induced ascites tumor growing intradermally, and the other one employed regrowth of methylcholanthrene-induced tumors after excision. In both systems poly A:U or theophylline, or a combination thereof, proved effective. Such treatment either prevented recurrence of tumor in the system where we were dealing with excision, or it retarded the growth of the RLV tumor as shown in the figure. You will note that the combination of poly A:U + theophylline was more effective than either agent alone. Initially we assumed that this effect was due to an appropriate stimulation of the immune response, but it turned out that the effects were just as good in irradiated animals that had received enough irradiation to knock out the immune response. We then proceeded to test the effects of pretreatment of tumor cells with poly A:U and/or theophylline, prior to implantation into syngenic hosts, on subsequent tumor growth in vivo, and found that this also resulted in an inhibition of growth. Consequently, we had to conclude that we are dealing not so much with altered immune responses, but rather we are dealing with a stimulation of endogenous cyclic AMP levels which has an inhibitory effect on tumor cell multiplication in vivo, the effects of the cyclic AMP stimulators being largely direct inhibitory effects on tumor cells.

RAMWELL: I am concerned about the action of PGA(2) in causing cell death in your tissue culture. You are going to make the cardiovascular people very unhappy, since they believe the PGA compounds have far less deleterious effects than the PGE compounds.

JOHNSON: I tested PGA(2) on L-929 cells, mouse embryo fibroblasts, and BHK-21 cells. In all three of these lines the cells rounded up and detached from the substratum within a few hours.

RAMWELL: You stated that there were some differences between the actions of prostaglandins and cyclic AMP in your cells. Is that correct?

JOHNSON: The majority of the responses we observed in our cells are identical to those observed by addition of dibutyryl-cyclic AMP. However, two responses are not. The L-929 cells treated with prostaglandin flatten at confluency and lose the spindly appearance. The same cells treated with dibutyryl-cyclic AMP retain this spindly appearance at confluency. The other response was that observed during treatment with PGA(2). Since this was not observed with dibutyryl-cyclic AMP, and PGA(2) does not stimulate adenylate cyclase activity, it seems likely that this is independent of cyclic AMP.

RASMUSSEN: You did not show any data on cell lines that were unresponsive to cyclic AMP in terms of their response to PGE(1) at 50 microgram/ml. Does PGE(1) at those concentrations have any effect on the growth rate or morphology of cells in which adenylate cyclase is unresponsive to PGE(1)?

JOHNSON: This is difficult to quantitate. It seems reasonable from my data that there is no effect in prostaglandin-unresponsive cells.

ANGGARD: Since we are here to consider what physiological role the prostaglandins might have in cell biology, it might be worthwhile to remember that the secretion rate of prostaglandin is such that the likely concentrations at the sites of action would be about a thousand times less than those you used.

WEISSMANN: This is a problem that one has to face not only with prostaglandins, but with any endogenously produced material which is degraded by the body. One wants really to calculate what its possible concentration might be at a critical site. If we consider all the prostaglandin substrates within the various cell membranes, how much of how many prostaglandins could be formed from all the substrate phospholipids or triglycerides that are available for prostaglandin biosynthesis in response to tissue injury or inflammation?

ANGGARD: It's difficult to make those types of calculations, because if you take the total amount of arachidonate esterified to phospholipids or to steroids, you may end up with wrong figures. All of that may not be accessible at the site of formation. The prostaglandins are probably localized within certain cells, and within those cells they probably act very close to the site of action, otherwise they would never get above the threshold concentration.

EDELMAN: The circulating level of prostaglandins may have no relationship to physiological effects. In general, hormones are made in discrete tissues or organs and exported to well-defined target tissues. In contrast, prostaglandins are made by a variety of tissues and would therefore resemble cyclic AMP, i.e., act at the site and within the cell of origin. The addition of prostaglandin to an extracellular medium may yield biological effects, but these responses do not give any information on the role of circulating levels of prostaglandins in physiological action. One may ask if the prostaglandins modulate cyclic AMP activity, or do they respond to intracellular cyclic AMP levels and constitute, therefore, a secondary manifestation of the primary effects of hormones on adenylate cyclase enzyme activity?

RAMWELL: There is still no convincing evidence that cyclic AMP stimulates prostaglandin synthesis. Our problem has been that the high concentrations of cyclic AMP tend to interfere with the bioassays. However, we have evidence in frog skin that cyclic AMP does not stimulate prostaglandin release, but that isoproterenol does. The question that tends to arise is whether one prostaglandin, say PGE(1), will stimulate other prostaglandin synthesis.

<u>ALLEN</u>: It turns out that 3 x 10^{-5} M PGE(2) results in lysis of red cells.

PROSTAGLANDIN RELEASE BY HUMAN CELLS *IN VITRO*

B. M. Jaffe, C. W. Parker
and G. W. Philpott

Departments of Surgery and Medicine
Wash. Univ. School of Med., St. Louis, Mo.

INTRODUCTION

Recent evidence suggests that prostaglandins may be involved in the regulation of cellular proliferation and differentiation. Kischer (1967) has shown that PGE compounds cause thickening, stratification and precocious keratinization of the epidermis as well as inhibition of feather development in the chick embryo in culture. Prostaglandins have been shown to markedly increase levels of cyclic AMP in phytohemagglutinin-stimulated human lymphocytes (Smith et al, 1971a), associated with the inhibition of incorporation of tritiated thymidine, uridine and leucine into DNA, RNA and protein (Smith et al, 1971b). In this Symposium, Johnson et al (1972) have demonstrated that PGE(1) *in vitro* alters L cell growth, motility, adhesiveness and morphology. These responses accompany an increase in the concentration of cyclic AMP, and are thus thought to be mediated by activation of the adenylate cyclase system.

Using a radioimmunoassay for prostaglandins developed in our laboratory, (Jaffe et al, 1971) we have measured concentrations of prostaglandins in media in which a variety of normal and neoplastic tissues had been cultured. The results of these studies form the basis of this report.

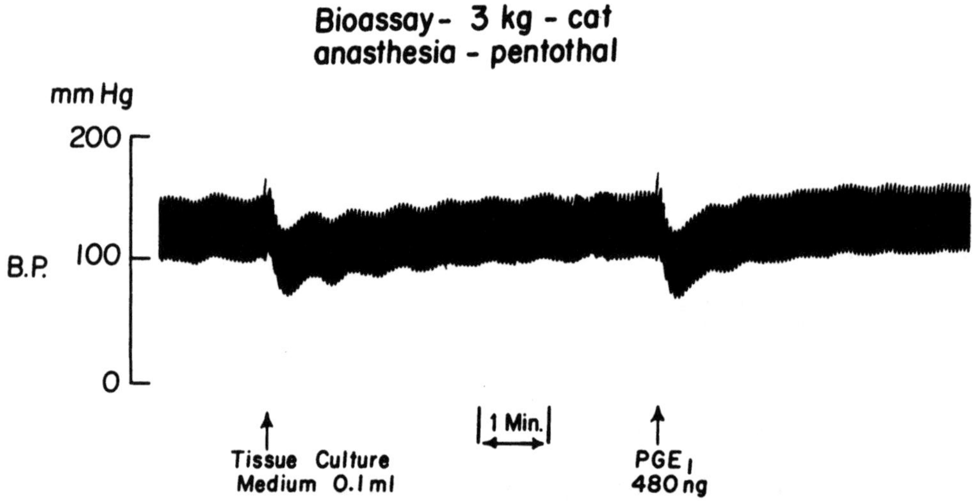

Figure 1. Kymographic recording of blood pressure in the anesthetized cat bioassay system. At the first arrow 0.1 ml of concentrated, extracted tissue culture medium from a colon carcinoma was injected. This is compared to the response to 480 nanograms of PGE(1) administered at the second arrow.

METHODS

Cultures of tumor tissue were established from sterile surgical specimens and maintained in matrix cultures by the method of Kalus et al (1968). The matrix support employed was a fibrin foam. Using media (MEM or 199) with up to 20% fetal calf serum, tumors have been maintained in culture for over one year with tissue growth into the interstices of the matrix. Normal gastrointestinal mucosa was studied in vitro using short term organ cultures as described by Browning and Trier (1969). HeLa and L cells were maintained as monolayers.

The method for prostaglandin radioimmunoassay has been previously described in detail (Jaffe et al, 1972). The antibodies to PGE(1) used in this study were elicited by immunizing a rabbit subcutaneously (in Freund's complete adjuvant) with PGE(1) conjugated to keyhole limpet hemocyanin with ethyl chloroformate (Oliver et al, 1968). After five to six immunizations over 15 months the serum contained antibodies which were suitable for use in a radioimmunoassay system. The maximal sensitivity of the immunoassay using [^{3}H]PGE(1) (New England Nuclear) as the radioactive marker is 10 picograms.

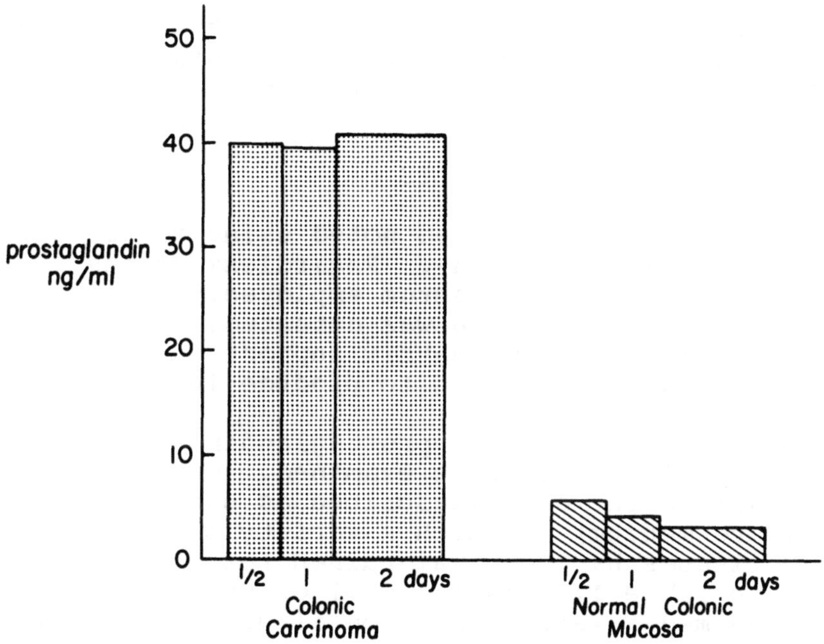

Figure 2. Concentrations of prostaglandin in tissue culture fluid in which normal colonic mucosa and colon carcinoma from the same surgical specimen were cultured. Media were changed every 6 hr and pooled as 0-12 hr, 12-24 hr, and 24-48 hr specimens for measurement of prostaglandins.

In order to measure prostaglandin in tissue culture medium, 0.1 ml of unprocessed medium was added to 0.05 ml of a 1:100 dilution of antibody globulin, 0.05 ml of carrier rabbit gamma globulin (10 mg/ml), and [^{3}H]PGE(1). After incubation for three days at 4°C, cold ammonium sulfate was added to a final concentration of 40% of saturation. One hr later, the reaction mixture was centrifuged at 2000 RPM and the supernatant was discarded. The precipitate was dissolved in 0.5 ml of Nuclear Chicago Solubilizer and counted in toluene scintillation solution. Concentrations of prostaglandins were determined from a calibration curve which was obtained by adding known concentrations of PGE(1) to measure displacement of antibody-bound [^{3}H]PGE(1).

In order to confirm that the compound measured in media was prostaglandin, media in which two colon carcinomas had been grown were pooled for two months, keeping all media frozen at -20°C. The pooled mixture was concentrated 120 fold by making an acidic-lipid extract (pH 3 - chloroform). This concentrated extract was tested in an anesthetized cat blood pressure bioassay (Kannegiesser and Lee, 1971) by Dr. James Lee.

RESULTS

There were no difficulties noted using tissue culture medium in the radioimmunoassay. Regardless of the concentration of fetal calf serum employed, no prostaglandin-like inhibition was detected. Media that had been incubated with the matrix support in the absence of cells for 48 hr also failed to produce inhibition.

Measurable quantities of prostaglandins were released into medium by all the tissues examined. Concentrations of prostaglandins in media (described as nanogram/ml of tissue culture medium) in which a variety of normal and neoplastic tissues had been maintained for at least 24 hr are listed in Table I. Since 2.0 ml of media were used in each dish and media were changed on alternate days, these figures approximate the amount of prostaglandin released in a given culture dish per day. The table shows marked concentration differences when epithelial and non-epithelial tissues are compared. Prostaglandin concentrations in the media of epithelial neoplasms were uniformly 15 nanogram/ml or higher whereas with non-epithelial tissues prostaglandin concentrations did not exceed 2.5 nanogram/ml. This difference between epithelial and non-epithelial tissues was not due only to differences in the rates of growth or the quantity of tissue present because both groups included rapidly growing as well as slowly growing tumors and the tissue culture sample sizes were comparable.

Confirmation of the presence of prostaglandin was obtained from the effect of concentrated, extracted medium (from colon carcinoma) in the cat blood pressure bioassay. There was a prompt fall in blood pressure (within 30 sec after infusion of the extract) followed by a slow return to control values over a 5 min period (Fig. 1). The shape of the curve strongly suggests the presence of PGE(1) and quite possibly PGA(1) as well. Calculating back to the original concentration of medium, the bioassay data yield an initial concentration of 40 nanogram/ml of PGE(1) which is quite consistent with the quantity of PGE(1) measured by radioimmunoassay. Activity in the bioassay is strong evidence that the compound released is not a degraded prostaglandin. This would have been unlikely in any case, since the most common prostaglandin degradation product is

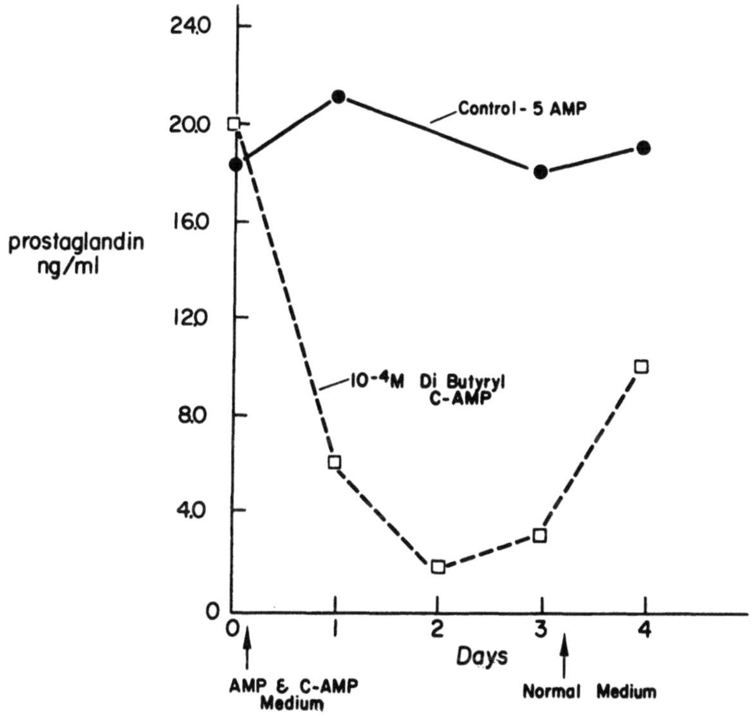

Figure 3. Prostaglandin release by colonic carcinoma in the presence of 10^{-4} M dibutyryl-cyclic AMP. After establishing baseline rates of prostaglandin release, test media were added at the first arrow. Media were restored to nucleotide-free media at the second arrow.

a 15-keto derivative and 15-keto prostaglandins cross react with anti-prostaglandin antibodies only to a negligible degree.

Since all of the matrix cultures of epithelial tumors contain stromal elements, it could be argued that they are the major source of the prostaglandins. To evaluate this possibility prostaglandin release was measured serially in a medullary carcinoma of the thyroid maintained for several weeks in tissue culture. Although initially epithelial components grew well, subsequently fibroblastic elements predominated. The release of prostaglandin was greater when the epithelial elements were still functioning (Table I). Some weeks later, when the culture had become a pure fibroblast line, it released less than 2.0 nanogram/day of prostaglandin into the culture medium.

Although the difference between epithelial and non-epithelial tissues was striking, there was one exception

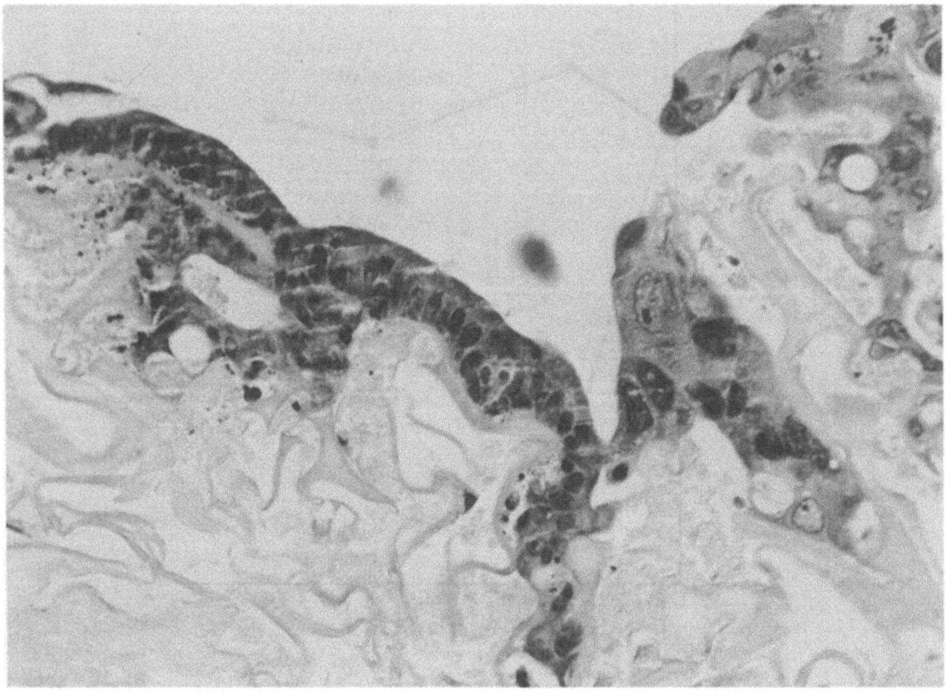

Figure 4. Photomicrograph of colon carcinoma maintained in culture for 8 months (x350). (Reduced 25% for reproduction.)

which merits mention. In monolayer cultures, HeLa cells (originally derived from an epidermoid carcinoma) released only 2 nanogram/day into the medium, which places it in the non-epithelial cell range. However, considering the long period of time in which this cell line has been maintained in tissue culture, marked alterations in cell behavior might have occurred. Thus it may not be truly representative of a malignant epithelial cell.

In another experiment, normal colonic mucosa and colon carcinoma from the same surgical specimen were maintained identically in short term cultures, changing the medium every 6 hr. Medium that had been removed from cultured cells was pooled (1-12, 12-24, 24-48 hr) and the prostaglandin concentrations measured. There was a marked difference in prostaglandin concentrations between media from the normal and neoplastic mucosa (Fig. 2) which did not appear to be due to differences in the number of epithelial cells in the two samples as judged by histological examination of serial sections. Prostaglandin release under these conditions was

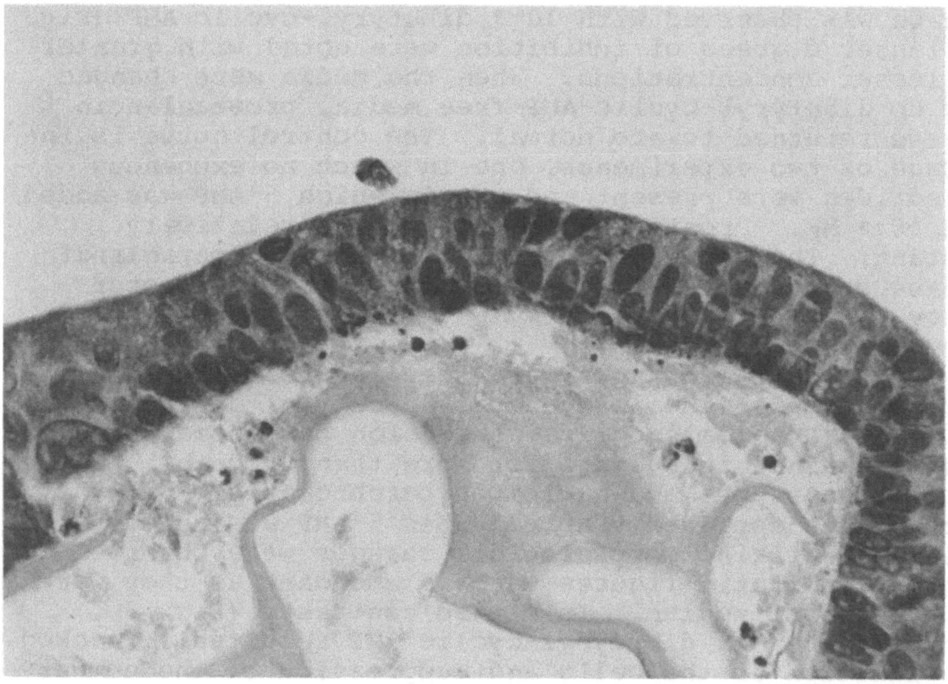

Figure 5. Photomicrograph of the same colon carcinoma after incubation in 5'AMP for 3 days (x600). (Reduced 25% for reproduction.)

relatively constant (Fig. 2). In another experiment with colon carcinoma we have documented increases in the amounts of prostaglandin found in the media after 8 as compared to 2 hr in culture, indicating that release is not a self-limited process restricted to the first 1-2 hr after the medium change.

A single preliminary experiment has been done to determine if the release of prostaglandin could be altered by culturing colonic carcinomas in exogenous dibutyryl-cyclic AMP. After determining baseline prostaglandin release rates for two colon carcinomas, these tumor cells were cultured in duplicate dishes in dibutyryl-cyclic AMP in concentrations of 10^{-3} M, 5 x 10^{-4} M, 10^{-4} M, and 5 x 10^{-5} M. The medium was changed daily, maintaining dibutyryl-cyclic AMP in the media at the desired concentrations. After 3 days, dibutyryl-cyclic-AMP-free medium was again utilized. Prostaglandin release was measured sequentially by radioimmunoassay. With each concentration of exogenous dibutyryl-cyclic AMP there was a significant decrease in the amount of prostaglandin released. The most striking inhibition of prostaglandin release was observed with 10^{-4} dibutyryl-cyclic AMP (Fig. 3); lesser degrees of inhibition were noted with greater and lesser concentrations. When the media were changed back to dibutyryl-cyclic-AMP-free media, prostaglandin release returned toward normal. The control curve is the average of two experiments, one in which no exogenous nucleotides were present and one in which 5'AMP was added (5×10^{-4} M). Prostaglandin release was relatively constant. The failure of 5'AMP to inhibit prostaglandin release suggests the importance of the phosphodiester moiety.

Histological changes were seen in the cultured carcinomas under the influence of dibutyryl-cyclic AMP. Fig. 4 is a photomicrograph of a colon carcinoma which had been maintained in culture for more than 8 months. There is cellular atypia and nuclear hyperchromatism, but most of the cells appeared viable. When 5'AMP was added to the medium no striking morphological changes were observed (Fig. 5). Mitotic figures were as abundant as they were in the control medium. In marked contrast, (Fig. 6), culture in 10^{-3} M dibutyryl-cyclic AMP resulted in marked vacuolization of the cells and suppression of the number of mitotic figures. This effect was reversible. When dibutyryl-cyclic AMP was subsequently omitted from the media, the cells grew normally.

DISCUSSION AND CONCLUSIONS

In these experiments we have applied the principle of radioimmunoassay to the measurement of prostaglandins in media in which a variety of normal and neoplastic tissues had been cultured. All of the tissues examined released significant amounts of prostaglandin into the media over a period of 24 to 58 hr. While the kinetics of prostaglandin release have not been studied in detail, it is already clear that in cultured colonic tumor cells prostaglandin release is sustained over a period of several hr despite the accumulation of nanogram amounts of prostaglandins in the medium. Detailed information as to the pattern of individual prostaglandin release is not available. However, the anti-PGE(1) antibody used in these studies has greater specificity for PGE(1) than for the other prostaglandins so that PGE(1) is the most likely source of the inhibition. In colonic carcinoma culture fluid the presence of large amounts of PGE was directly substantiated by the results of the cat bioassay. Moreover, the quantity of PGE(1) measured by bioassay was

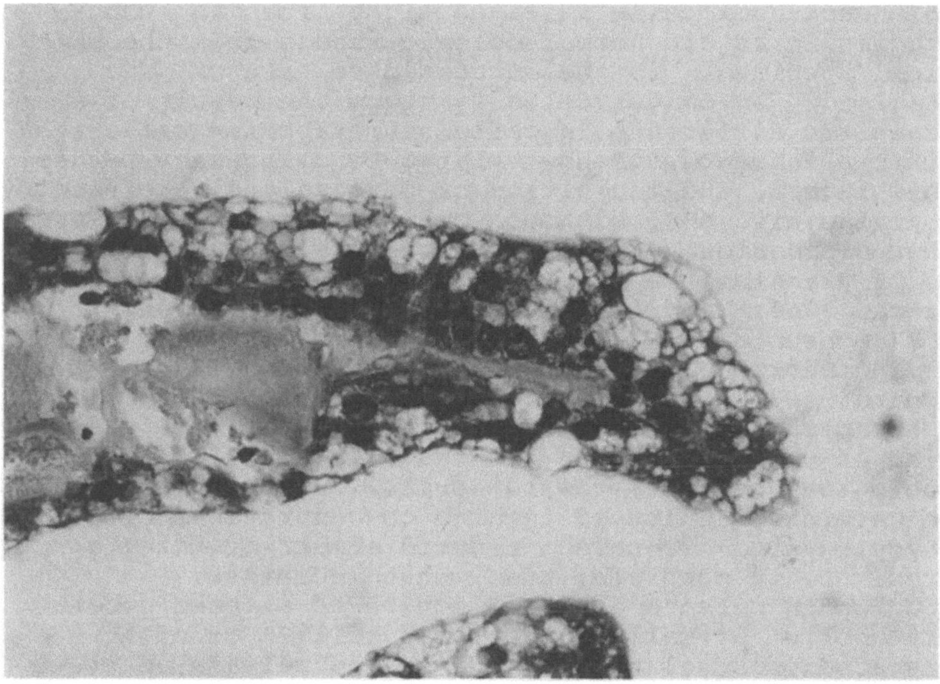

Figure 6. Photomicrograph of the same colon carcinoma after incubation in dibutyryl-cyclic AMP for 3 days. Pronounced vacuolization of the cells has occurred (x600). (Reduced 25% for reproduction.)

in good agreement with the value obtained by radioimmunoassay. Thus, in colon cancer cell media at least, non-prostaglandin long chain fatty acids which cross react immunologically with PGE(1) were not contributing in any major way to the immunoassay results.

The difference in the amounts of prostaglandin released by epithelial as compared to non-epithelial tissues is quite interesting. The greater rate of release from epithelial cell cultures cannot be ascribed to differences in the growth rates. Furthermore, a medullary carcinoma of the thyroid released large amounts of prostaglandin when epithelial elements predominated, but only minimal amounts after the culture had become fibroblastic in character. In view of the limited number of tissues examined we cannot exclude the possibility of significant overlap between these two groups. Indeed HeLa cells, epithelial in origin, released only small amounts of prostaglandin.

Of even greater interest is the observation that colonic carcinoma cells released eight times as much prostaglandin as did normal colonic mucosa from the same surgical specimen. Is the increased release of prostaglandin by cancer cells in vitro the result of a non-specific difference in cellular metabolism or is it a specific phenomenon somehow related to malignancy per se? In this regard, the inhibition of prostaglandin release by dibutyryl-cyclic AMP, although only preliminary, deserves further discussion. Data with a variety of tumor cell lines [F 1 amnion, Hep-2, L cells and HeLa cells (Ryan and Heidrick, 1968; Heidrick and Ryan, 1970), L-929 and XC cells (Johnson et al, 1971), Chinese hamster ovary cells (Hsie and Puck, 1971), and 3T6 and PyV 3T3 cells (Sheppard, 1971)] suggests that culture in high concentrations of cyclic and dibutyryl-cyclic AMP causes slowing of the growth rate and restoration of morphological features toward normal. We have found that colon carcinomas cultured in high concentrations of dibutyryl-cyclic AMP have a reduced number of mitotic figures. Is it possible, then, that increased prostaglandin release is a reflection of altered cyclic AMP metabolism by malignant colonic cells? Or might accelerated prostaglandin synthesis and release be a primary event with secondary changes in intracellular metabolism favoring unrestrained growth? Clearly, answers to these and other questions require extensive further investigation in regard to prostaglandin synthesis, release and catabolism in malignant and normal cells. Of equal importance will be additional studies with the same cells on the interactions of intra- and extracellular prostaglandins with the adenylate cyclase system in vitro.

TABLE I

CONCENTRATIONS OF PROSTAGLANDINS IN TISSUE CULTURE MEDIUM

Epithelial cells	Prostaglandin (nanogram/ml)
Colon carcinomas (13)	30 - > 50
Carcinoid	17
Esophageal carcinoma	22
Duodenal carcinoma	17
Papillary carcinoma of thyroid	18
Medullary carcinoma of thyroid	30
Insulinoma	44
Dog renal cell carcinoma	> 50
Ovarian carcinoma	> 30
Endometrial carcinoma	> 30
Pancreatic carcinoma	> 30
Non-epithelial cells	
Human fibroblasts (4)	< 2.5
L cells	0.3
Lymphosarcoma	2.0
Liposarcoma	0.2
Neuroblastoma	0.5

ACKNOWLEDGEMENTS

The authors gratefully acknowledge Dr. J. Lee and H. Kannegiesser for having performed the cat bioassay experiment, Drs. J. Pike and W. Magee of the Upjohn Corporation for supplying the prostaglandins, and Miss D. Collier for her skilled technical assistance.

The work described in this paper was supported by Training Grant NIGMS 371.

REFERENCES

Browning, T. H., and Trier, J. S., 1969, Organ culture of mucosal biopsies of human small intestine, J. Clin. Invest. 48:1423.

Heidreck, M. L., and Ryan, W. L., 1970, Nucleotides on cell growth *in vitro*, Cancer Res. 30:376.

Hsie, A. W., and Puck, T. T., 1971, Morphological transformation of chinese hamster ovary cells by dibutyryl-adenosine 3',5'-monophosphate and testosterone, Proc. Nat. Acad. Sci. 68:358.

Jaffe, B. M., Smith, J. W., Newton, W. T., and Parker, C. W., 1971, Radioimmunoassay for prostaglandins, Science 171:494.

Jaffe, B. M., and Parker, C. W., 1972, Radioimmunoassay for prostaglandins in "Radioassays in Clinical Medicine", (R. Donati and W. T. Newton, eds.) Charles Thomas, Springfield, Ill.

Johnson, G. S., Friedman, R. M., and Pastan, I., 1971, Restoration of several morphological characteristics of normal fibroblasts in sarcoma cells treated with adenosine 3',5'-cyclic monophosphate and its derivatives, Proc. Nat. Acad. Sci. 68:425.

Johnson, G. S., Pastan, I., Peery, C. V., Otten, J., and Willingham, M., 1972, The role of prostaglandins in the regulation of growth and morphology of transformed fibroblasts, in the proceedings of this Symposium.

Kalus, M., Ghidoni, J., and O'Neal, R., 1968, The growth of tumors in matrix cultures, Cancer 22:507.

Kannegiesser, H., and Lee, J. B., 1971, Difference in hemodynamic response to prostaglandins A and E, Nature 229:498.

Kischer, C. W., 1967, Effects of specific prostaglandins on development of chick embryo skin and down feather organ *in vitro*, Dev. Biol. 16:203.

Oliver, G. C., Parker, B. M., Brasfield, D. L., and Parker, C. W., 1968, The measurement of digitoxin in human serum by radioimmunoassay, J. Clin. Invest. 47:1035.

Ryan, W. L., and Heidreck, M. L., 1968, Inhibition of cell growth *in vitro* by adenosine 3',5'-monophosphate, Science 162:1484.

Sheppard, J. R., 1971, Restoration of contact-inhibited growth to transformed cells by dibutyryl-adenosine 3',5'-cyclic monophosphate, Proc. Nat. Acad. Sci. 68:1316.

Smith, J. W., Steiner, A. L., Newberry, W. M., and Parker, C. W., 1971a, Cyclic adenosine 3',5'-monophosphate in human lymphocytes. Alterations after phytohemagglutinin stimulation, J. Clin. Invest. 50:432.

Smith, J. W., Steiner, A. L., and Parker, C. W., 1971b, Human lymphocyte metabolism: Effects of cyclic and noncyclic nucleotides on stimulation by phytohemagglutinin, J. Clin. Invest. 50:442.

DISCUSSION

JOHNSON: I didn't understand what you said in the beginning of your paper about serum and the way it affects measurement of prostaglandin.

JAFFE: There is no question that prostaglandin does bind to albumin (see Shio and Ramwell, this Conference), and we were afraid that this would complicate the determination of prostaglandin in tissue culture medium. Indeed, if you take straight human serum, it is impossible to measure prostaglandin with the radioimmunoassay system directly, presumably because of the albumin present. We have found, however, that in serum diluted 1:10, you can measure prostaglandin with little difficulty. More importantly, measurement of prostaglandin in tissue culture medium with up to 25% fetal calf serum, did not impose any problems. Calibration curves in buffer and tissue culture media with varying amounts of fetal calf serum were very similar. There was no non-specific inhibition of the reaction between labeled prostaglandin and antibody by medium alone, which suggests that there are insignificant amounts of prostaglandin in control media. Furthermore, we have avoided just the problem you alluded to in your talk, that is, non-specific binding to proteins, enough to invalidate measurement of prostaglandins in tissue culture media.

JOHNSON: We have tested five epithelial cell lines and we saw essentially no morphological change in the presence of dibutyryl-cyclic AMP.

JAFFE: At what concentrations?

JOHNSON: 1 mM, the same as you used.

JAFFE: Were these normal or cancer cells?

JOHNSON: Hela, monkey kidney, and human choreocarcinoma cells. We have seen no changes similar to what we have seen in our fibroblast lines. In regard to the vacuoles that you saw, we have actually found one line, an SP40 transformed 3T3, which shows extensive vacuolization in the control, but after treatment there was very little. This is essentially just the opposite of what you have seen.

ELLIS: I would like to know how you interpret your results in terms of Dr. Johnson's findings. You report that cancer cells produce prostaglandins, while Dr.

Johnson indicates that some prostaglandins stop cellular growth in tissue culture. Do you feel that prostaglandins enhance cellular growth of cancer cells, or have the cancer cells you studied lost the mechanism whereby prostaglandins may inhibit growth?

JAFFE: It was no accident that I didn't make any inference at all. The initial problem is to determine whether production of large amounts of prostaglandin by tumors is a primary or a secondary factor. I don't know if there is, as a primary factor, an alteration in the adenylate cyclase system resulting in some sort of alteration in metabolism that results in unrestrained growth (manifested by production of large amounts of prostaglandins) or if a prostaglandin abnormality is the primary factor. It is also possible that the tissue produces prostaglandins in an attempt to limit its own growth. I realize all of this is pure speculation. I think we can consider at this point, from our studies and those of Dr. Johnson, that cancer cells in culture are not totally independent and do respond to stimuli which can be measured and quantitated. I think it would be inappropriate to say any more than that.

JOHNSON: I think it's important to emphasize that we have to really specify the cell type when discussing data like these. Your data deal primarily with epithelial cells. Have you also compared prostaglandin effects in normal fibroblasts to see how that would compare to the L cells?

JAFFE: Yes, they were very much the same. We have measured prostaglandin release by four normal fibroblast lines and L cells and they were the same, that is in the 2 nanogram/day range.

KORNBERG: There seems to be some difference in results. I wonder if you are aware of and concerned about the pH of the medium and its influence on the response of cells. Harry Eagle, in his recent work, showed that the normal contact inhibition is not observed if one controls the pH. Is this something that you have looked at closely?

JAFFE: Not closely, but there is no way that you can successfully maintain tissue in culture without being conscious of pH. Among other things, there is a marker in the medium. You invariably must know if there has been a significant change in pH.

KORNBERG: Yes, but isn't it invariably indicating that the medium is too acid or alkaline? It was my impression that when one sees the phenol red go yellow, everyone ignores it. Eagle has added strong buffers to the bicarbonate medium and finds remarkable effects; he obtains cell densities severalfold greater than are seen even with tumor cell lines.

JAFFE: I can't answer that specifically, but we have measured pH sequentially and it doesn't change significantly. Consequently, we haven't specifically attempted to buffer this out or change the medium. We used two mediums during our experiments: one of them is the MEM, essential medium, and the other a commercial medium, 199. They were used uniformly and there were no media-related differences and the pH remained constant.

RAMWELL: In a previous conversation, you alluded to the use of synthetase inhibitors. Could you comment on that?

JAFFE: That is something we intend to look at shortly, but first we have to document and measure synthesis using labeled precursors. It would of course be unrealistic to assume that there has been no synthesis of prostaglandin by these cultured tumors. We have measured prostaglandin release by two colon cancers over the course of a year. The concentrations of prostaglandin that these tumors make now is about the same that they were a year ago. Incidentally, these are the same ones incubated in dibutyryl-cyclic AMP. I think it must be obvious that if they are still making prostaglandin a year later, they must be synthesizing it, but we have no data to document the specific synthesis. Once we have measured the rates of prostaglandin synthesis, then we will look at specific inhibitors, in particular some of the substituted long chain fatty acids and substituted prostaglandin analogs.

EDELMAN: I am curious about some of the temporal aspects of the relationship between exogenous cyclic AMP content and prostaglandin release in the dibutyryl-cyclic AMP experiments. 24 hr after the removal of the dibutyryl-cyclic AMP from the medium, you still had a substantially inhibited level. In the systems that I am familiar with, cyclic AMP tends to be metabolized very rapidly, and the physiological effects disappear in less than 1 hr.

JAFFE: There are several possible explanations. First, this particular concentration was the only one in which we did not see prostaglandin release return to baseline levels. When we used the other concentrations, which were both higher and lower, we did return to the initial rate of release. Experimentally, at the end of each day, when we change the medium, the dibutyryl-cyclic AMP was washed out with three changes of normal media and then the dibutyryl-cyclic AMP-containing medium was replaced to maintain the dibutyryl-cyclic AMP concentration. It is only an assumption at this point, that the diburyryl-cyclic AMP must be taken up intracellularly to cause the effects we have described. Second, this is not a physiological situation. I have no idea what enzymes are present and functioning in these media, which all contain fetal calf serum. After the colon cancers have been out of the body 8 months, I can't tell you for sure which enzymes are operative and which are not. Third, this study was done with matrix cultures, in which tumors are grown on a fibrin foam deliberately arranged with interstices for the tumor to grow into. One of the obvious answers is that there is no way that we could have gotten these interstices completely free of whatever we put in the media. I presume one of the reasons we still saw a late effect was failure to completely wash out all the dibutyryl-cyclic AMP. So, although I can't answer your question directly, I have explained some of the possible places where we might have had problems.

GOODSON: Do your media contain arachidonic acid?

JAFFE: The media do not, but it is possible that the fetal calf serum does contain a small amount. Additionally, arachidonic acid does cross react with the radioimmunoassay, and we didn't measure any cross-reacting activity in new media, as we discussed before.

RAMWELL: To what degree does arachidonic acid cross-react?

JAFFE: It depends on the concentration. At 10^{-5} M it's about 70%, but at a concentration of unlabeled prostaglandin at 10^{-5} M we have total inhibition. At as low as 10^{-7} M, there's virtually no cross-reactivity, and at 10^{-7} M prostaglandin, we still have total inhibition. At the concentrations we are measuring, about 100 picoM/ml, cross-reactivity is not a problem.

LEE: If you have demonstrated synthesis over a period of a year, there must be some supply of precursor.

JAFFE: I assume that's true. The experiments have been done in fetal calf serum, and we expect that the fetal calf serum contains some precursors. I don't think it's necessary that it be arachidonic.

RAMWELL: Since you're measuring prostaglandin E(1), the immediate precursor is homo-gamma-linolenic acid.

JAFFE: That is true, but nonetheless homo-gamma-linolenic acid had to have come from precursors; primarily smaller fatty acids in the presence of malonyl CoA. Although the media may not have contained homo-gamma-linolenic acid, fetal calf serum contains smaller fatty acids which are then increased in size to become the 20 carbon precursors which are then cyclized to form prostaglandins. The question of whether or not a direct precursor is in media is relevant only in relation to the bioassay. Dr. Lee, do arachidonic and homo-gamma-linolenic acids cross-react in the cat bioassay?

LEE: In the rat blood pressure bioassay and, I presume, in the cat assay, they do not. We originally looked very closely at homo-gamma-linolenic as well as arachidonic assay in the course of isolating prostaglandins from the kidneys some years ago. We did find that if we took a commercial preparation of arachidonic acid, injected it into the rat and measured blood pressure, we observed a fall in pressure with a curve that was identical to our renomedullary extracts. After much effort we found out that commercial preparation of these unsaturated fatty acids are often contaminated with peroxides of fatty acids. Peroxides of fatty acids will give you a fall in blood pressure with characteristics very similar to those observed with prostaglandins. In fact, we originally thought that the vasodepressor activity of the renal medulla was due to such peroxides. In any case, if you chromatograph arachidonic acids on alumina, fatty acid peroxides are absorbed and it is then clear that pure arachidonic acid is devoid of any blood pressure lowering effects. I suspect your tumors are not making peroxides of fatty acids, but indeed are making prostaglandins and that you are not dealing with an unsaturated fatty acid precursor. The remote possibility remains, however, that these tumors contain peroxides of fatty acids.

BEHRMAN: Dr. Jaffe, you have raised antibodies to other prostaglandin conjugates. Have you assayed the media in which you cultured explants of cancerous tissue for prostaglandins other than PGE? It is possible that these abnormal tissues may not increase total prostaglandin biosynthesis but rather produce an increase in PGE at the expense of some other prostaglandin.

JAFFE: That is obviously something that has intrigued us, but which we haven't completely evaluated. It looked like the prostaglandin released is primarily PGE(1), but what you are saying may be right. We now have antibodies to PGE(1) and PGE(2), PGA(1) and PGA(2), and PGF(1-alpha), so that we ought to be able to answer that question. We are actually right now in the process of characterizing which prostaglandins are indeed being synthesized, and although I can't tell you the answer now, we will have it shortly. The good agreement between the bioassay and the radioimmunoassay data strongly suggests that we are measuring prostaglandin, and likely PGE(1), but we have not fully characterized the media to find out which other prostaglandins may also be released.

ANGGARD: May I elaborate, please? We have tried similar experiments with bacteria and we are concerned with the possibility that certain species of bacteria may be producing prostaglandin-like substances which may be responsible for certain inflammatory conditions. In our growth of batch cultures, upon extraction we originally found what we called prostaglandin activity by bioassay on a guinea pig ileum. When we tried this in the rat uterus preparation, it was not effective. I think this illustrates the fact that this is a very difficult question to answer until you have really gotten several different measurement methods to coincide.

RAMWELL: I would like to substantiate Dr. Anggard's statement. Working with sensitized leukocytes provided by Dr. Lichtenstein and Dr. Bourne, which were immunogenically challenged, we found a rat fundus-contracting material in ethyl acetate extracts of the cells and fluid. But in subsequent experiments, using a more specific extraction procedure and bioassay preparation (i.e., the rat uterus), we observed little activity that could be ascribed to the known prostaglandins.

JAFFE: We would like nothing better than to be able to further document our observations by still another technique. The problem is that we are dealing with such small concentrations. In order to collect enough medium so that we could document the changes by converting PGE to PGB and reading it in a spectrophotometer, we would have to collect media for several weeks. That involves still more problems. I recognize the problem you are facing. The good agreement between bioassay and radioimmunoassay data makes it almost sure that this is prostaglandin. Epithelium lining the gastrointestinal tract makes prostaglandin *in vivo*. There is no reason to think that prostaglandin is not released *in vitro*.

RAMWELL: Vogt et al have already shown that isolated intestine synthesizes prostaglandins, and Pamela Davison in our laboratories at Stanford finds that prostaglandin plays a significant role in maintaining prostaglandin tone in guinea pig ileum both *in vivo* and *in vitro*. Moreover, a synthetase inhibiter, 5,8,11,14-tetraynoic acid abolishes prostaglandin synthesis and tone.

RELEASE AND ACTIONS OF PROSTAGLANDINS IN INFLAMMATION AND FEVER: INHIBITION BY ANTI-INFLAMMATORY AND ANTIPYRETIC DRUGS

A. L. Willis, P. Davison, P. W. Ramwell, W. E. Brocklehurst,* and B. Smith+

Department of Physiology, Medical School
Stanford University, Cal.

INTRODUCTION

When studying the pharmacological mediation of an inflammatory reaction, several difficulties are encountered. These lie in the multiplicity of factors involved and the complexity of their interactions. However, there are rigid criteria which ought to be satisfied before a substance can be definitely classified as a mediator of inflammation; these are listed below:
(1) The putative mediator of inflammation should induce some or all of the signs of inflammation.
(2) During the inflammatory reaction, the putative mediator should be released in amounts whose local concentrations in the tissue are capable of inducing these signs; ideally its appearance should correlate with development of the inflammatory reaction.
(3) Release or actions of the putative mediator should be reduced by known anti-inflammatory drugs and, as a corollary, drugs which reduce the release or actions of the mediator should possess anti-inflammatory properties.

---

*Lilly Research Centre, Windlesham, Surrey, England.
+Royal College of Surgeons, London, England.

Suppression of carrageenin-induced paw swelling in rats is probably the most widely used screening test for anti-inflammatory drugs. Indeed this test played an important part in the introduction of indomethacin (Winter et al, 1962, 1963). Evidence presented here will show that PGE(2) has a role in carrageenin-induced inflammation and that the non-steroidal anti-inflammatory drugs may owe their efficacy in this test to inhibition of PGE(2) production in the inflamed tissue. Furthermore, a similar mode of action in brain provides an explanation for the anti-pyretic properties of these drugs.

PHLOGISTIC ACTIONS OF PROSTAGLANDINS

Evidence is accumulating to show that PGE(1) and PGE(2) are well suited to mediate inflammation, thus satisfying criterion (1), above.

Increased Vascular Permeability and Vasodilatation

Erythema and increased vascular permeability are invariably seen during acute inflammation. The ability of prostaglandins to induce these effects is well documented, and these findings are summarized in Table I. In general, PGE(1) and PGE(2) induce vascular permeability in skin, but the corresponding PGF compounds and PGA(1) are much less effective. PGE(1) can induce hind paw edema in rats following continuous intravenous infusion (Weeks and Lawson, unpublished, quoted by Bergstrom et al, 1968) or following subcutaneous injection into the plantar surface (Arora et al, 1970). The latter finding has been extended to other species for PGE(1), PGE(2), and PGA(2) (see Glenn, this Symposium).

In most species (including man) permeability effects of the prostaglandins appear to be due to a direct action on the microvasculature and not through release of other endogenous substances such as histamine or serotonin. However, the rat appears to be a special case. There is clear evidence through the use of antagonist drugs (mepyramine and methysergide) and pretreatment with the mast cell disrupter, compound 48/80, that PGE compounds (at least in doses of up to 100 nanograms) act through release of mast cell histamine and serotonin, the effect attributable to the former being largest (Crunkhorn and Willis, 1969, 1971a). Similar results for PGE(1) were reported by Arora et al (1970) who also published photomicrographs demonstrating mast cell disruption

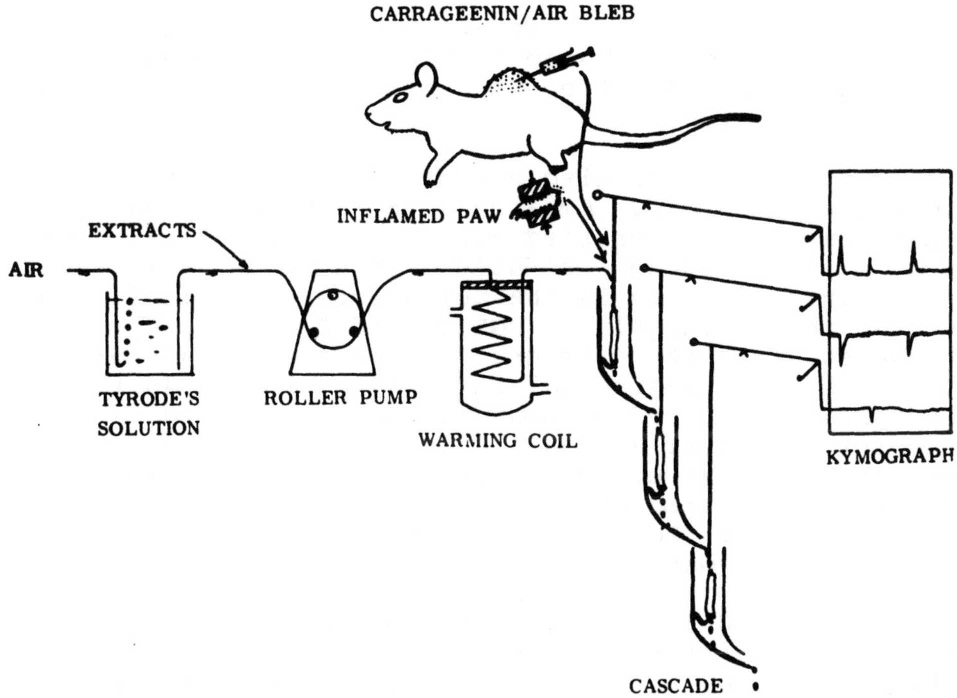

Figure 1. Outline of the approach used in examination of pharmacological activity present in inflammatory exudates. Preliminary qualitative examination or simultaneous parallel assay of extracts was carried out on several selectively sensitive isolated smooth muscle preparations superfused in cascade with Tyrode's solution. Crude exudate squeezed from excised inflamed paws or from carrageenin-air blebs was injected straight onto the tissues. This was to avoid loss of labile activity or formation of activity by storage. Extracts, reconstituted in Tyrode's solution, were injected into the self-sealing silicone rubber tubing used in the roller pump (Willis, 1971).

induced <u>in vivo</u> by PGE(1). on the other hand, the findings of Kaley and Weiner (1971) are different. Their work indicated that prostaglandins have a direct action on the microcirculation (notably the venules). These discrepancies could be due to the different colony of Wistar rats used by Kaley and Weiner. Harris and West (1963) reported profound differences between rats of the same strain in their edematous reaction when injected with dextran or egg white; furthermore the "anaphylactoid" reaction induced by dextran may be mediated through liberation of mast cell amines (Parrat and West, 1958).

Possibly there is always a direct component in the permeability effects induced by "large" (microgram) doses of prostaglandins in rat or when prostaglandin is formed during inflammation.

PGE compounds are also potent inducers of edema and vasodilatation in man. Ambache (1961) reported that chromatographically purified "irin" (a hydroxy fatty acid extract of iris tissue) produced erythematous weals when injected intradermally in the human forearm; irin is now known to consist largely of PGE(2) and PGF(2-alpha) (Ambache et al, 1966). Later, Bergstrom et al (1965) observed that intra-arterial infusion of PGE(1) in the human forearm produced pronounced flushing and edema which lasted for well over an hour, and Solomon et al (1968) showed that similar effects were seen locally following intradermal injection of PGE(1). The comparative effects of PGE(1), PGE(2), PGF(1-alpha) and PGF(2-alpha) given intradermally in man have been investigated (Crunkhorn and Willis, 1969, 1971a; Juhlin and Michaelsson, 1969; Michaelsson, 1970); and the latter workers demonstrated that the permeability and vasodilator effects were not greatly reduced by antihistamines or pretreatment of the skin with compound 48/80.

Leukotaxis and Granuloma Formation

During the "delayed phase" of acute inflammation, the inflamed tissue is invaded by leukocytes (polymorphs and macrophages) and fibroblasts lay down granuloma tissue (Spector and Willoughby, 1968). Prostaglandins might play some part in these processes, for PGE(1) has been shown to be chemotactic *in vitro* for rabbit polymorphs (Kaley and Weiner, 1971) and to increase the weight of chronically implanted cotton pellets (Arora et al, 1970). PGE(2) should be tested in these systems, since this is the principal prostaglandin released during carrageenin inflammation in rat (Willis, 1970).

Production of Pain

Sites of inflammation are commonly sensitive to touch and may be painful, but evidence for a role of prostaglandins in the generation of pain is so far scanty. Horton (1963) reported that PGE(1) failed to produce pain when applied to human blister bases and likewise Crunkhorn and Willis (1969, 1971a) showed that no pain was produced when PGE(1), PGE(2), PGF(1-alpha) or PGF(2-alpha) were

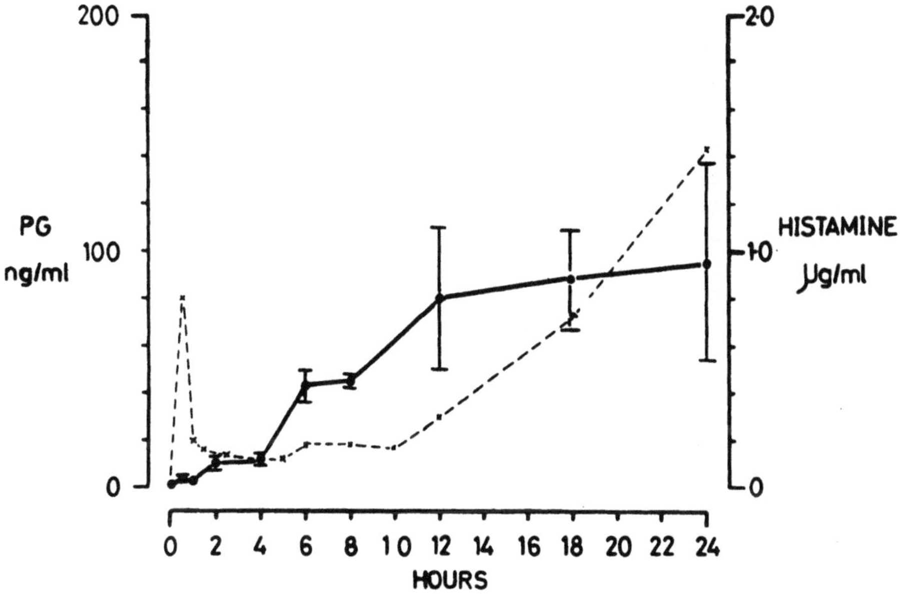

Figure 2. Mean concentration (±S.E.) of prostaglandin activity extracted from carrageenin-air bleb fluid up to 24 hr after injection of 2% carrageenin suspension; each value is the mean of concentrations in bleb fluid from 5-12 animals, except in the case of "time 0" when three samples of carrageenin suspension were extracted. The solid line shows activity in ethyl acetate extracts of acidified exudate, as nanogram/ml of PGE(2), assayed on the rat stomach strip. The broken line shown mean concentrations of histamine (Willis, 1971).

injected intradermally in the forearm. However in similar experiments, Solomon et al (1968) showed that when microgram doses of PGE(1) were injected, the erythmatous area was sensitive to touch and this finding was confirmed by Juhlin and Michaelsson (1969). Thus prostaglandins may exert a facilitatory effect on sensory nerve endings to stimulation by another mediator such as bradykinin, for a well known effect of prostaglandins is to decrease the threshold for stimulation in a number of excitable tissues (Bergstrom et al, 1968). Alternatively, as suggested by Collier (1971), prostaglandins may not be significantly involved in the mediation of pain in skin, where aspirin has little analgesic effect.

However, there is evidence that prostaglandins can induce pain directly. PGE(2) and PGF(2-alpha) produce irritation when applied topically to the eye (Bethel and Eakins, 1971) and we have recently shown that PGE(1) produces "writhing" when injected intraperitoneally in mice (Willis and Ramwell, unpublished data). This effect was seen in eight mice injected with PGE(1) (1-20 micrograms) but in 14 mice which received PGE(2) (5-100 micrograms) only two of the mice responded (doses of 5 and 10 micrograms, respectively). PGF(2-alpha) had no effect (5 microgram doses, n = 4) and the vehicle (0.5 ml of 0.9% NaCl solution) was also inactive. This result appears to exclude a role for PGE(2) in pain production in mice, indeed it has been reported to inhibit writhing induced in this species by acetylcholine (Holmes, 1968).

In addition to the fever which often accompanies an inflammatory reaction (see "PROSTAGLANDINS AND FEVER"), headache is often produced, which is alleviated by aspirin (now known to inhibit prostaglandin synthesis) and intravenous infusion of PGE compounds in man produces headache (Bergstrom et al, 1968).

PRODUCTION OF PROSTAGLANDINS DURING INFLAMMATION

General Review

Prostaglandins are released during inflammation engendered by various stimuli in several species, including man. The earliest implication of prostaglandins in inflammation was by Ambache et al (1965, 1966), who showed that mechanical irritation of rabbit eyes evoked release of "irin" [PGE(2) and PGF(2-alpha)] into perfusates of the anterior chamber; this finding forms a background for subsequent studies on the inflammatory role of prostaglandins in the eye (Waitzman and King, 1967; Solomon et al, 1968; Beitch and Eakins, 1969; Bethel and Eakins, 1971).

PGE(2) is released during carrageenin-inflammation in rat (Willis, 1969a, 1970), as well as histamine, kinin and possibly serotonin (Willis, 1969b, 1971). PGE(2) is known to be released into lymph from scalded dog paws (Anggard and Jonsson, 1971). Finally, a mixture of smooth muscle stimulating prostaglandins are released into perfusate from the skin of human subjects with allergic contact dermatitis, although the less inflammatory PGF compounds predominate (Greaves et al, 1971).

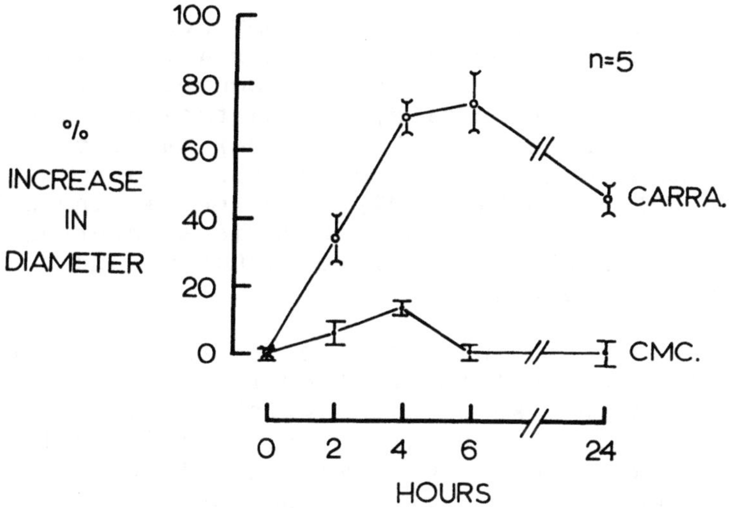

Figure 3A. A comparison of the inflammatory and prostaglandin-producing properties of carboxymethylcellulose (CMC) and carrageenin (CARRA). Lack of paw swelling on injection of CMC, compared to swelling induced by injection of carrageenin. Shown are effects on diameter of rat hind paws after injection of 1% suspensions of carrageenin or CMC. Results are expressed as mean increases in diameter (±S.E.) of the paws, each individual result being expressed as percent increase in diameter of the CMC or carrageenin-treated paws compared with diameter of the contralateral paws injected only with 0.1 ml of saline.

Prostaglandin Release During Carrageenin Inflammation in Rat

The approach used in studying occurrence of pharmacologically active substances in exudates from carrageenin inflamed tissue is shown in Fig. 1. Rapid preliminary identification of pharmacologically active material in the exudates was possible using a cascade superfusion method similar to that introduced by Vane (1964). Several isolated tissues, chosen for their selective sensitivity to various endogenously occurring substances were superfused in cascade with Tyrode's solution and selectivity of the tissues could be further enhanced by the use of blocking agents.

In some early experiments, edema fluid was squeezed from carrageenin-inflamed paws in a manner similar to that described by Bonta and De Vos (1965). However, the most obvious drawback inherent in this approach was that it was impossible to be sure how much of this activity was due to the physical trauma of excising and squeezing the paws. To obviate these difficulties, the carrageenin-air bleb technique was developed (Willis, 1969a) based on the granuloma pouch method of Selye (1954). A suspension of carrageenin (2%) in NaCl (0.9%) solution was injected into subcutaneous air blebs in rats at various intervals; 0.1-0.2 ml samples of bleb fluid were withdrawn and immediately bioassayed.

Evidence was obtained for the presence of histamine and kinin, but the most interesting finding was detection of a prostaglandin-like material which appeared about 2-3 hr after injection of the carrageenin. This active material was extracted at pH 3 with ethyl acetate and on partition between aqueous ethanol and petroleum ether, it remained in the ethanol phase, i.e. it behaved as a hydroxylated fatty acid (Samuelsson, 1963; Shaw and Ramwell, 1969). When submitted to two-stage thin layer chromatography (TLC) in Green and Samuelsson's (1964) AII and AI solvent systems, this material appeared to consist almost entirely of PGE(2) (Willis, 1970). Finally, it was found that the PGE(2)-like material eluted from the plates, not only behaved like authentic PGE(2) in parallel smooth muscle bioassays, but it also induced increased vascular permeability in rat skin which (as for PGE compounds) was inhibited by mepyramine maleate (Willis, 1971).

Next, a relationship was established between concentrations of prostaglandin in the exudate and time after carrageenin injection. Groups of rats were injected with carrageenin at various time intervals up to 24 hr before harvesting the exudate. All the available exudate from each rat was then withdrawn and extracted for prostaglandins. Some of the material was not extracted but was stored at -20°C until assayed for histamine or for its capacity to synthesize prostaglandins from arachidonic acid. In addition, small samples were routinely taken for histological and bacteriological examination or were pooled and, after centrifuging, the cell-free supernatant was examined for its content of free lysosomal enzyme activity, using the methods of Anderson (1970).

Fig. 2 shows the time course for histamine and prostaglandin content in the bleb fluid up to 24 hr. After approximately 2.5 hr, prostaglandin levels began to rise and reached a maximum between 12 and 24 hr. Levels of histamine rose at a time when those for prostaglandin

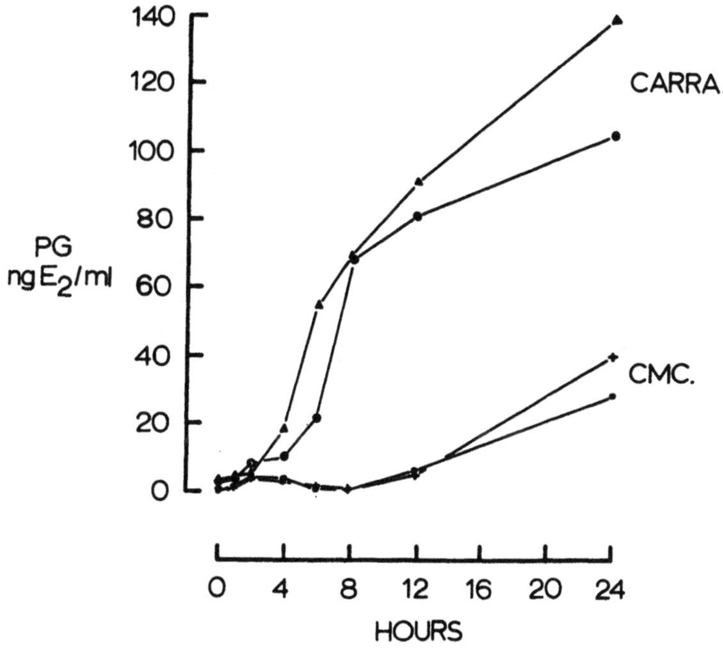

Figure 3B. A comparison of the inflammatory and prostaglandin-producing properties of carboxymethylcellulose (CMC) and carrageenin (CARRA). Concentrations of prostaglandin-like activity (as PGE(2) assayed on rat stomach strip) in bleb fluid of individual rats up to 24 hr after injection of 2% suspensions of CMC or carrageenin (Willis and Ramwell, unpublished data).

were also high and this could have been partly due to histamine release induced by the PGE(2) (see "Increased Vascular Permeability and Vasodilatation").

Relationship Between Appearance of Prostaglandins and Development of the Inflammatory Response

In order to completely satisfy criterion 2 (see "INTRODUCTION"), the amount of prostaglandin in the exudate should appear as a consequence of the irritant nature of the carrageenin and should correlate with development of the inflammatory response. Possibly, the PGE(2) occurring in carrageenin-air blebs may have accumulated as "overflow" from normal prostaglandin turnover in the surrounding tissue and was not released as a consequence of the irritant nature of carrageenin. However, no evidence was obtained for this possibility. Early in this study (Willis, 1971), it was shown that a 1%

suspension of carrageenin produced less prostaglandin (and exudate) than a 2% suspension, and that the occasional occurrence of pathogenic bacteria in the exudate did not contribute significantly to the prostaglandin production. However, the ideal result would be a greatly diminished appearance of prostaglandin in the bleb fluid when a non-irritant fluid is injected instead of carrageenin. Carboxymethylcellulose (CMC) did not provoke an inflammatory response (Fig. 3A), and when it was injected into the subcutaneous air blebs (Fig. 3B), little prostaglandin appeared up to 12 hr after injection, less than 5 nanogram/ml as PGE(2), although in control animals injected with carrageenin, the characteristic rise in prostaglandin was seen. At 24 hr, when prostaglandin levels in the carrageenin-injected animals were maximal, the prostaglandin in the CMC blebs had increased to about 30 nanogram/ml [as PGE(2)], and this probably represents accumulation of spontaneously released prostaglandin.

In addition to prostaglandins, histamine, kinin and possibly serotonin are also found in carrageenin-air bleb exudate (Willis, 1969b, 1971) and so any component of paw swelling induced by endogenous prostaglandin production could well be masked by effects of these other mediators. Elegant work by Di Rosa et al (1971) has exposed for the first time a delayed phase of paw swelling induced by carrageenin and which may be due to PGE(2). These workers pretreated the rats with compound 48/80 (to deplete mast cells of histamine and serotonin) and cellulose sulphate (to deplete kininogen). By combining both treatments, it was shown that onset of carrageenin-induced paw swelling was delayed until 2.5 hr, when swelling developed rapidly, reaching maximal levels at 4-5 hr. The delay in onset of swelling is similar to that for appearance of prostaglandin in the carrageenin-air blebs.

Appearance of prostaglandins in carrageenin-air bleb exudate is paralleled by infiltration of the exudate with leukocytes, predominantly polymorphs (Willis, 1971; Anderson et al, 1971). These findings were similar to those of DiRosa and Willoughby (1971), who showed that the "prostaglandin" phase of swelling in carrageenin-inflamed paws was well correlated with infiltration of the tissue by leukocytes (mainly polymorphs), and moreover that only the "prostaglandin phase" of swelling was reduced by therapeutic doses of non-steroidal anti-inflammatory drugs, a finding of considerable relevance to the recently reported ability of these drugs to inhibit prostaglandin production (see "Aspirin and Other Non-Steroidal Anti-Inflammatory Drugs").

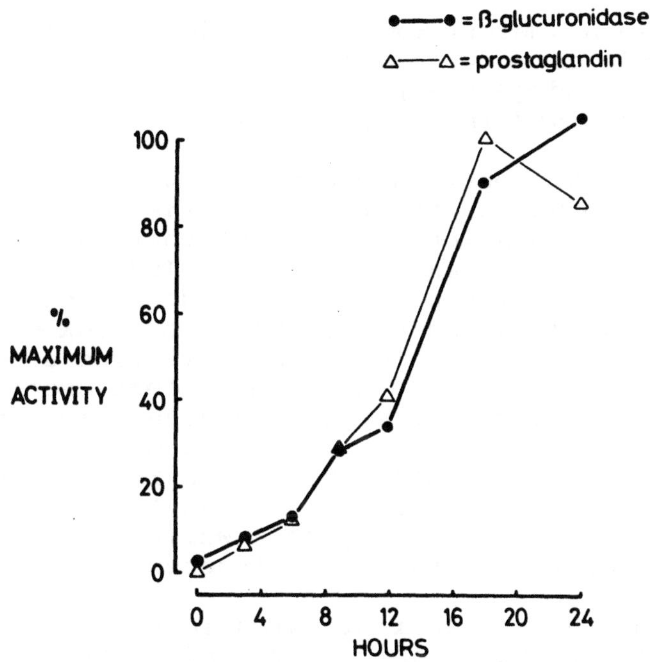

Figure 4. Levels of beta-glucuronidase and prostaglandin, compared directly in batches of pooled carrageenin-air bleb exudate. For comparison purposes, values are expressed as percentage of the maximum levels seen throughout the period up to 24 hr after injection of carrageenin. Maximum concentration of prostaglandin activity (as PGE(2) assayed on rat stomach strip) was 127 nanogram/ml and that of beta-glucuronidase was 120 microgram/ml (expressed as weight of phenolphthalein released by 1 ml of exudate from phenolphthalein-beta-glucuronide substrate using the assay method described by Anderson, 1970). From: Anderson, Brocklehurst and Willis, 1971; reproduced by permission of Academic Press, New York.

Possible Mechanisms Involved in Production of Prostaglandins During Acute Inflammation

Formation of prostaglandins is thought to involve liberation by phospholipase of unsaturated fatty acids which are then cyclized and converted to prostaglandins by the prostaglandin-synthetase enzyme system. For instance, PGE(2) and PGF(2-alpha) are synthesized from arachidonic acid which can be cleaved from phospholipids by phospholipase A (Bergstrom et al, 1964; Bartels et al, 1970). Little is known about the source of phospholipase A involved in endogenous prostaglandin production, or the mechanisms involved in its activation.

There are two obvious mechanisms which might be involved in production of prostaglandins during carrageenin inflammation: the first postulates activation of complement (Giroud and Willoughby, 1970), and the second requires participation of lysosomal phospholipase, possibly released during phagocytosis by leukocytes (Anderson et al, 1971). These possibilities are not mutually exclusive, for activation of complement, which appears to be a primary step in the development of acute inflammation (Willoughby et al, 1969), may initially generate factors (including prostaglandins), which are chemotactic for leukocytes, which then liberate lysosomal enzymes, including phospholipase A while engaged in phagocytosis of the carrageenin (Anderson et al, 1971).

Lysosomes have been implicated in the pathogenesis of inflammation (see Weissmann, 1967) and there is now evidence for the occurrence of phospholipase A in lysosomes of liver, adrenal gland, polymorphonuclear leukocytes and macrophages (Waite et al, 1969; Fowler and De Duve, 1969; Smith and Winkler, 1968; Elsbach and Rizack, 1963; Elsbach, 1966) and levels of free acid phosphatase (a typical lysosomal enzyme) are elevated in carrageenin-inflamed rat paws (Coppi and Bonardi, 1968). Evidence presented here indicates that lysosomal phospholipase A may be involved in prostaglandin formation during inflammation engendered by carrageenin.

When the relationship between levels of prostaglandin in carrageenin-air bleb exudate and those of lysosomal enzymes was examined, a similarity was observed (Anderson et al, 1971). Indeed, when this comparison was carried out in the same samples of exudate there was a correlation coefficient of 0.98 between concentrations of free beta-glucuronidase and prostaglandin withdrawn at increasing time intervals after carrageenin injection (Fig. 4). In these experiments both prostaglandin and beta-glucuronidase concentrations were similar to the degree of infiltration of the exudate by leukocytes, suggesting that they might be involved in prostaglandin production. However, interpretation was made difficult by the possible leukotaxic effects of PGE(2) (see "Leukotaxis and Granuloma Formation") and the variety of cell types which could contribute to the production of prostaglandins in the bleb fluid.

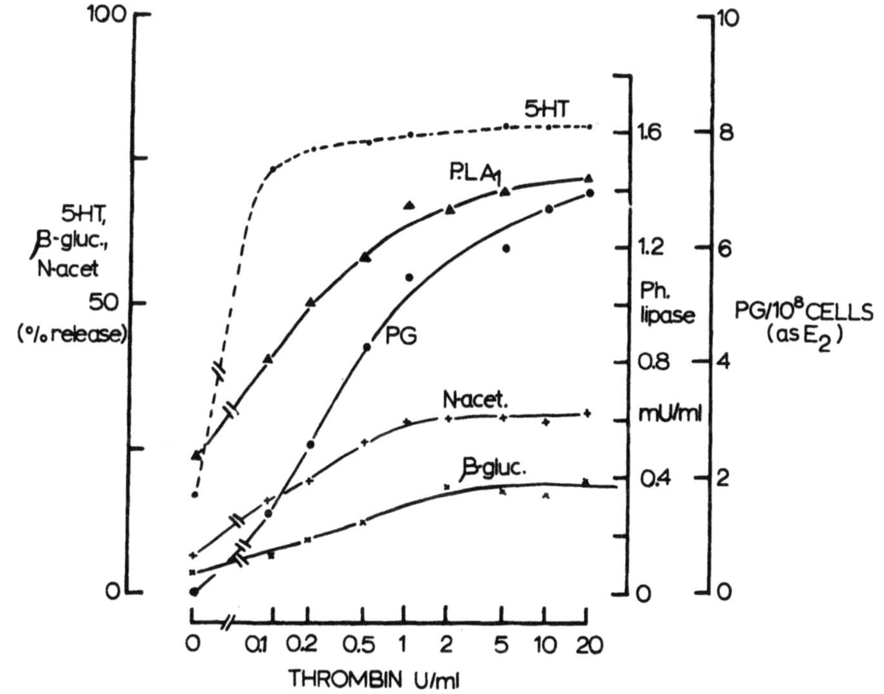

Figure 5. Thrombin-induced release of prostaglandins and platelet constituents from human platelets. Platelet-rich plasma (from two donors) was incubated for 1 hr with ^{14}C serotonin and the platelets isolated, washed and resuspended in Tris-saline (pH 7.4). The platelet suspension (90 ml, platelet count 13 x 10^8 cells/ml) was divided into 6 ml aliquots and incubated with thrombin (0.1-20 units/ml); one tube received no thrombin. After incubation (5 min at 37°C), the reaction was terminated by cooling and centrifugation in the cold to remove the platelets from contact with the supernatant. Some (4.5 ml) of the supernatant was acidified (pH 3) and extracted for prostaglandins with ethyl acetate. The remainder was examined for release of ^{14}C serotonin (5-HT) and lysosomal enzymes, including a phospholipase A(1) with a pH optimum of 4.0 [PLA(1)]. Release of serotonin, beta-glucuronidase (beta-gluc) and N-acetyl-beta-glucosaminidase (N-acet) are expressed as percentage of the total activity liberated from sonicated platelets in the absence of thrombin. Prostaglandin activity was assayed in terms of PGE(2) on rat stomach strip and is expressed here in nanogram/10^8 platelets. Phospholipase activity is expressed as activity in the supernatant in arbitrary units, because the platelet content of this enzyme is difficult to assess (there is pool dilution of the labelled substrate by unlabelled endogenous substrate from the platelets) (Willis and Smith, unpublished data).

These difficulties were overcome by using human platelets as the source of prostaglandin production and directly measuring the liberation of phospholipase A. When incubated with thrombin, isolated human platelets, resuspended in an artificial saline solution produce PGE(2) and PGF(2-alpha) and release them into the supernatant solution (Smith and Willis, 1970); moreover adenine nucleotides, serotonin and lysosomal enzymes are released (Holmson and Day, 1970) and also phospholipases A(1) and A(2) with acid pH optima (Silver et al, 1971). There was a similarity between release of lysosomal enzymes and prostaglandins from the platelets (Willis and Smith, unpublished data) both on a temporal basis and with ascending concentrations of thrombin, the similarity between release of phospholipase and prostaglandin was marked (Fig. 5). However, there was no correlation between prostaglandin production and release of serotonin or adenine nucleotides.

In order to further establish the relationship betwen prostaglandin production and lysosomal phospholipase A, it was necessary to use a drug which inhibited the release of lysosomal enzymes. Known anti-inflammatory drugs have been reported to reduce levels of free lysosomal enzyme activity in inflamed tissue (Coppi and Bonardi, 1968; Anderson, 1970) and therefore may be expected to act similarly on isolated platelets. However, neither the release of lysosomal enzymes or the other platelet constituents (serotònin and adenine nucleotides) was altered by indomethacin, aspirin or hydrocortiscne, although, as discussed in section "Aspirin and Other Non-Steroidal Anti-Inflammatory Drugs", prostaglandin production was suppressed by the former two drugs. Indeed, hydrocortisone, in spite of its reported lysosomal stabilizing properties (see Weissmann and Thómas, 1964) failed to reduce release of lysosomal enzymes (including acid phospholipase) even when incubated at a concentration of 1 mg/ml with platelet-rich plasma for 5 hr (Willis and Smith, unpublished data).

These results with platelets were not dissimilar to those obtained in carrageenin inflammation (section "Inhibition of Prostaglandin Production During Carrageenin-Induced Inflammation"), suggesting that the platelet may serve as a useful model in the study of inflammatory mechanisms in vitro. In fact, a thrombin-like mechanism has been implicated in carrageenin inflammation in rat (Wiseman and Chang, 1968); this points to a possible link between lysosomes and the complement system, which when activated, liberates enzymes with some of the characteristics of thrombin (Muller-Eberhard, 1969).

MECHANISM OF ACTION OF ANTI-INFLAMMATORY DRUGS

We now examine the ability of prostaglandins to fulfill the third criterion, i.e. that the production or the inflammatory actions of the prostaglandins are inhibited by anti-inflammatory drugs.

Aspirin and Other Non-Steroidal Anti-Inflammatory Drugs

Willoughby's group (see "Relationship Between Appearance of Prostaglandins and Development of the Inflammatory Response") showed that the "prostaglandin phase" of carrageenin inflammation was selectively inhibited by non-steroidal anti-inflammatory drugs. These workers suggested that non-steroidal anti-inflammatory drugs act by inhibiting cell immigration. In view of this work and the finding that aspirin does not inhibit increased vascular permeability induced by PGE(2) (Michaelsson, 1970), this class of drug might suppress prostaglandin production by interfering with release or activity of lysosomal phospholipase, or by inhibition of prostaglandin-synthetase (Willis, 1971). Evidence is available for the latter possibility.

Inhibition of Prostaglandin Production in Vitro. Piper and Vane (1969) discovered that low concentrations of indomethacin and aspirin inhibited release of a "rabbit aorta contracting substance" (RCS) during anaphylactic shock in isolated perfused guinea-pig lungs; this was the first report that these drugs could act *in vitro* in concentrations of the same order of magnitude as those found in the body during anti-inflammatory therapy (Smith and Dawkins, 1971). In these studies, prostaglandins were also released, but the inhibitory effect of the anti-inflammatory drugs upon their release was not noticed, although upon later re-examination of the data there did appear to be some inhibition (Vane, 1971).

Subsequently, it was shown that RCS is liberated from perfused lungs by infusion of arachidonic acid (Vargaftig and Dao, 1971; Piper and Vane, unpublished data). Using cell-free homogenates of guinea-pig lung, Vane (1971) observed that synthesis of PGE(2) and PGF(2-alpha) from arachidonic acid was inhibited by small concentrations of

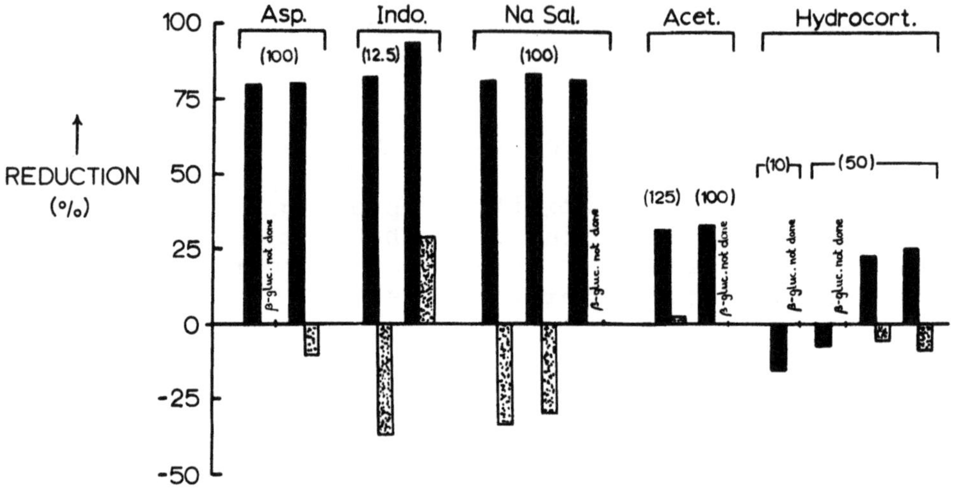

Figure 6. In vivo effects of anti-inflammatory and anti-pyretic drugs on prostaglandin content of carrageenin-air bleb exudate. The drugs used or the vehicle in which they were dissolved were injected intraperitoneally, 50 min prior to the carrageenin. At a fixed time interval between 6 and 9 hr later, exudate was withdrawn, pooled and analyzed. In the case of hydrocortisone, in all but the first experiment, an additional dose was given 6 hr before injection of the carrageenin. In the third of these experiments, the rats were given additional doses, twice daily for 2 days beforehand. Shown are the effects of prostaglandin content and free (cell-free supernatant) beta-glucuronidase. Results are expressed as percentage reduction in levels in exudate from the drug-treated animals compared to those in the controls. Figures in parentheses refer to the doses of each drug in mg/kg. Solid bars: effects on prostaglandin levels. Dappled bars: effects on free beta-glucuronidase levels.

indomethacin and aspirin, the former being about 43 times more potent on a molar basis (which correlates well with findings in several anti-inflammatory tests). The potency of the drugs in this system can be judged by the fact that there was an 80% reduction in prostaglandin synthesis with indomethacin at only 1 microgram/ml. However, sodium salicylate was considerably less active and hydrocortisone, morphine and mepyramine had little effect in concentrations up to 50 microgram/ml.

Prostaglandin production in thrombin-treated human platelets was also inhibited *in vitro* by non-steroidal anti-inflammatory drugs and with similar orders of potency to those in lung homogenates. The selectivity of this inhibition for release of adenine nucleotides, serotonin and lysosomal enzymes was not affected by any of the drugs tested (Smith and Willis, 1971 and unpublished data). Relevance of these findings to the clinical situation was enhanced when it was shown that oral ingestion of aspirin (600 mg) or indomethacin (100 mg) virtually abolished prostaglandin production in platelets, subsequently isolated and treated with thrombin. This effect was not seen, however, in a similar experiment in which codeine (60 mg) was taken orally. Phospholipase release from the platelets was not diminished by the non-steroidal anti-inflammatory drugs, which therefore must have inhibited the prostaglandin-synthetase system, a conclusion complementary to that of Vane (1971).

In the isolated, perfused spleen of dog, aspirin and indomethacin inhibited prostaglandin production (the latter drug being exceptionally potent), but no inhibition was seen with sodium salicylate in concentrations of up to 40 microgram/ml, and once more hydrocortisone had no effect (Ferreira et al, 1971). Thus, aspirin and indomethacin might inhibit prostaglandin production in a variety of tissues and species, but there may be differences in the relative ability of different non-steroidal anti-inflammatory drugs to inhibit prostaglandin production in various systems.

Inhibition of Prostaglandin Production During Carrageenin-Induced Inflammation. Using the carrageenin-air bleb technique, non-steroidal anti-inflammatory drugs have been shown to inhibit prostaglandin production induced by carrageenin. Rats were injected intraperitoneally with the drug under investigation or else the same volume of the vehicle in which the drug was dissolved (0.9% saline, 2% ethanol in saline or 2% neutralized sodium carbonate solution in saline). Carrageenin blebs were prepared 50 min later, and at a fixed time interval between 6-9 hr later (i.e. during the rising phase of prostaglandin production), exudate from each rat was withdrawn, pooled and examined for its content of prostaglandin and free beta-glucuronidase. When effects of hydrocortisone were examined, multiple doses were given for up to 2 days beforehand. Results are shown in Fig. 6.

Aspirin (100 mg/kg) and indomethacin (12.5 mg/kg) dramatically reduced (about 80%) prostaglandin content of the exudate. Surprisingly (in view of the usual poor effect seen *in vitro*), sodium salicylate (100 mg/kg) also

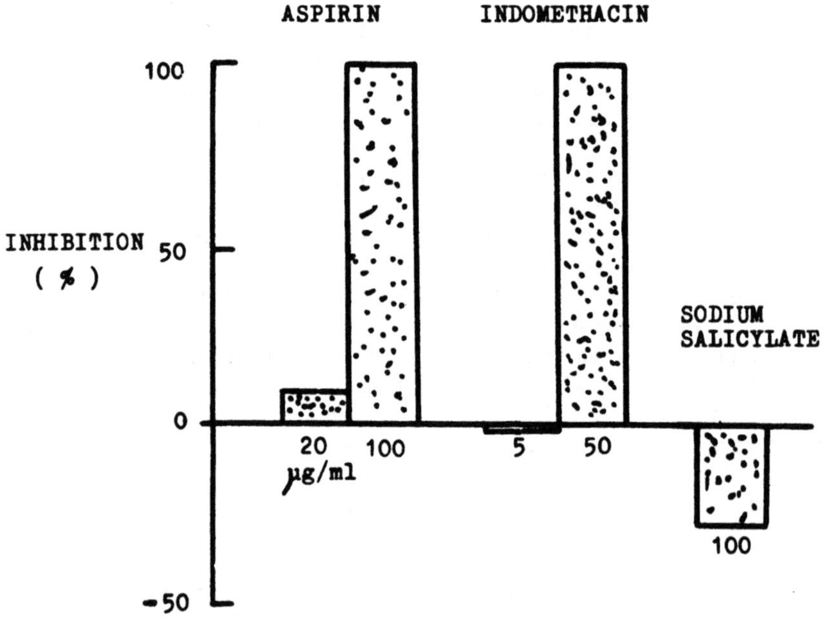

Figure 7. Prostaglandin synthetase activity in 8 hr carrageenin-air bleb exudate. Effects are shown of aspirin, indomethacin and sodium salicylate on the ability of a broken cell suspension (in Tyrode solution) to synthesize acid lipid prostaglandin-like activity from exogenous arachidonic acid (20 microgram/ml). Added at 5 min before the arachidonic acid, aspirin (100 microgram/ml) and indomethacin (50 microgram/ml) completely suppressed the synthetase activity, although five- to tenfold lower concentrations were ineffective. Sodium salicylate (100 microgram/ml) produced no decrease in the production of prostaglandin-like activity, indeed there was some stimulation; whether this might involve liberation of exogenous arachidonic acid in this crude preparation remains to be seen (Willis, Davison and Ramwell, unpublished data).

produced a diminution of prostaglandin production in the exudate, indicating that this in vivo situation is more predictive for the anti-inflammatory activity invariably seen with the salicylates. Although there were usually increases (about 30%) in concentrations of free beta-glucuronidase, exceptions were seen and further investigation is required.

The potency of sodium salicylate in these experiments raised the possibility that the prostaglandin synthetase involved in carrageenin inflammation is different from that in lung, spleen and platelets. In a preliminary examination of this possibility (Willis, Davison, and Ramwell, unpublished data) the effect of sodium salicylate on prostaglandin synthetase in the exudate was examined. It can be seen (Fig. 7) that the ability of a broken cell suspension of 8 hr carrageenin-air bleb exudate to synthesize prostaglandin was totally inhibited by aspirin (100 microgram/ml) or indomethacin (50 microgram/ml) although five- to tenfold lower concentrations of these drugs were ineffective. However, no reduction of prostaglandin production was seen with sodium salicylate (100 microgram/ml); indeed there was some stimulation (also seen in preparations of mouse brain; see "PROSTAGLANDINS AND FEVER"). We have yet to examine effects on prostaglandin synthesis in the skin and connective tissue surrounding the blebs (other potential sources of prostaglandin in the exudate). However, taken together with other *in vitro* results (see "Inhibition of Prostaglandin Production *in Vitro*") it appears that sodium salicylate may be converted *in vivo* to an active metabolite.

Possible Mechanisms of Action of Hydrocortisone in Carrageenin Inflammation

Lack of Effect on Release of Prostaglandins and Lysosomal Enzymes. Although hydrocortisone is a potent inhibitor of carrageenin inflammation (Winter et al, 1962), it does not alter *in vitro* production of prostaglandins in lung, platelets or spleen (see "Inhibition of Prostaglandin Production *in Vitro*") but the possibility remained that it might produce this effect *in vivo*. However, using the carrageenin-air bleb technique it was shown that hydrocortisone (50 mg/kg intraperitoneally) failed to markedly alter carrageenin-induced prostaglandin production and there was no indication that there had been a swing of synthesis from PGE(2) to PGF(2-alpha) (shown by differential responses on the isolated tissues used for bioassay of the prostaglandins). This finding was obtained even when the hydrocortisone was given in repeated doses for up to 2 days previously. In addition, this drug failed to significantly alter levels of free beta-glucuronidase in the exudate. Both these findings are remarkably similar to those obtained *in vitro* with platelets (see "Inhibition of Prostaglandin Production *in Vitro*") and show that the mode of action of hydrocortisone in acute inflammation (at least that engendered by carrageenin) does not involve a

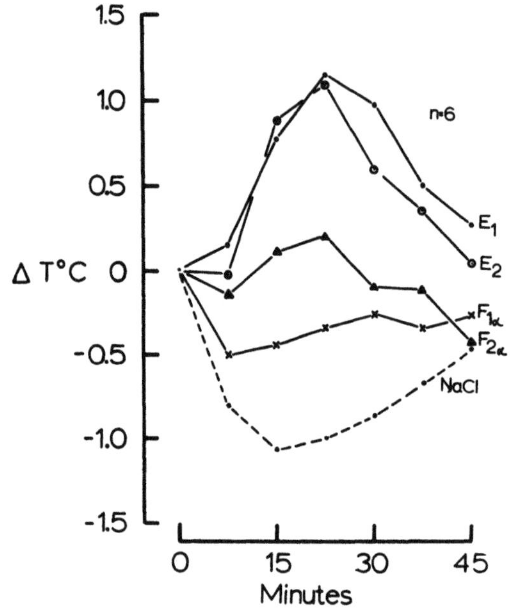

Figure 8. Comparative effects of prostaglandins E(1), E(2), F(1-alpha), F(2-alpha) and saline, injected into the cerebral ventricles of conscious mice. Albino mice (19-21 g) were injected intracerebrally (I.C.) with the prostaglandins (600 nanograms) in 20 microliters of 0.9% sterile pyrogen-free sodium chloride solution. Rectal temp was measured with a thermistor probe at intervals before and after the I.C. injections. Results are expressed as change in temp from initial (delta T°C), initial temp being the mean of three consecutive readings taken before the I.C. injections. Each point is the mean of values obtained in six mice. Standard errors (not shown) were of the order of 0.1-0.2°C (Willis, Davison, and Ramwell, 1972).

decrease in release of prostaglandins or lysosomal enzymes.

<u>Other Possible Mechanisms</u>. It is possible that hydrocortisone could interfere with release of a mediator other than prostaglandin or that it protects the microvasculature from increases in permeability induced by prostaglandins and/or other vaso-active substances.

There is ample evidence for a role of kinins in carrageenin inflammation in rat (van Arman et al, 1965; DiRosa et al, 1971) but there is conflicting evidence for inhibition of kinin-formation by hydrocortisone. Although the *in vitro* work of Melmon and Cline (1967) indicated that formation of kinins by human granulocytes is inhibited by low concentrations of hydrocortisone, these conclusions were not supported by Eisen et al (1968) using kallikrein from a variety of sources. However the possibility of interference with the kinin forming system cannot yet be ruled out completely for Michaelsson (1970) has shown that repeated ingestion of prednisone markedly reduced weals induced in man by intradermal injection of kallikrein, although not altering responses to histamine and bradykinin.

Weals induced by PGE(2) in rat skin are diminished by prior repeated dosing with hydrocortisone (50-100 mg/kg, intrapitoneally) although permeability effects due to other agents (histamine, serotonin or bradykinin) are also somewhat reduced (Willis, unpublished data). These findings are not unprecedented, for Michaelsson (1970) has shown that ingestion of prednisone causes some reduction in diameter of weals induced by PGE(1) in human skin and a similar though much more marked effect has been reported for topically applied fluocinolone acetamide (Juhlin and Michaelsson, 1969). This compound, applied under occlusion, promptly eliminated the hyperemia induced by topical application of PGE(1) (V. Place, private communication).

Thus, ability of the anti-inflammatory steroids to reduce carrageenin inflammation may, to some extent, involve a suppression of the increased vascular permeability induced by PGE(2).

PROSTAGLANDINS AND FEVER

The possibility of a PGE compound being involved in the pathogenesis of fever was first put forward by Milton and Wendlandt (1970), who showed that small doses of PGE(1) injected into the cerebral ventricles of conscious cats induced a pronounced hyperthermic response with a rapid onset and which was not reduced by intraperitoneal injection of acetaminophen, although this drug did reduce hyperthermia induced by intracerebrally injected serotonin. As serotonin induces release of prostaglandins into the perfused cerebral ventricles of dog (Holmes, 1970), it was concluded that acetaminophen and other antipyretic drugs might act by interfering with production of prostaglandin in brain (Milton and Wendlandt, 1970). Recently, these workers have extended their study to other prostaglandins, showing that PGE(1) and PGE(2) are potent

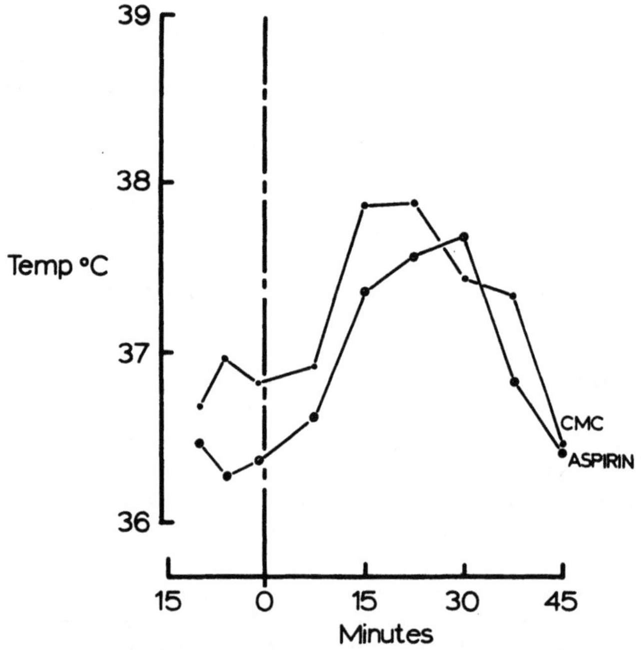

Figure 9. Inability of aspirin to inhibit pyrexia induced in mice by intracerebral injection of PGE(2). No difference was seen between PGE(2)-induced rises in temp in three mice given aspirin (50 mg/kg orally) and three mice given only the vehicle (0.5% carboxymethylcellulose (CMC) in 0.9% sodium chloride solution); a similar result was seen in a pair of mice in which the hyperthermia was induced by PGE(1). Shown is an example of the results obtained. At -15 min, one mouse was given 0.25 ml of 0.5% CMC orally, while the other received aspirin (50 mg/kg) suspended in the same volume of CMC). At time "0", PGE(2) (600 nanograms, 20 microliters) was injected into the cerebral ventricles of both mice; the usual rapid rise in rectal temp was seen in both (Willis, Davison and Ramwell, 1972).

inducers of hyperthermia in cats although PGF(1-alpha), PGF(2-alpha) and PGA(1) are much less active (Milton and Wendlandt, 1971); hyperthermic effects of PGE(1) have also been demonstrated in rabbits (Feldberg and Saxena, 1971).

As aspirin and indomethacin are now known to interfere with prostaglandin production (see "Aspirin and Other Non-Steroidal Anti-Inflammatory Drugs"), we evaluated the hypothesis of Milton and Wendlandt. Mice were used in our study because Cashin and Heading (1968) have developed an inexpensive and simple method for injecting drugs intracerebrally in groups of mice, without the need for stereotaxic apparatus. Furthermore, these workers have characterized the relative effectiveness of several drugs inhibiting hyperthermia induced by intracerebral injection of "E" pyrogen.

PGE(1) and PGE(2) (600 nanograms in 20 microliters) induced hyperthermia when injected intracerebrally in mice, but the corresponding PGF compounds were less active (Fig. 8). As reported by others (Brittain, 1966; Cashin and Heading, 1968), intracerebral injection of 0.9% sodium chloride solution (20 microliters) induced a transient fall in temperature.

If antipyretic drugs act by inhibiting prostaglandin production, then they need not necessarily reduce prostaglandin-induced hyperthermia. In fact, we found that acetaminophen (50 mg/kg, orally) given 15 min before intracerebral injection of PGE(2) (600 nanograms) potentiated the ensuing temp rise. The mean maximum temp rise (±S.E.) in control animals (given 0.25 ml of 0.5% CMC in saline orally) was 1.02 ± 0.11°C, but that for the mice which received acetaminophen was 1.62 ± 0.17°C. Aspirin (50 mg/kg) also failed to reduce hyperthermia induced by PGE(2) or PGE(1) (600 nanograms) and an example from one of these results is shown in Fig. 9.

In addition, we obtained evidence indicating that some antipyretic drugs may act by inhibition of prostaglandin synthesis in brain. Mouse or gerbil brains were frozen in liquid nitrogen and crushed into a powder which was suspended in Tyrode's solution at 37°C with arachidonic acid (final concentration 20 microgram/ml). Aspirin (20 microgram/ml) or indomethacin (5 microgram/ml) added 5 min before arachidonic acid, markedly inhibited the substrate stimulated production of PGE(2) (identified by TLC bioassay, and combined gas chromatography/mass spectroscopy). The degree of inhibition of prostaglandin synthesis by these drugs ranged from 50-100% (two or three experiments with each drug in brain of each species). The most interesting finding, however, was that acetaminophen inhibited synthesis of prostaglandin in mouse brain, a concentration of 20 microgram/ml causing a mean reduction in prostaglandin of about 50% (range 35-80%) in five experiments). The result for sodium salicylate (20 microgram/ml) was similar to that seen for prostaglandin synthetase in exudate (see Fig. 7); no inhibition of prostaglandin production was produced, but some stimulation (mean of 25%).

How do these results compare with the relative antipyretic potencies of these drugs? In the study by Cashin and Heading (1968), the antipyretic potency of indomethacin was about 2.5 times that of aspirin, a finding not inconsistent with the relative ability of these drugs to inhibit prostaglandin production from arachidonic acid (albeit with fixed concentrations). These workers also showed that acetaminophen was about equipotent with indomethacin, whereas in our studies on prostaglandin production indomethacin was more potent. Furthermore, although sodium salicylate has similar antipyretic properties to aspirin (Woodbury, 1970), it actually stimulated prostaglandin production in vitro. The minor discrepancy for acetaminophen might be due to a higher sensitivity of hypothalamic prostaglandin synthetase to this drug (we used whole brain). Alternatively factors might be involved concerning in vivo distribution, metabolism and fate of the drugs. This is certainly a possibility for sodium salicylate which is capable of suppressing prostaglandin production in vivo (see Fig. 6); further investigation is obviously required.

Thus, while we have yet to show that pyrogens release PGE compounds in mouse brain, our results so far indicate that this may be expected.

CONCLUSIONS

Evidence discussed here indicates that PGE compounds are involved in the pathogenesis of acute inflammation and perhaps also fever. Certainly, PGE(2) now fulfills all the necessary criteria (see "INTRODUCTION") to enable it to be classed as a mediator of carrageenin inflammation in the rat. Future confirmatory work will obviously entail the use of potent and specific prostaglandin antagonists. We also need to increase our knowledge of mechanisms involved in prostaglandin production during inflammation and their interrelationships with other mediator substances.

To these conclusions, there are several far reaching consequences. Not the least of these is the probability that PGE compounds may be involved in the acute inflammatory stage of rheumatoid arthritis, which is ameliorated by indomethacin, aspirin and other non-steroidal anti-inflammatory drugs which inhibit prostaglandin synthesis (see Ringold et al, and Ham et al, this Symposium). Furthermore, as PGE(1) is chemotactic for leukocytes (Kaley and Weiner, 1971) and since repeated injection of carrageenin into rabbit knee joints leads to an experimental "arthritis" (Gardner, 1960), the possibility arises that prostaglandins might participate in the transition from acute to chronic inflammation.

Assuming some validity for these hypotheses, we have a logical basis for the search for new anti-inflammatory, and perhaps anti-pyretic drugs. These drugs could interfere with any of the processes involved in production of prostaglandins or in their phlogistic effects. There are also other exciting possibilities: in general PGF compounds do not have marked pro-inflammatory properties (see Table I). Indeed in rats, PGF(2-alpha) has local anti-inflammatory properties (Crunkhorn and Willis, 1971b). It is therefore possible that drugs capable of directing prostaglandin synthesis *in vivo* from PGE(2) to PGF(2-alpha) might form a new class of anti-inflammatory drugs.

TABLE I

ABILITY OF PROSTAGLANDINS TC INDUCE INCREASED VASCULAR PERMEABILITY IN SKIN

Species	Site	Prostaglandin						References
		E(1)	E(2)	F(1-alpha)	F(2-alpha)	A(1)	A(2)	
Rat	Skin	+++	+++	0	0	+	N.T.	Kaley and Weiner, 1968, 1971; Crunkhorn and Willis, 1969, 1971*
Rat	Paw	+++	N.T.	N.T.	N.T.	N.T.	N.T.	Arora et al, 1970*
Rabbit	Skin	++	N.T.	N.T.	0	N.T.	N.T.	Solomon et al, 1968
Rabbit	Eye	++	+++	+	+	+	N.T.	Solomon et al, 1968; Waitzman and King, 1967; Beitch and Eakins, 1969**
Guinea pig	Skin	++	N.T.	N.T.	0	N.T.	N.T.	Horton, 1963; Solomon et al, 1968
Man	Skin	+++	+++	+	+	N.T.	N.T.	Ambache, 1961; Bergstrom et al, 1965***; Solomon et al, 1971; Juhlin and Michaelsson, 1969; Crunkhorn and Willis, 1969, 1971

* Appears to act mainly through histamine release
** Blocked by pclyphloretin phosphate
*** Intra-arterial infusion

ACKNOWLEDGEMENTS

We wish to thank Ken Diamond, Mary Heidger, and Gordon Ringold for technical assistance at Stanford. Synthetic prostaglandins were obtained from Dr. N. Weinshenker, Alza Corp., Palo Alto, Ca.

This work was supported in part by USPHS Grant 3-R01-NS09585 and ONR Contract No. N00014-67A.

REFERENCES

Ambache, N., 1961, Prolonged erythema produced by chromatographically purified irin, J. Physiol. (London) 160:3.

Ambache, N., Brummer, M. C., Rose, S. G., and Whiting, J., 1966, Thin layer chromatography of spasmogenic unsaturated hydroxy acids from various tissues, J. Physiol. (London) 185:77.

Ambache, N., Kavanagh, L., and Whiting, J., 1965, Effect of mechanical stimulation on rabbits' eyes: release of active substances in anterior chamber perfusates, J. Physiol. (London) 176:378.

Anderson, A. J., 1970, Lysosomal enzyme activity in rats with adjuvant-induced arthritis, Ann. Rheum. Dis. 29:307.

Anderson, A. J., Brocklehurst, W. E., and Willis, A. L., 1971, Evidence for the role of lysosomes in the formation of prostaglandins during carrageenin-induced inflammation in the rat, Pharm. Res. Commun. 3:13.

Anggard, E., and Jonsson, C. E., 1971, Efflux of prostaglandins in lymph from scalded tissue, Acta. Physiol. Scand. 81:440.

Arora, S., Lahiri, P. K., and Sanyal, R. K., 1970, The role of prostaglandin E(1) in inflammatory process in the rat, Int. Arch. Allergy, 39:186.

Bartels, J., Kunze, H., Vogt, W., and Wille, G., 1970, Prostaglandin liberation from and formation in frog intestine, Nuanyn-Schmiedeberg's Arch. Pharmak. Exp. Path. 206:199.

Beitch, B. R., and Eakins, K. E., 1969, The effects of prostaglandins on the intraocular pressure of the rabbit, Br. J. Pharmac. 37:158.

Bergstrom, S., Carlsson, L. A., Ekelund, L. G., and Oro, L., 1965, Cardiovascular and metabolic response to infusions of prostaglandin E(1) and to simultaneous infusions of noradrenaline and prostaglandin E(1) in man, Acta. Physiol. Scand. 64:332.

Bergstrom, S., Carlsson, L. A., and Weeks, J. R., 1968, The prostaglandins: A family of biologically active lipids, Pharmac. Rev. 20:1.

Bergstrom, S., Danielsson, H., and Samuelsson, B., 1964, The enzymatic formation of prostaglandin E(2) from arachidonic acid, Biochim. Biophys. Acta. 90:207.

Bethel, R. A., and Eakins, K. E., 1971, Antagonism by polyphloretin phosphate of the intraocular pressure rise induced by prostaglandins and formaldehyde in the rabbit eye, Fed. Proc. 30:626.

Bonta, I. L., and De Vos, C. J., 1965, Presence of a slow contraction inducing material in fluid collected from the rat paw edema induced by serotonin, Experientia, 21:34.

Brittain, R. T., 1966, The intracerebral effects of noradrenaline and its modification by drugs in the mouse, J. Pharm. Pharmac. 18:621.

Cashin, C. H., and Heading, C. E., 1968, The assay of antipyretic drugs in mice, using intracerebral injection of pyretogenins, Br. J. Pharmac. 34:148.

Collier, H. O. J., 1971, Prostaglandins and aspirin, Nature (London) 232:17.

Coppi, G., and Bonardi, G., 1968, Effect of two non-steroidal anti-inflammatory agents on alkaline and acid phosphatases of inflamed tissue, J. Pharm. Pharmac. 20:661.

Crunkhorn, P., and Willis, A. L., 1969, Actions and interactions of prostaglandins administered intradermally in rat and man, Br. J. Pharmac. 36:216.

Crunkhorn, P., and Willis, A. L., 1971a, Cutaneous reactions to intradermal prostaglandins, Br. J. Pharmac. 41:49.

Crunkhorn, P., and Willis, A. L., 1971b, Interaction between prostaglandins E and F given intradermally in the rat, Br. J. Pharmac. 41:507.

DiRosa, M., Giroud, J. P., and Willoughby, D. A., 1971, Studies of the mediators of the acute inflammatory response induced in rats in different sites by carrageenin and turpentine, J. Path. 104:15.

DiRosa, M., and Willoughby, D. A., Screens for anti-inflammatory drugs, J. Pharm. Pharmac. 23:297.

Eisen, V., Greenbaum, L., and Lewis, G. P., 1968, Kinins and anti-inflammatory steroids, Br. J. Pharmac. 34:169.

Elsbach, P., 1966, Phospholipid metabolism by phagocytic cells, I. A comparison of conversion of [^{32}P]-lysolecithin to lecithin by homogenates of rabbit polymorphonuclear leukocytes and alveolar macrophages, Biochim. Biophys. Acta 125:510.

Elsbach, P., and Rizack, M. A., 1963, Acid lipase and phospholipase activity in homogenates of rabbit polymorphonuclear leukocytes, Am. J. Physiol. 205:115.

Feldberg, W., and Saxena, P. N., 1971, Fever produced in rabbits and cats by prostaglandin E(1) injected into the cerebral ventricles, J. Physiol. 215:23.

Ferreira, S. H., Moncada, S., and Vane, J. R., 1971, Indomethacin and aspirin abolish prostaglandin release from spleen, Nature New Biology, 231:237.

Fowler, S., and De Duve, C., 1969, Digestive activity of lysosomes. III. The digestion of lipids by extracts of rat liver lysosomes, J. Biol. Chem. 244:471.

Gardner, D. L., 1960, Production of arthritis in the rabbit by the local injection of the mucopolysaccharide carrageenin, Ann. Rheum. Dis. 19:369.

Giroud, J. P., and Willoughby, D. A., 1970, The interrelations of complement and a prostaglandin-like substance in acute inflammation, J. Path. 101:241.

Glenn, M. E., 1972, Pro-inflammatory and anti-inflammatory effects of certain prostaglandins, in "Prostaglandins in Cellular Biology and the Inflammatory Process" (B. B. Pharriss and P. W. Ramwell, eds.), Plenum Press, New York (in press).

Greaves, M. W., Sondergaard, J., and McDonald-Gibson, W., 1971, Recovery of prostaglandins in human cutaneous inflammation, Br. Med. J., 2:258.

Green, K., and Samuelsson, B., 1964, Prostaglandins and related factors: XIX. Thin layer chromatography of prostaglandins, J. Lipid. Res. 5:117.

Harris, J. M., and West, G. B., 1963, Rats resistant to the dextran anaphylactoid reactions, Br. J. Pharmac. 20:550.

Holmes, S. W., 1968, Concentrations of prostaglandins in the nervous system, Ph. D. Thesis in the University of London.

Holmes, S. W., 1970, The spontaneous release of prostaglandins into the cerebral ventricles of the dog and the effect of external factors on this release, Br. J. Pharmac. 38:653.

Holmsen, H., and Day, H. J., 1970, The selectivity of the thrombin-induced platelet release reaction: subcellular localization of released and retained constituents, J. Lab. Clin. Med. 75:840.

Horton, E. W., 1963, Action of prostaglandin E(1) on tissues which respond to bradykinin, Nature (London) 200:892.

Juhlin, S., and Michaelsson, G., 1969, Cutaneous vascular reactions to prostaglandins in healthy subjects and in patients with urticaria and atopic dermatitis, Acta. Derm-Vener. (Stockholm) 49:251.

Kaley, G., and Weiner, R., 1968, Microcirculatory studies with prostaglandin E(1), in "Prostaglandin", symposium of the Worcester Foundation for Experimental Biology, (P. W. Ramwell and J. E. Shaw, eds.), Interscience, New York, pp. 321-328.

Kaley, G., and Weiner, R., 1971, Prostaglandin E(1): a potential mediator of the inflammatory response, Ann. N. Y. Acad. Sci. 180:338.

Melmon, K. L., and Cline, M. J., 1967, Interaction of plasma kinins and granulocytes, Nature, (London) 213:90.

Michaelsson, G., 1970, Effects of antihistamines, acetylsalicylic acid and prednisone on cutaneous reactions to kallikrein and prostaglandin E(1), Acta. Derm. Vener. (Stockholm) 30:31.

Milton, A. S., and Wendlandt, S., 1970, A possible role for prostaglandin E(1) as a modulator for temperature regulation in the central nervous system of the cat, J. Physiol. Lond. 207:768.

Milton, A. S., and Wendlandt, S., 1971, J. Physiol. (London) 218:325.

Muller-Eberhard, H. J., 1969, Complement, Ann. Rev. Biochem. 38:389.

Parrat, J. R., and West, G. B., 1958, Inhibition by various substances of edema formation in the hind paw of the rat induced by 5-hydroxytryptamine, histamine, dextran, eggwhite, and compound 48/80, Br. J. Pharmac., 13:65.

Piper, P. J., and Vane, J. R., 1969, Release of additional factors in anaphylaxis and its antagonism by anti-inflammatory drugs, Nature (London) 225:29.

Samuelsson, B., 1963, Isolation and identification of prostaglandins from human seminal plasma, J. Biol. Chem. 238:3229.

Selye, H., 1954, An experimental model illustrating the pathogenesis of the diseases of adaptation, J. Clin. Endocrin., 14:997.

Shaw, J. E., and Ramwell, P. W., 1969, Separation, identification and estimation of prostaglandins, Meth. Biochem. Analysis, 17:325.

Silver, M. J., Smith, J. B., and Webster, G. R., 1971, Phospholipase activity in human platelets, Pharmacologist, 13(2):276, abs. 474.

Smith, A. D., and Winkler, H., 1968, Lysosomal phospholipases A(1) and A(2) of bovine adrenal medulla, Biochem. J., 108:867.

Smith, J. B., and Willis, A. L., 1970, Formation and release of prostaglandins by platelets in response to thrombin, Br. J. Pharmac. 40:545.

Smith, J. B., and Willis, A. L., 1971, Aspirin selectively inhibits prostaglandin production in human platelets, Nature New Biology, 231:235.

Smith, M. J. H., and Dawkins, P. D., 1971, Salicylate and enzymes, J. Pharm. Pharmac. 32:729.

Solomon, L. M., Juhlin, L., and Kirschbaum, M. B., 1968, Prostaglandin on cutaneous vasculature, J. Invest. Derm. 51:280.

Spector, W. G., and Willoughby, D. A., 1968, The pharmacology of inflammation, English Universities Press Ltd. (London).

van Arman, C. G., Begany, A. J., Miller, L. M., and Pless, H. H., 1965, Some details of the inflammations caused by yeast and carrageenin, J. Pharmac. Exp. Ther. 150:328.

Vane, J. R., 1964, The use of isolated organs for detecting active substances in the circulating blood, Br. J. Pharmac. Chemother. 23:360.

Vane, J. R., 1971, Inhibition of prostaglandin synthesis as a mechanism of action from aspirin-like drugs, Nature New Biology, 231:232.

Vargaftig, B. B., and Dao, N., Pharmacology, (in press).

Waite, M., Scherphof, G. L., and Van Deenen, L. L. M., 1969, Differentiation of phospholipases A in mitochondria and lysosomes of rat liver, J. Lipid Res. 10:411.

Waitzman, M. B., and King, C. D., 1967, Prostaglandin influences on intraocular pressure and pupil size, Am. J. Physiol. 212:329.

Weissman, G., 1967, The role of lysosomes in inflammation and disease, Ann. Rev. Med. 18:97.

Weissman, G., and Thomas, L., 1964, The effect of corticosteroids upon connective tissue and lysosomes, Recent Progr. Horm. Res. 20:215.

Willis, A. L., 1969a, Parallel assay of prostaglandin-like activity in rat inflammatory exudate by means of cascade superfusion, J. Pharm. Pharmac. 21:126.

Willis, A. L., 1969b, Release of histamine, kinin and prostaglandins during carrageenin-induced inflammation in the rat, in "Prostaglandins Peptides and Amines", (P. Mantegazza and E. W. Horton, eds.), Academic Press, London. pp. 31-38.

Willis, A. L., 1970, Identification of prostaglandin E(2) in rat inflammatory exudate, Pharmac. Res. Commun. 2:297.

Willis, A. L., 1971, Prostaglandins: their release and inter-relationship with biogenic amines, Ph. D. Thesis in the University of London.

Willis, A. L., Davison, P., and Ramwell, P. W., 1972, An action of antipyretic drugs in mice through inhibition of prostaglandin synthesis in brain, in "Prostaglandins" (in press).

Willoughby, D. A., Coote, E., and Turk, J. L., 1969, Complement in acute inflammation, J. Path. 97:295.

Winter, C. A., Risley, E. A., and Nuss, G. V., 1962, Carrageenin-induced edema in hind paws of the rat as an assay for anti-inflammatory drugs, Proc. Soc. Exp. Biol. Med. 111:544.

Winter, C. A., Risley, E. A., and Nuss, G. V., 1963, Anti-inflammatory and antipyretic activities of indomethacin, 1(p-chlorobenzoyl)-5-methoxy-2-methyl-indole-3-acetic acid, J. Pharmac. Exp. Ther. 141:369.

Wiseman, E. H., and Chang, Y., 1968, The role of fibrin in the inflammatory response to carrageenin, J. Pharmac. Exp. Ther. 159:206.

Woodbury, D. M., 1970, Analgesic-antipyretics, anti-inflammatory agents and inhibitors of uric acid synthesis, in "Pharmacological Basis of Therapeutics", 4th edn. (L. S. Goodman and A. Gilman, eds.), The MacMillan Company, London and Toronto, pp. 314-347.

DISCUSSION

KLIGMAN: It seems to me that prostaglandins are in a state of chaos. Whether they are pro-inflammatory or anti-inflammatory, inhibitors or stimulators, they seem to have the same antithetic effects on observers. There are contradictions in the experimental data. Moreover, the amount of proof that quells skepticism is very different for the experimenter and his critics. I think we are going to hear one hell of a lot of ingenious rationalizations this morning. I shall simply put out my own generalizations concerning inflammatory models. I note that non-dermatologists may share similar disabilities, namely too much dependance on their eyes. We have been proselytizing for some time now, "don't rely on what you see". It can be very misleading. For example, what is a weal? Are all weals similar? In rats or mice, as you know, weals don't present themselves to the naked eye. You have to visualize them by extravasation of circulating dyes. One can produce a weal with histamine that looks like a very good weal, similar to the kind produced by antigen-antibody reactions or bradykinin or prostaglandin. Yet these weals are different in their time course, severity and also in their histopathology, that is, their content of cells. The antigen-antibody weal has many eosinophils in it. Weals produced by histamine do not. There are weals that are simply an extravasation of fluid. Others have many neutrophils. There are weals which come up late and weals which come up early. The point is that models have to be characterized in their entirety and not simply by appearance.

One can stray very badly by relying simply on anti-inflammatory assays. For example, would you predict that anti-inflammatory drugs would be highly effective in the sunburn reaction? Can you block the sunburn reaction with steroids? Physicians think so and have treated people that way. It turns out that this is not the case. Steroids are in fact rather feeble in moderating sunburn reactions, topically or orally (excluding the vasoconstrictive effect). If steroid is placed on the skin and then irradiated, the sunburn reaction will not occur. If the skin is irradiated before the steroid, there is little suppression. The explanation for this is that steroids absorb erythemic radiation and act as sunscreens. This is an example of a mistaken interpretation of the effect of anti-inflammatory agents.

Curiously, non-steroids such as aspirin seem to curtail the sunburn reaction quite well.

Is histamine pro-inflammatory or anti-inflammatory? It just depends on how you set up the experiment. This is part of the tactics of investigation. If you inject histamine along with tuberculin, you can diminish the expected reaction. One would not expect that histamine would intensify the reaction, since it contributes to the early phase of the inflammatory response. In this case histamine acts by increasing the blood flow and removing the allergen from the site so this is less present to elicit the reaction. Just one point, Dr. Willis - you pointed out that prostaglandin synthesis is not inhibited by cortisone. Are you suggesting that hydrocortisone does not diminish the inflammatory infiltrate? Isn't it an anti-inflammatory agent? And hasn't it been shown many times that it can, in fact, suppress the inflammation? Could you please discuss the reaction of cortisone on tissue.

WILLIS: Firstly, hydrocortisone does reduce carrageenin-induced inflammation in rat; it does not interfere with the formation of prostaglandins, but it does reduce their ability to induce increases in vascular permeability. Secondly, your point as to whether prostaglandins are pro- or anti-inflammatory: Any inducer of inflammation, whether acting directly or indirectly, can inhibit inflammation induced by itself or another phlogistic agent if it is given to a different site in the body. This is the well known "counter-irritant" effect and is seen with carrageenin itself (Atkinson and Hicks, 1971, Br. J. Pharmac. 41:487), and one certainly would not suggest that carrageenin is not a phlogistic agent! PGE compounds are the most potent known inducers of increased vascular permeability in man and certain animals, in fact, just a nanogram of PGE(2) can induce a reaction in rat skin (Crunkhorn and Willis, 1971, Br. J. Pharmac. 41:49). Furthermore, aspirin and similar drugs inhibit prostaglandin production at very low concentrations, and they have never worked in any other system in such a potent manner. For instance, if one takes only two aspirin tablets, then the ability of his platelets to produce prostaglandins is virtually abolished (Smith and Willis, 1971, Nature New Biology, 231:235). Taken together, I think this evidence is very convincing.

KALEY: I do not believe that steroids are vasoconstrictors. In some isolated experiments they may potentiate the effects of certain vasopressor agents (S.

Kalsner, 1969, Circ. Res., 24:383), but on their own they are certainly not vasoconstrictors. I would be curious to know how Dr. Willis reconciles his experimental results with those of many others who have shown that the anti-inflammatory effects of steroids are based on their ability to stabilize lysosomal membranes, and how in his view hydrocortisone affects the PGE(2)-induced vascular permeability change? Are other agents, which increase vascular permeability, also affected by hydrocortisone?

STRAUSS: In the dermatological literature there is ample documentation that when steroids are applied to the skin there is a vasoconstrictor response. This response has been used to develop a standard assay for corticosteroids.

RINGOLD: In our hands, the ability of non-steroidal anti-inflammatory agents to block the biosynthesis or release of prostaglandins correlates with their anti-inflammatory effects. The work of Willis and of Vane prompted us (Tomlinson, R. V., Ringold, H. J., Qureshi, M. C., and Forchielli, E., 1971, Biochem. Biophys. Res. Comm., submitted for publication) to study the effect of a number of agents upon the isolated bull seminal vesicle system. This microsomal system converts the requisite unsaturated fatty acid precursors into PGE(1) and PGE(2) in high yields and with mg throughput, which permits actual isolation and physical characterization of the products. Thus, yields and inhibition can be readily quantitated either via isotope techniques or spectrophotometrically by the UV absorption at 278 nanometers following alkaline conversion to PGB.

In a first experiment, aspirin, sodium salicylate, indomethacin, meclofenamate and naproxen (the d-isomer of 2-[6'-methoxy-2'-naphthyl]-propionic acid) at the relatively high concentration of 0.6 mg/ml were shown to cause 50-100% inhibition of the biosynthesis of both PGE(1) and PGE(2) when eicosatriencic or eicosatetraenoic acids were the substrates at a concentration of 0.33 mM. In further experiments, dose-response curves for the inhibition of PGE(2) biosynthesis were studied with indomethacin, d-naproxen, the l-enantiomer of naproxen and aspirin. The concentration of anti-inflammatory agent found to cause a 50% inhibition were 7×10^{-6} M, 1×10^{-4} M, 7×10^{-3} M and 1.5×10^{-2} M respectively. Since the arachidonic precursor was found unchanged, the blockade was at an early stage, which suggests the possibility of direct competitive inhibition.

Table I lists the activities of these four compounds in some conventional assays for non-steroidal anti-inflammatory agents as determined in our laboratories. It may be noted that the ability of these agents to inhibit the biosynthesis of PGE(2) parallels, in general, their potencies in the various assays. In particular, it is striking that naproxen is 70-fold more potent as an inhibitor of PGE(2) biosynthesis than its l-enantiomer, a change involving the asymmetry about only one carbon atom. This correlates with the markedly reduced anti-inflammatory, anti-pyretic and analgesic activity of the l-compound compared to naproxen and suggests that the blockade of prostaglandin biosynthesis or release is intimately related to the pharmacological properties of non-steroidal anti-inflammatory agents.

WEISSMANN: I think the data on aspirin are impressive. I'd like to say something in response to Dr. Kligman's comment on the mode of action of steroids in skin. In 1956, together with Dame Honor B. Fell, we produced "sunburn in a dish" (Weissmann, G., and Fell, H. B., 1962, J. Exp. Med. 116:365). We took mouse skin and floated it in rats and exposed it to UV light in the absence of intact vasculature. Cortisol successfully suppressed most of the signs of UV induced tissue damage; subsequent histochemical examination showed that the intracellular organelles looked better in the cortisol-treated explants. At least some of the effects of cortisol in living skin in vitro are therefore unrelated to vasculature. Secondly, Dr. Willis, I am deeply troubled by investigators who aspirate subcutaneous tissue and measure beta-glucuronidase, claiming this has something to do with lysosomes. This procedure may produce nothing but a biochemical cell count, because what one does is to measure influx of white cells into the carrageenin or other induced granulomata at that phase. A number of people have supported the idea that cortisone does not entirely suppress the appearance of white cells in certain kinds of inflammation. Certainly, in most systems where proper biochemistry has been done and total cell enzymes and subcellular fractions have been studied rather than the appearance of materials in the supernatants (which indicate live, dead and dying cells), cortisone does stabilize the membranes of lysosomes. Inability to show an effect upon platelet release of beta-glucuronidase is in complete keeping with our studies which show that cortisone does not prevent the release of packets of preformed granules either in the macrophage, the polymorph or the platelet. However, the failure to find an effect upon the appearance of beta-glucuronidase in the tissue simply means that there are not as many dead or dying white cells in the area as existed before cortisone. So I don't think that these data can be used either to support or deny an action of cortisone on lysosomes. Finally, putting all our eggs in the basket of one mediator may lead one into deep trouble. Basically, polymorphs have a large number of mediators within them. They have proteases which can cleave complement components. Tissue injury will provoke hemotactic factors, and it may be the polymorphs themselves which elaborate these as well as many of the factors that you have studied, not to speak of the permeability-inducing proteins, as well as collagenases, elastases, and neutral proteases (Weissmann, G., and Dukor, P., 1970, Advances in Immunology, 12:283; Davies, P., Krakauer, K., and Weissmann, G., 1971, Biochem. J., 123:559). If you are to put all of your emphasis on prostaglandins as mediating all kinds of inflammation, you may be in trouble. Unless these experiments are repeated in leukocyte-depleted animals, complement-depleted animals, or with animals genetically lacking discrete complement components, one may have difficulty extrapolating acute inflammation in the carrageenin system to other sorts of inflammation.

WILLIS: My objection to the lysosomal stabilizing theory is that the concentrations of hydrocortisone used *in vitro* to show a lysosomal stabilizing effect are very high, whereas the effects observed *in vivo* for carrageenin inflammation may be more real. In our platelet work we even took platelet-rich plasma and incubated it for 5 hr with hydrocortisone (1 mg/ml) and found no reduction in release of lysosomal enzymes. As Dr. Weissmann says, hydrocortisone does not influence release of "packets" of lysosomal enzymes. However, these "packets" are presumably lysosomal granules, and one would therefore expect hydrocortisone to stabilize them, if the "stabilizing" theory is correct.

IGNARRO: In relation to Dr. Weissmann's statement, I think that the choice of a particular model of inflammation is very important, e.g., I think that one ought to look at chronic models of inflammation as well as acute models. Something like adjuvant-induced arthritis in the rat would certainly be more akin to rheumatoid arthritis. You might find a difference in the spectrum of prostaglandin activity when you compare adjuvant-arthritis to carrageenin edema. We have shown last year that certain anti-inflammatory drugs such as phenylbutazone, indomethacin and hydrocortisone can certainly alter lysosome membrane integrity in adjuvant polyarthritis. In polyarthritis, liver lysosomes become more fragile, and this condition is associated with a great elevation in the plasma levels of lysosomal enzymes. There is an increase in the local concentration of lysosomal enzymes in the inflamed paws, and the time course follows that of the delayed hypersensitivity. Drugs that are effective in attenuating inflammation, such as hydrocortisone, are quite effective in altering the integrity of the lysosomal membrane. Such drugs stabilize the liver lysosomes, reduce the plasma levels of lysosomal enzymes, and reduce the local concentration of lysosomal enzymes to normal. So I think that the choice of a model here is important. I believe that the steroids certainly can function by enhancing the stability of the lysosomal membrane in chronic models of inflammation. I don't know if this can be demonstrated for acute models of edema.

WILLIS: The real testing point is that we believe that lysosomal phospholipase A is involved in prostaglandin production in the exudate or in thrombin-treated platelets. Hydrocortisone definitely does not inhibit prostaglandin production *in vitro* or *in vivo* (Vane, 1971, Nature New Biology, 231:232; Smith and Willis, 1971, Ibid, 231:235; Ferreira, Moncada and Vane, 1971, Ibid, 231:237). Now, if we can definitely establish a link between lysosomes and prostaglandin production, then a direct stabilizing effect of hydrocortisone on lysosomes is not likely to be its mode of action, at least in acute inflammation.

IGNARRO: I agree, but I do think you really ought to look at the entire picture also in a chronic model of arthritis, i.e., after 14 or 21 days of disease and not several hours.

KUEHL: As a result of the observations of Vane et al, we decided to study prostaglandin synthetase and its possible relation to the action of non-steroidal anti-inflammatory agents. In these studies we used sheep seminal vesicles as the enzyme source and examined the effect of compounds upon the conversion of labeled arachidonic acid to [^{3}H]-PGE(2). The [^{3}H]-PGE(2) was separated by extraction and thin layer chromatography, and measured by scintillation counting. Utilizing this method, we examined about 200 compounds, including fenamates, phenylbutazone, aspirin, indocin and compounds structurally related thereto, for their effect upon prostaglandin synthetase activity. In general, there appears to be a quite good correlation between the inhibitory activity of these compounds upon prostaglandin synthetase and their activity in the rat carrageenin foot edema assay. This is particularly true within groups of related structures. There is, of course, not absolute agreement, but I don't think one should expect this. I am sure that there are sites prior to and subsequent to the action of prostaglandin involved in the inflammatory processes, as well as completely unknown areas that could be additional targets for anti-inflammatory agents. In any event, the most potent argument we have favoring the Vane concept is the finding that in a number of optically active pairs, of which only the d-isomer is active in the foot edema assay, perfect correlation was noted between this parameter and inhibitory activity in the prostaglandin-synthetase inhibition assay. Kinetic studies with indocin itself reveal that it is a competitive inhibitor of the synthetase.

DI PASQUALE: One comment I would like to make is that cortisone is apparently a better stabilizer of lysosomes than cortisol or dexamethasone, yet physiologically it is believed that cortisone has to be converted to cortisol for its action. Similarly, in animal models and also clinically, cortisol and dexamethasone are many times more active than cortisone. Also Sachs and DeDuve (1962) correlated a cortisol-induced thymolysis to an elevated acid hydrolase activity, thereby indicating that cortisol labilizes instead of stabilizing thymic lysosomes. These facts, plus the aspect of dual localization of lysosomal marker enzymes, differential release of lysosomal enzymes, and the present controversy on the effect of non-steroidal anti-inflammatory compounds on lysosomal membranes, bring up some important questions.

WILLIS: I would like to return to the adjuvant polyarthritis question. It has been shown that PGE(1) is chemotactic for polymorphs (Kaley and Weiner, 1971, Ann. N. Y. Acad. Sci., 180:338), and this might turn out to be true for PGE(2) also. This, PGE compounds could conceivably be involved in the change from acute inflammation to chronic inflammation. During adjuvant-induced arthritis in rats, PGE compounds are present in the inflamed paws. Levels in the more inflamed "primary reaction" paws are about six times those in the secondary reaction paws, and these results are similar to those for free lysosomal enzyme activity and leukocyte population (Anderson, 1970, Ann. Rheum. Dis. 29:307). PGE(2) might be released early in the inflammatory reaction and then contribute towards the attraction of leukocytes, which produce more prostaglandins. Thus, a positive feedback chain reaction is initiated, leading to chronicity of the inflammation. There is little evidence which suggests this; for instance, regular injection of carrageenin into rabbit knee joints leads to a chronic condition resembling rheumatoid arthritis (Gardner, 1960, Ann. Rheum. Dis. 19:369). The possibility of PGE(2) being involved in such a situation could be tested by repeating these experiments but injecting PGE(2) into the knee joints instead of carrageenin.

TABLE I

COMPARISON OF THE RELATIVE ANTI-INFLAMMATORY, ANTI-PYRETIC AND ANALGESIC POTENCIES OF INDOMETHACIN, NAPROXEN, ITS L-ENANTIOMER AND ASPIRIN AND THEIR ABILITY TO INHIBIT PGE(2) SYNTHESIS

	Animal Models				BSVM System
	Anti-inflammatory				Inhibitor Activity
Compound	Adjuvant Induced Arthritis Activity**	Carrageenin	Anti-pyretic	Analgesic	Anti-PGE(2) Synthesis
Indomethacin	2000**	48	18	58	2140
Naproxen (d-)	200**	33	22	7	150
Naproxen enantiomer (l-)	--	≥1.5	1.5	≥0.5	2
Aspirin	1*	1	1	1	1

*Relative activity of aspirin arbitrarily set at 1.0.

**As determined on days 14-28 of the adjuvant induced arthritis.

FORMATION OF PROSTAGLANDINS IN THE SKIN FOLLOWING A BURN INJURY

E. Anggard and C.-E. Jonsson

Dept. of Pharmacology, Karolinska Inst., Stockholm
and the Burn Centre, Dept. of Plastic Surgery,
University Hospital, Uppsala, Sweden

Two phases are discerned in the vascular response to a burn injury, one immediate, lasting 5-10 min, and one delayed starting after about 30 min and lasting several hours to days. Histamine has been suggested as a mediator of the first phase of the inflammatory response, and vasoactive polypeptides have been implied in the delayed phase (Spector and Willoughby, 1968). In the following we shall describe results showing that prostaglandin E(2) is formed in the skin following a burn injury, and may thus be an additional mediator in the delayed phase of the tissue response to a burn injury.

RELEASE OF PGE(2) INTO LYMPH FOLLOWING SCALDING

A polyethylene cannula was inserted in one of the paw lymphatics of an anesthetized dog. Lymph was collected from the cannula for immediate bioassay on isolated smooth muscle organs (guinea pig ileum, rat colon, fundus and uterus). Scalding was effectuated by immersion of the paw in boiling water for 10 sec. For details of the experimental procedure see Jonsson (1971).

The paws developed an edema which usually reached a maximum after 1-2 hr. This was paralleled by a 10-15 fold increase in the rate of lymph flow (Fig. 1). The peak lymph flow was typically reached after one hr and then gradually subsided.

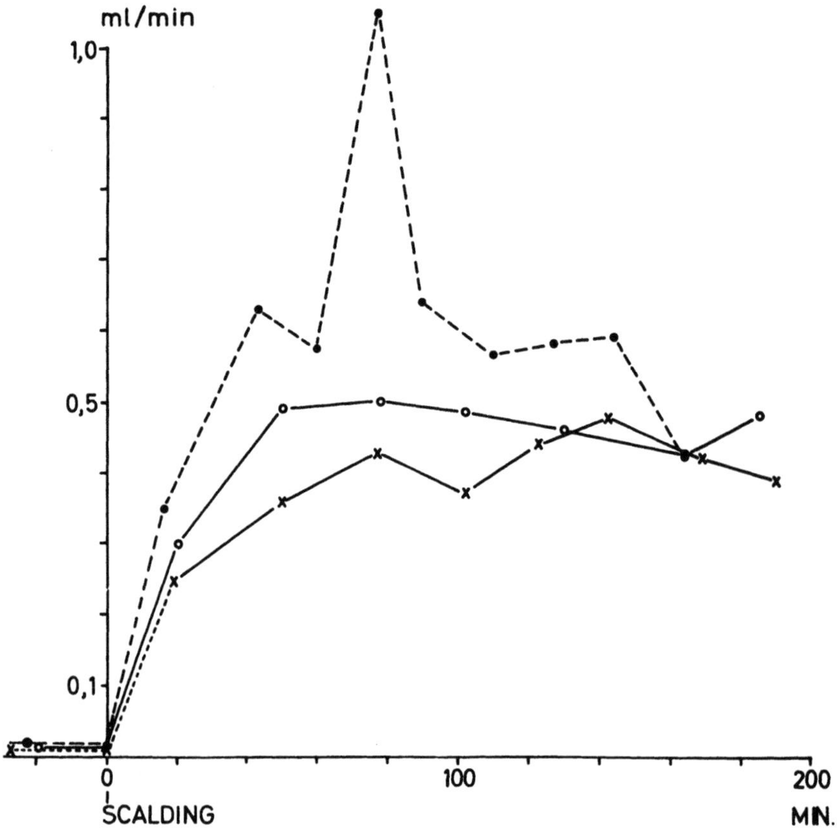

Figure 1. Lymph flow (ml/min) in paw lymphatics from three dogs.

Prior to the scalding no effect of the lymph on the smooth muscle bioassay was observed. After scalding, the direct addition of lymph to the isolated organ elicited a slow reacting response, which was unaffected by the presence of atropine, 2-bromo-d-lysergic acid and mepyramine in the bathing fluid. This is illustrated in Fig. 2.

The time course of the appearance of smooth muscle stimulating activity roughly paralleled that of the increase in lymph flow (Fig. 3). Most of the smooth muscle stimulating material was extractable from lymph into organic solvents, after acidification. It therefore appeared possible that the compound could be a prostaglandin. The type of contraction obtained on isolated organs was also similar to that caused by PGE(2).

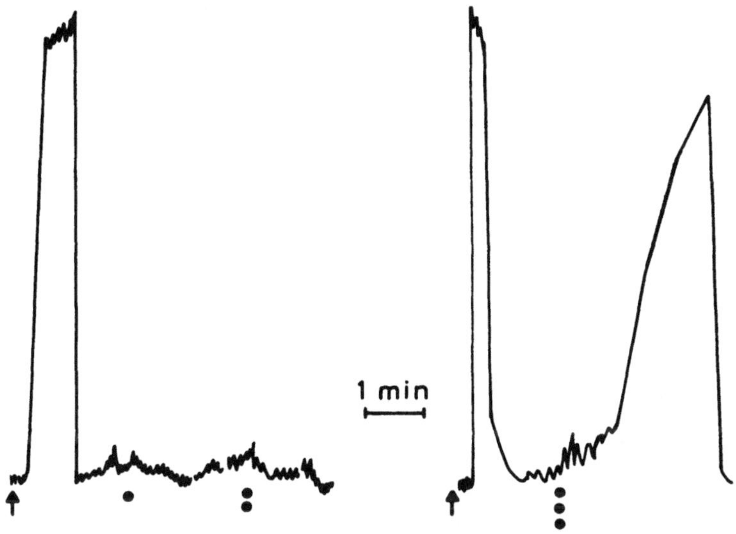

Figure 2. Smooth muscle stimulating activity in paw lymph after scalding. Direct bioassay on isolated guinea pig ileum, bath volume 5 ml, containing Tyrode solution with atropine sulfate (7 x 10^{-7} M), 2-bromo-d-lysergic acid (5 x 10^{-7} M) and mepyramine maleate (1 x 10^{-6} M). (Arrow) bradykinin 10^{-7} M; • 0.5; •• 1.0 ml lymph before scalding; ••• 1.0 ml lymph 45 min after scalding.

To investigate this possibility further, 0.15 microcuries of ^{3}H-labeled PGE(2) (4.5 curie/mM) was added to pooled lymph collected in ethanol. The ethanol was filtered and evaporated ("crude extract"). The aqueous residue was extracted with heptane, acidified to pH 3 and extracted with ether. Acidic compounds were then extracted back into phosphate buffer, pH 8, and then taken over into ether again at an acid pH. The ether was evaporated and the dry residue dissolved in Tyrode solution ("acidic lipid extract"). The biological activity of the "crude extract" and that of the "acidic lipid" extract was assayed using the rat fundus strip. It was found that the mean recovery of biological activity in the "acidic lipid fraction" was over 90% (93, 85, and 94%, n = 3) after correction for losses through the extraction procedure. Thus almost all of the smooth muscle stimulating material present in the lymph seemed to be of acidic lipid character.

By a combination of chromatographic, biological and enzymatic methods it was possible to identify PGE(2) as

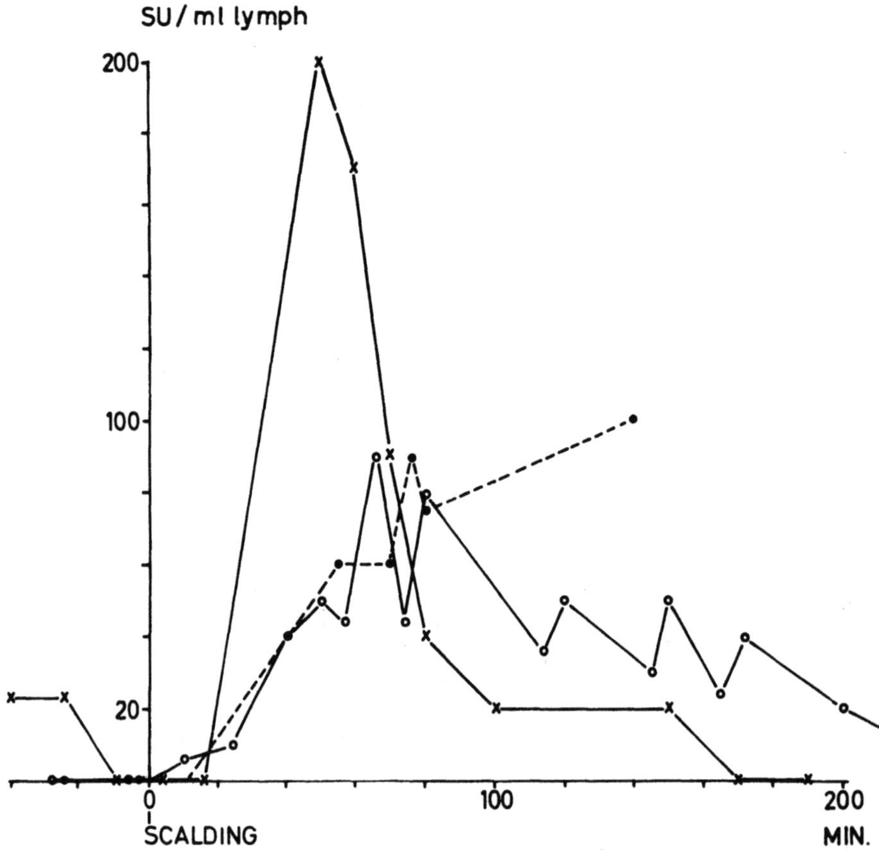

Figure 3. Smooth muscle stimulating activity on isolated rat colon before and after scalding. Direct bioassay of lymph from three dogs. SU = spasmogenic units, 1 SU being equivalent to 1 nanogram of PGE(1).

the main component of the extract. For experimental details see Anggard and Jonsson (1971).

Chromatography of the acidic lipid fraction on silicic acid in a system which affords group separation of the prostaglandins (Bygdeman and Samuelsson, 1967) showed that nearly all biological activity co-chromatographed with the ^{3}H-labeled PGE(2) (Fig. 4). Using reversed phase partition chromatography, which separates the individual prostaglandins (Bergstrom et al, 1963), most of the spasmogenic activity appeared together with PGE(2) with small amounts at the expected position of PGE(1). Furthermore, the biologically active material isolated from lymph by extraction and by silicic acid

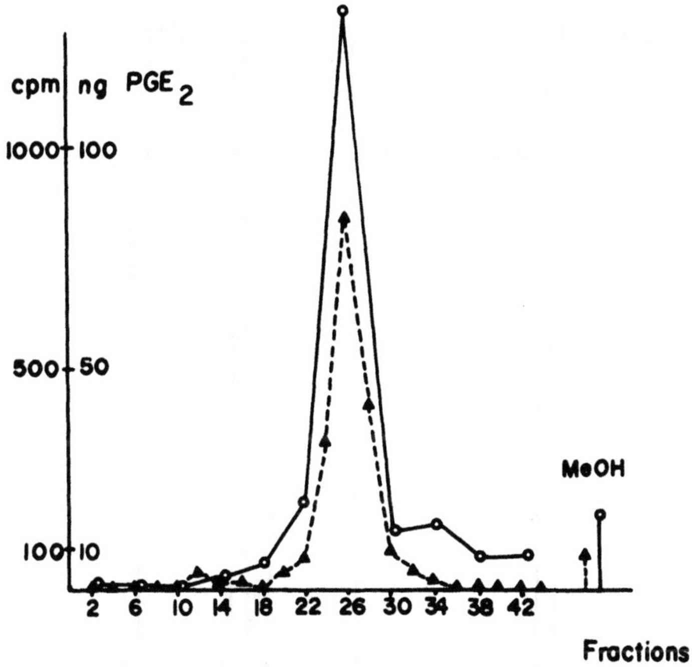

Figure 4. Silicic acid chromatography of lipid extract prepared from lymph collected after scalding. Silicic acid 1 g. Linear gradient of ethyl acetate: benzene in volumetric ratios from 1:9 to 9:1. Fractions 1.5 ml. Smooth muscle assay (circles); radioactivity (triangles).

chromatography and reversed phase partition chromatography was analyzed by gas liquid chromatography with electron capture detection (Jouvenaz et al, 1970). A peak was observed with the same retention time as that of the same derivative of PGE(2). The chromatogram is shown in Fig. 5.

Finally the active material was subjected to a recent enzymatic identification procedure. 15-hydroxyprostaglandin dehydrogenase (PGDH), highly purified from swine lung, is specific for the 15(S)-hydroxy group in the prostaglandins (Anggard and Samuelsson, 1966; Nakano et al, 1969; Vonkeman et al, 1969; Shio et al, 1970). The equilibrium of the reaction favors the oxidation of the prostaglandins. The resulting 15-keto metabolites possess only a few percent or less of the activity of the parent compound (Anggard 1966; Pike et al, 1967). When the smooth muscle stimulating material isolated from lymph was incubated with PGDH and NAD^+,

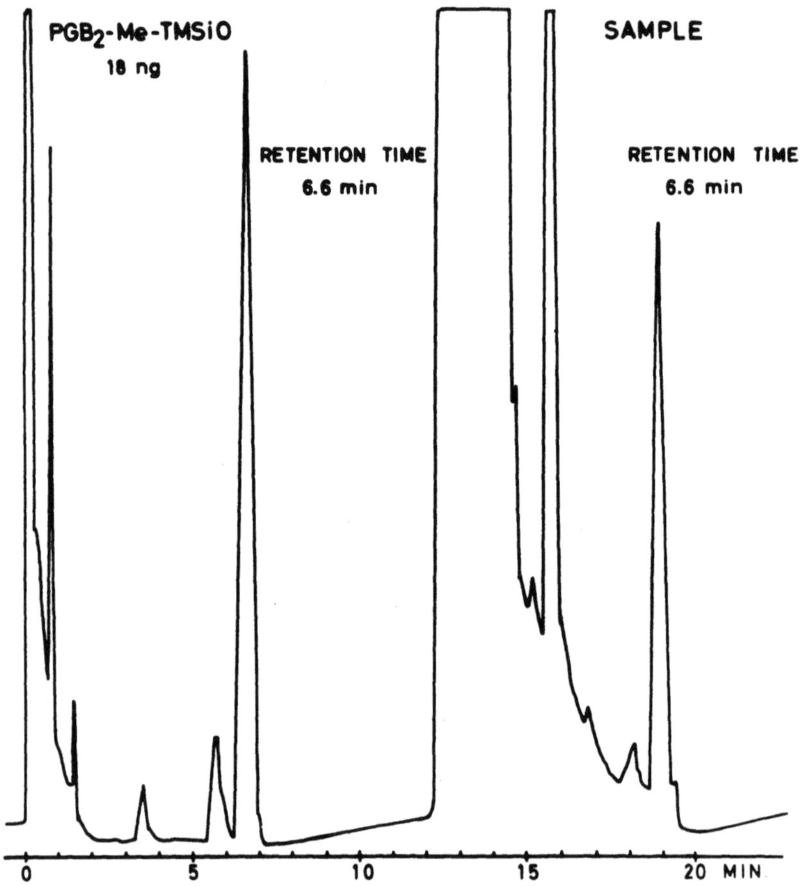

Figure 5. Gas chromatography of material isolated from dog lymph after scalding. The reference PGE(2) and the sample were treated with alkali and resulting PGB-compound converted to methyl ester trimethylsilyl ether. Conditions: 1% OV-1, temp 225°C, ^{63}Ni electron capture detector.

complete biological (rat fundus strip) inactivation was observed (Table I). Control incubations omitting either the enzyme or the cofactor retained full activity. It is evident therefore that the smooth muscle stimulating compounds were a substrate for the PGDH. Together the data establish that the major part of the biological activity appearing in peripheral lymph following a scalding injury is due to PGE(2).

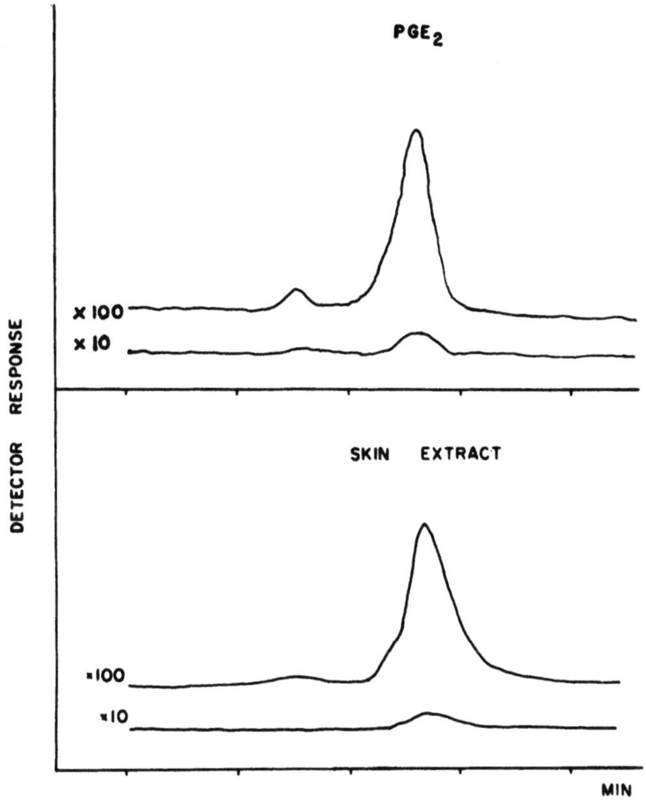

Figure 6. Identification of PGE(2) from human skin homogenates as the methyl ester, methoxime, acetate (Me-MO-Ac) derivative (Green, 1969) by combined gas chromatography-mass spectrometry. The mass spectrometer was focused on fragments having a m/e value of 328. This fragment is formed by the elimination of a methoxy group and two acetate groups from the PGE(2)-Me-MO-Ac. Column: 4 ft 1% SE-30, temp 235°C, the energy of the electrons 22,5 eV. Responses to authentic PGE(2)-Me-MO-Ac-derivative (upper curve) and the same derivative prepared from a skin extract (lower curve).

FORMATION AND METABOLISM OF PROSTAGLANDINS IN HUMAN SKIN

As a part of our studies on the role of the prostaglandins in the burn syndrome seen in humans, we decided to investigate various aspects of the prostaglandin system in human skin. PGE(2) has recently been identified in rat skin (Jouvenaz et al, 1970; Ziboh and Hsia, 1971). Also Hansson and Samuelsson (1965) in an autoradiographic study had noted a high uptake of radioactivity in skin after intravenous injection of ^{3}H-labeled PGE(1) in mice.

Skin was obtained from ten subjects during breast reduction (Skoog, 1963). The skin, free from fat and without its deep dermal layer, was frozen on dry ice immediately after excision and stored at -80°C before use. A 10% (weight/volume) homogenate was prepared using a rotating knife homogenizer operated at 0°C for 2 min. The medium was 0.15 M potassium phosphate buffer (pH 7.4) containing 1.5 mM reduced glutathione and 30 mM EDTA. In the biosynthesis experiments ^{3}H-labeled arachidonic acid was added and the mixture incubated at 37°C with shaking. The acidic lipids were extracted by standard procedures. Radioactive prostaglandins were isolated by silicic acid chromatography. For experimental details see Jonsson and Anggard (1972).

When a human skin homogenate was incubated at 37°C appearance of smooth muscle (gerbil colon) stimulating material was noticed. Using chromatographic, enzymatic and mass spectrometric methods it could be established that essentially all biological activity was due to PGE(2). The analysis by combined gas liquid chromatography and mass spectrometry (GLC-MS) is shown in Fig. 6. If the frozen skin was homogenized directly in ethanol to halt biosynthetic processes, the levels of PGE(2) were found to be relatively low, about 15 nanogram/g (= mean, n = 7, range 0-35 nanogram/g). When the tissue was homogenized in buffer at 0°C the levels of PGE(2) were 25 nanogram/g (= mean, n = 9, range 20-35 nanogram/g). Upon incubation of the homogenate at 37°C the yield of PGE(2) was increased 10-15 fold. The time course of the appearance of PGE(2) in the incubation fluid is shown in Fig. 7.

When a homogenate of human skin was incubated with ^{3}H-labeled arachidonic acid, a 1.5-3.2% (n = 3) conversion into ^{3}H-PGE(2) was observed. When the radioactive PGE(2) was isolated from the incubated homogenate and the radioactivity and biological activity determined, it was found that a 200-fold decrease in specific activity had occurred. This shows that the major part of the PGE(2) had originated from endogenous unlabeled arachidonic acid.

The relative abundance of prostaglandin precursors of phospholipids in human skin was determined in separate experiments. In agreement with Vroman, Nemecek and Hsia (1969) we found that arachidonic acid was the major C20 fatty acid, representing about 15% of the total

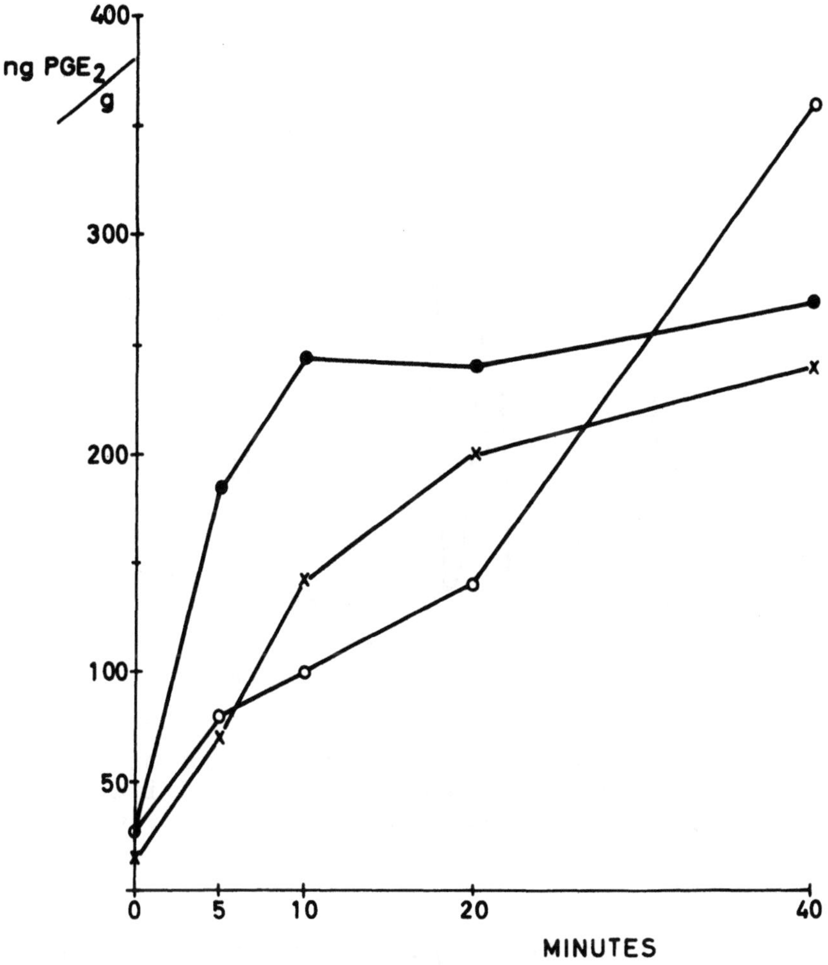

Figure 7. Formation of PGE(2) in human skin homogenates. Skin from three patients was homogenized in 0.15 M potassium phosphate buffer, pH 7.4, and incubated at 37°C with 1.5 mM reduced glutathione and 30 mM EDTA. Acidic lipid extracts were prepared and bioassayed on gerbil colon with PGE(2) as reference.

phospholipid fatty acids. In addition small amounts of bis-homo-gamma-linolenic acid were found. A representative chromatogram is shown in Fig. 8.

We have also obtained evidence for the metabolism of PGE(2) in human skin via the 15-hydroxy-dehydrogenase pathway. Following incubation of a skin homogenate with ^{3}H-PGE(2) and NAD^{+} a 50-70% conversion into 11-alpha-hydroxy-9,15-diketoprost-5-enoic acid [15-keto-dihydro-

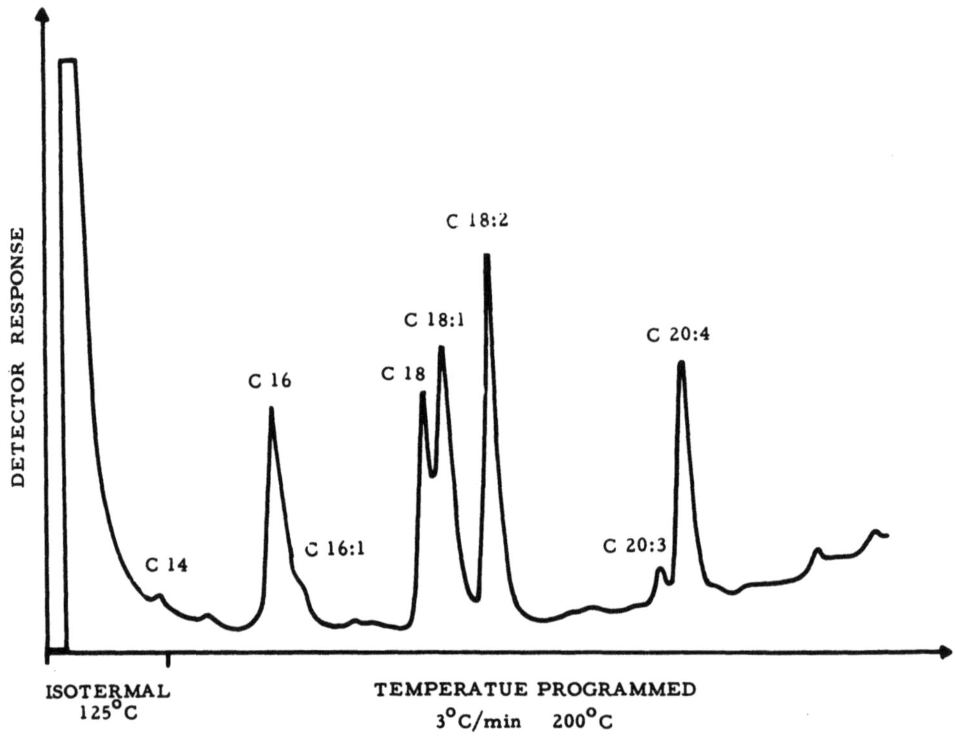

Figure 8. Gas chromatogram of methylesters of skin phospholipids. Column: 8% EGSS-X, operated isothermally at 125°C for 5 min, then temp programmed at 3°C/min to a final temp of 210°C.

PGE(2)] was observed (Fig. 9). The identity of the metabolite was confirmed by a mass fragmentographic technique.

DISCUSSION

The present results clearly demonstrate that human skin possesses substrates and enzymes for the formation and metabolism of prostaglandins. As in the rat skin (Jouvenaz et al, 1970; Ziboh and Hsia, 1971), PGE(2) was found to be the major prostaglandin. The levels of PGE(2) which may be present normally in the skin are probably fairly low. The lowest levels found by us, about 15 nanogram/g, were observed when the rapidly excised and frozen skin was homogenized in ethanol to interrupt enzymatic processes. It is however not unlikely that some of the PGE(2) might have been formed during the surgical procedures and before the tissue was frozen. At any rate

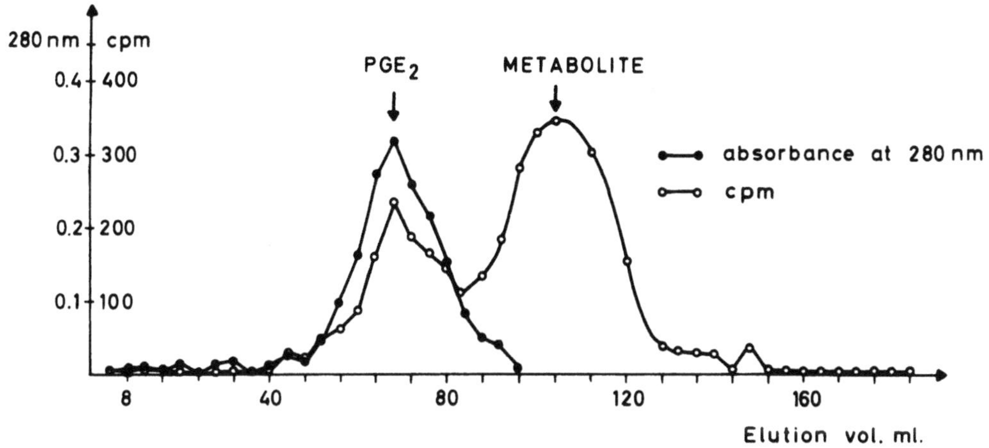

Figure 9. Reversed phase partition chromatography of skin metabolite of ^{3}H-PGE(2). A skin homogenate was incubated for one hr at 37°C, in 0.1 M potassium phosphate buffer, pH 8.0, containing 10 mM NAD^{+} and 10 mM NADH. After extraction of the radioactivity, 0.5 mg of PGE(2) was added and chromatography was performed using reversed phase partition chromatography on hydrophobic supercel (Bergstrom et al, 1963; Anggard et al, 1965), using the C-47 system. The unlabeled reference PGE(2) was detected by determining the absorbance at 280 nanomicrons after treatment with alcoholic 0.5 N NaOH. Radioactivity: open circles. Absorbance: closed circles.

the endogenous level is probably low under normal circumstances.

Data obtained from other tissues (Hopkin et al, 1968; Anggard et al, 1972) indicate that prostaglandins are not stored within any subcellular compartment but formed upon physiological demand, probably to act close to the site of synthesis (Anggard, 1970, 1971). In some tissues the prostaglandins may be inactivated by the ubiquitous (Anggard et al, 1971) prostaglandin dehydrogenase prior to their release into the circulation. This concept is illustrated by the scheme shown in Fig. 10a.

Let us then consider what might happen when the tissue is damaged by a thermal or mechanical injury (Fig. 10b). Firstly cellular membrane functions would predictably be damaged. This, like most processes which disturb membrane function (Gilmore et al, 1969), would activate the hydrolysis of arachidonic acid from phospholipids, the rate-limiting step in prostaglandin biosynthesis (Samuelsson, 1970). With an excess of

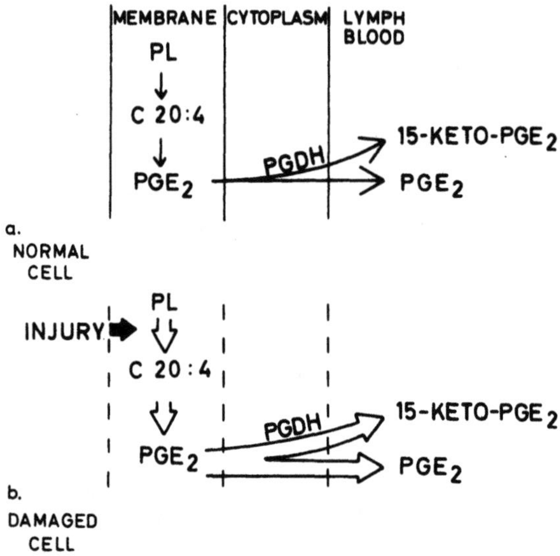

Figure 10. Hypothetical scheme describing prostaglandin formation in (a) normal skin and (b) following a burn injury. PL = phospholipid, C20:4 = arachidonic acid, PGDH = 15-hydroxy-prostaglandin dehydrogenase.

arachidonic acid at the site of the prostaglandin synthetase much more PGE(2) would be formed than normally and the cell, the interstitial fluid, the lymph, and the blood would be flooded with this highly active compound. The amount of prostaglandin dehydrogenase would not be sufficient to inactivate the non-physiological excess of PGE(2), which would be free to influence capillary function. PGE(2) has been demonstrated to give rise to long-lasting erythema and edema when injected into human skin (Solomon et al, 1968; Juhlin and Michaelsson, 1969; Crunkhorn and Willis, 1971). It thus seems possible that PGE(2) could play a role in the pathophysiology of the human burn syndrome.

TABLE I

INACTIVATION OF SMOOTH MUSCLE STIMULATING COMPOUNDS WITH 15-HYDROXY PROSTAGLANDIN DEHYDROGENASE (PGDH)

Additions			
Lymph Extract	PGDH 0.15 mU	NAD^+ 5 mM	Bioassay nanograms PGE(2)
+	+	+	0
+	-	+	24
+	+	-	20
+	-	-	24
-	+	+	0

Incubation of lipid extract from post-scalded paw lymph, corresponding to 24 nanograms PGE(2), was carried out at 37°C for 30 min in 0.2 ml 0.1 M Tris buffer pH 8.0, containing 5 mM NAD^+, 0.15 mU of PGDH. Bioassay was performed on rat fundus strip.

REFERENCES

Anggard, E., 1966, The biological activities of three metabolites of prostaglandin E(1), Acta Physiol. Scand. 66:509.

Anggard, E., 1970, Chairman's report of the discussion, in: Proceedings of the Fourth International Congress of Pharmacology, Vol. IV, Schwabe and Co., Basel-Stuttgart, p. 54.

Anggard, E., 1971, Studies on the analysis and metabolism of the prostaglandins, Ann. N. Y. Sci. 180:200.

Anggard, E., Bohman, S. O., Griffin, J. E., Larsson, C., and Maunsbach, A. B., 1972, Prostaglandins in renal papilla, Acta Physiol. Scand. (in press).

Anggard, E., Green, K., and Samuelsson, B., 1965, Synthesis of tritium-labeled prostaglandin E(2) and studies on its metabolism in guinea pig lung, J. Biol. Chem. 240:1932.

Anggard, E., and Jonsson, C.-E., 1971, Efflux of prostaglandins in lymph from scalded tissue, Acta Physiol. Scand. 81:440.

Anggard, E., Larsson, K., and Samuelsson, B., 1971, The distribution of 15-hydroxy prostaglandin dehydrogenase and prostaglandin delta-13-reductase in tissues of the swine, Acta Physiol. Scand. 81:396.

Anggard, E., and Samuelsson, B., 1966, Purification and properties of a 15-hydroxy prostaglandin dehydrogenase from swine lung, Arkiv f. Kemi. 25:293.

Bergstrom, S., Ryhage, R., Samuelsson, B., and Sjovall, J., 1963, The structures of prostaglandins E(1), F(1-alpha), and F(1-beta), J. Biol. Chem. 238:3555.

Bygdeman, B., and Samuelsson, B., 1966, Analysis of prostaglandins in human semen: prostaglandins and related factors, 44, Clin. Chim. Acta. 13:465.

Crunkhorn, P., and Willis, A. L., 1971, Cutaneous reactions to intradermal prostaglandins, Brit. J. Pharmacol. 41:49.

Gilmore, N., Vane, J. R., and Wyllie, J. H., 1969, Prostaglandin release by the spleen in response to an infusion of particles, in: Prostaglandins, Peptides and Amines (P. Mategazza and E. W. Horton, Eds.), Academic Press, London, p. 21.

Green, K., 1969, Gas chromatography-mass spectrometry of O-methyloxime derivatives of prostaglandins, Chem. Phys. Lipids 3:254.

Hansson, E., and Samuelsson, B., 1965, Autoradiographic distribution of ^{3}H-labeled prostaglandin E(1) in mice. Prostaglandins and related factors. 31, Biochim. Biophys. Acta, 106:379.

Hopkin, J. M., Horton, E. W., and Whittaker, V. P., 1968, Prostaglandin content of particulate and supernatant fractions of rabbit brain homogenates, Nature (London) 217:71.

Jonsson, C.-E., 1971, Smooth muscle stimulating lipids in peripheral lymph after experimental burn injury, Scand. J. Plast. Reconstr. Surgery, 5:1.

Jonsson, C.-E., and Anggard, E., 1972, Biosynthesis and metabolism of prostaglandin E(2) in human skin, Scand. J. Clin. Lab. Investig. 29: (in press).

Jouvenaz, G. H., Nugteren, D. H., Beerthuis, R. K., and van Dorp, D. A., 1970, A sensitive method for the determination of prostaglandins by gas chromatography with electron-capture detection, Biochim. Biophys. Acta 202:231.

Juhlin, L., and Michaelsson, G., 1969, Cutaneous vascular reactions to prostaglandins in healthy subjects and in patients with urticaria and atopic dermatitis, Acta Derm.-Venerol. 49:251.

Nakano, J., Anggard, E., and Samuelsson, B., 1969, 15-hydroxy-prostanoate dehydrogenase: Prostaglandins as substrates and inhibitors, Europ. J. Biochem. 11:386.

Pike, J. E., Kupiecki, F. P., and Weeks, J. R., 1967, Biological activity of the prostaglandins and related analogs, Nobel Symp. II. Prostaglandins. (S. Bergstrom and B. Samuelsson, Eds.), Almqvist and Wiksell, Stockholm, pp. 161-171.

Samuelsson, B., 1970, Biosynthesis and metabolism of prostaglandins, in: Proceedings of the Fourth

International Congress on Pharmacology, Vol. IV, Schwabe and Co., Basel-Stuttgart, p. 12.

Shio, H., Ramwell, P. W., Andersen, N. H., and Corey, E. J., 1970, Stereo-specificity of the prostaglandin 15-dehydrogenase from swine lung, Experientia, 26:355.

Skoog, T., 1963, A technique of breast reduction, Acta Chir. Scand. 126:453.

Solomon, L. M., Juhlin, L., and Kirschenbaum, M. B., 1968, Prostaglandin on cutaneous vasculature, J. Invest. Dermatology 51:280.

Spector, W. G., and Willoughby, D. A., 1968, The pharmacology of inflammation. The English University Press, Ltd., London.

Vonkeman, H., Nugteren, D. H., and van Dorp, D. A., 1969, The action of prostaglandin 15-hydroxy-dehydrogenase on various prostaglandins, Biochim. Biophys. Acta 187:581.

Vroman, H. E., Nemecek, T. A., and Hsia, S. L., 1969, Synthesis of lipids from acetate by human preputial and abdominal skin in vitro, J. Lipid Res. 10:507.

Ziboh, V. A., and Hsia, S. L., 1971, Prostaglandin E(2): biosynthesis and effects on glucose and lipid metabolism in rat skin, Arch. Biochem. Biophys. 146:100.

DISCUSSION

LEE: Dr. Anggard's presentation illustrates again that much of the chaos in this field can be alleviated by adhering to strict chemical identification of prostaglandins. The word prostaglandin should really not be used if one is identifying by bioassay or superperfused organ, and using chromatography, since there are many things that will co-chromatograph with prostaglandins. In this instance it is probably much wiser to use the word, "prostaglandin-like". Dr. Anggard, did you recover any PGA(2)-like activity in the skin?

ANGGARD: We used assay organs which were insensitive to PGA(2). However, in the biosynthesis experiments where we used tritiated precursors and followed the results chromatographically, there was a small amount of radioactivity associated with the PGA(2) position. However, this might have been formed non-enzymatically during the isolation.

KESSLER: Is the 15-keto metabolite the only product you find in the skin? I ask this question because in some of our other experiments we found the presence of some delta-13,14-reductase activity.

ANGGARD: The major metabolite was the 15-keto-dihydro-PGE(2) which is formed from two enzymatic steps, one involving reduction at the 13 position.

KLIGMAN: What percentage of your preparation was dermis?

ANGGARD: We used skin without the deep dermal layer.

HSIU: We have not found prostaglandin synthetase activity in the microsomal portions of human epidermis.

ZIBOH: Although the prostaglandins have been shown to exert effects on smooth muscle of the uterus and blood vessels or mediating inflammatory reactions, there is increasing evidence that this class of compounds may play an important role in the epithelia. In an interesting study, Kirscher (Kirscher, C. W., 1967, Develop. Biol. 16:203) reported that two prostaglandins, PGE(1) and PGE-278 [now PGB(1)] enhanced maturation of skin by increasing the layers of epidermal cells and causing precocious keratinization in chick skin in culture. Jouvenaz et al (Jouvenaz, G. H., Nugteren, D. H., Beerthuis, R. K., and Van Dopp, D. A., 1970, Biochem. Biophys. Acta 202:231)

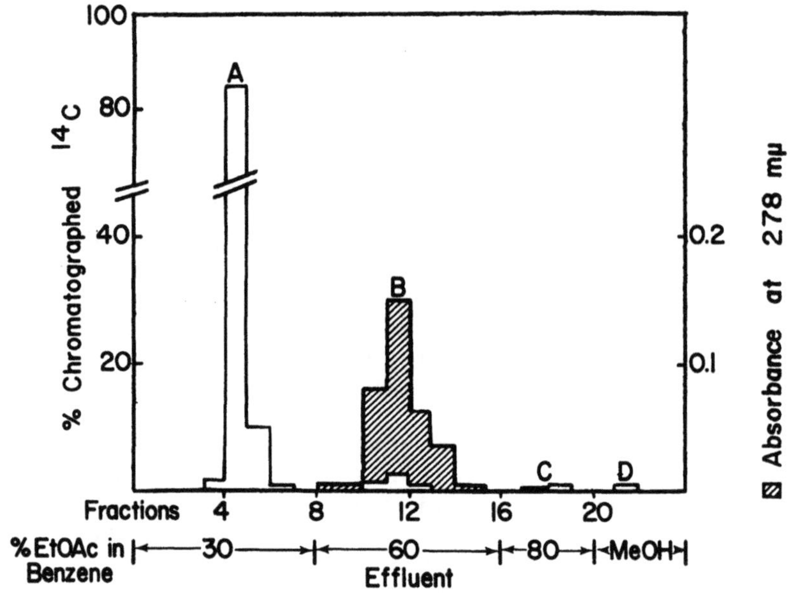

Figure 1. Silicic acid chromatography of unlabeled PGE(2) and the product obtained after incubating 1-^{14}C-arachidonic acid with homogenate of rat skin. The composition of the eluent is given as percentage of ethyl acetate in benzene. 10-ml fractions were collected. The heights of the open bars from the base line to the top indicate the percentages of chromatographed ^{14}C; the crossed bars indicate the absorbance at 278 nanometers of the product formed after treatment with NaOH.

found unusually high levels of PGE(2) in the epidermis of the rats. Furthermore, the most visible and consistent symptoms of essential fatty acid (EFA) deficiency in the rat is the condition of skin, which is characterized by severe scaling of the dorsum and plantar surfaces of the limbs and tail. This abnormality can be reversed by the administration of linolenic or arachidonic acid to the EFA-deficient animals. The demonstration that all cis C20 polyunsaturated fatty acids can be converted to prostaglandins suggests the possible involvement of prostaglandins in the function of essential fatty acids.

In view of a previous study in our laboratory which showed that approximately 9% of the total fatty acids in human skin is a 20:4 fatty acid (Vroman, H. E., Nemecck, R. A., and Hsia, S. L., 1969, J. Lipid Res. 10:507), it was therefore of interest to test as a first step whether this essential fatty acid is transformed to prostaglandins

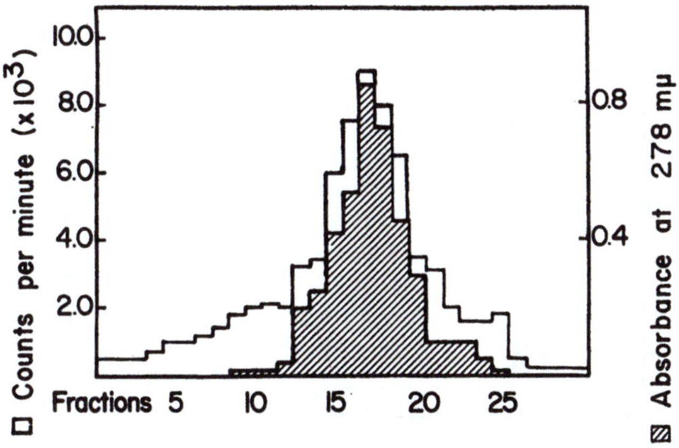

Figure 2. Reverse phase partition chromatography of fractions 9-16 from Fig. 1. Stationary phase was chloroform-isooctanol (15:15), and moving phase, methanol-water (135:165). The heights of the open bars from the base line to the top indicate the percentages of chromatographed ^{14}C; the crossed bars indicate the absorbance of dehydrated PGE(2) at 278 nanometers.

by enzyme preparations from rat and human skin. Furthermore, we have (1) examined whether the cutaneous lesions induced in rats by a diet deficient in EFA can be reversed by PGE(2), and (2) tested the effects of PGE(2) on carbohydrate and lipid metabolism.

For the biosynthetic studies, ^{14}C labeled arachidonic acid was incubated aerobically with a homogenate of rat skin prepared in modified phosphate buffer, as reported previously (Ziboh, V. A., and Hsia, S. L., 1971, Arch. Biochem. Biophys. 146:100); isolation, purification and characterization of radioactive products were achieved by chromatography on silicic acid and hydrophobic Hyflo Supercel. Results of these studies are shown in Fig. 1, 2, and 3. Our results demonstrated that arachidonic acid is transformed to PGE(2) by homogenates of rat skin. This capability of the skin to transform arachidonic acid to PGE(2) may be an important aspect of the function of arachidonic acid in its maintenance of normal skin. It was therefore interesting to investigate if prostaglandins play any role in the syndrome of EFA deficiency.

Essential fatty acid deficiency was produced in rats with a basal diet supplemented with hydrogenated coconut oil. Symptoms of EFA deficiency appeared after 6 to 8 wk with increasing severity in all animals fed the basal diet

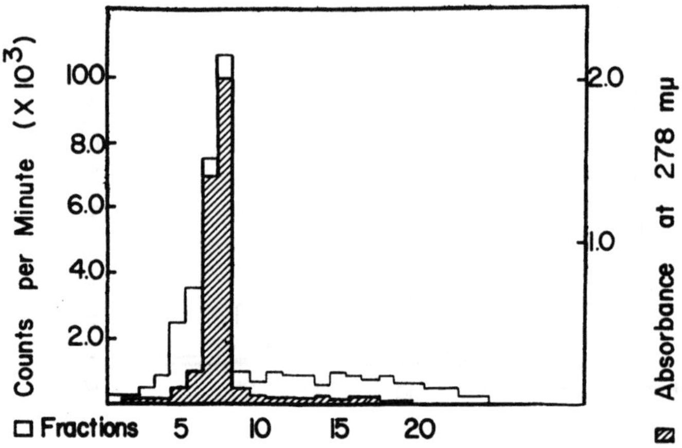

Figure 3. Reverse phase partition chromatography of the product obtained after treatment of radioactive fractions indicated in Fig. 2 with 0.5 N NaOH. Stationary phase was chloroform-heptane (45:5), and moving phase was methanol-water (165:135). The heights of the open bars from the base line to the top indicate the percentages of chromatographed ^{14}C; the solid bars indicate the absorbance of dehydrated PGE(2) at 278 nanometers.

supplemented with hydrogenated coconut oil. The control rats fed the diet containing safflower oil maintained a steady growth and did not exhibit any of the symptoms associated with deficiency of EFA.

The fatty acid composition of skin lipids under the two dietary regimens was determined with gas liquid chromatography. The changes in EFA deficiency included marked increases in monoenoic acids (16:1 and 18:1) and in a fatty acid with chromatographic characteristics of eicosatrienoic acid (20:3). There are also marked decreases in dienoic (18:2) and tetraenoic (20:4) acids. Topical application of PGE(2) which caused the disappearance of the scaly lesions in the skin of the EFA deficient rats did not alter appreciably the profile of fatty acids in the skin of these animals.

The possibility that PGE(2) may be involved in the EFA-deficiency syndrome was tested. Table I shows the effects of PGE(2) on scaly lesions by various modes cf administration. It is significant to mention that intraperitoneal administration of PGE(2) did not clear the cutaneous lesions. The inability of PGE(1) to clear the cutaneous lesions when infused into EFA deficient rats has been reported previously (Kupiecki, F. P., Sekhar, N. C., and Weeks, J. R., 1968, J. Lipid Res. 9:602) and (Gottembos, J. J., Beerthuis, R. K., and Van Dopp, D. A., 1967, in Nobel Symposium 2: Prostaglandins, S. Bergstrom and B. Samuelsson, eds., Almqvist and Wiksell, Stockholm; Interscience, New York, pp. 57-62.) The lack of effect is probably due to rapid systemic metabolism of PGE(2) and PGE(1) to biologically inactive metabolites (Ferreira, I. M., and Vane, J. R., 1967, Nature 216:868). It is interesting to point out, however, that topical treatment of PGE(2) did clear the scaly skin lesions. In another experiment, each EFA deficient rat was treated topically with PGE(2) dissolved in the vehicle on one leg and with the vehicle alone on the other leg. The results in Fig. 4 showed that the scaly lesions cleared in the limb (A and B) treated with PGE(2), but not in the limb (C and D) treated with vehicle. These results demonstrated that the action of PGE(2) was localized in the area of application, and did not exert its action on the other limbs via the systemic circulation. This is also consistent with the view expressed by Dr. Anggard that the action of prostaglandins may be a local one.

In order to gain more understanding of the mechanism of action of PGE(2) in the skin of EFA-deficient rats, *in vitro* experiments were carried out to test the effects of PGE(2) on carbohydrate and lipid metabolism in skin specimens from normal and EFA-deficient rats. Incubations were carried out in Krebs-Ringer bicarbonate buffer, pH 7.4, containing skin biopsy specimens from normal or EFA-deficient rats, with uniformly-labeled glucose. PGE(2) (0.5 mg/ml) dissolved in propylene glycol was added to the incubation mixture. Skin of EFA-deficient rats showed a marked increase in the incorporation of glucose carbon into lipid fractions. This increase is particularly marked in the neutral lipids. Fractionation of the neutral lipids carried out on a Florisil column (Carrol, K. K., 1961, J. Lipid Res. 162:155) revealed a 4-fold increase in the sterol ester fraction from EFA-deficient rats. The effect of PGE(2) on biosynthesis of sterol esters from glucose was then tested *in vitro*. Our results showed that PGE(2) inhibited sterol esterification in both normal and EFA-deficient rats, although this inhibitory effect was approximately 70% in the skin of the EFA deficient rats. This finding is in harmony with recent reports describing the inhibition of cholesterol ester synthetase by PGE(1) in the liver (Schweppe, J. S., and Jungman, R. A., 1970, Proc. Soc. Exptl. Biol. Med. 133:1307) and PGF(2-alpha) in the corpus luteum (Behrman, H. R., Lipids, in press).

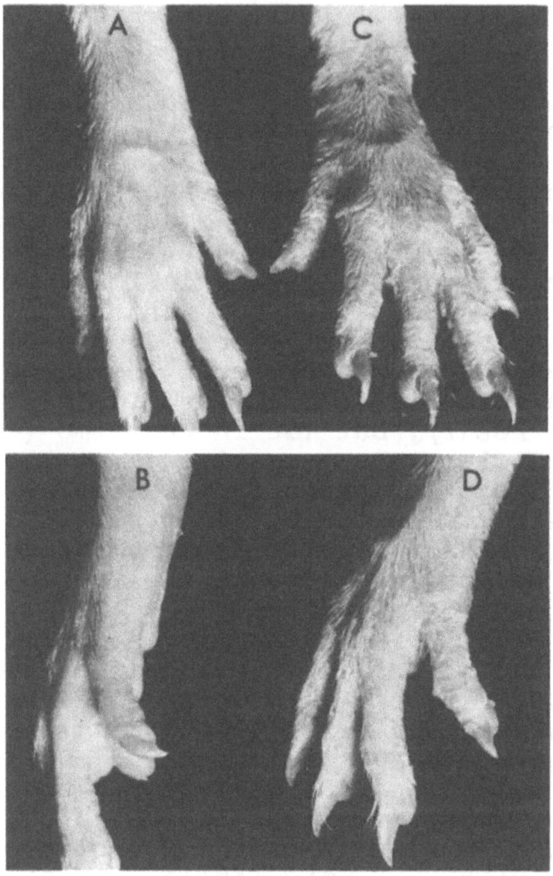

Figure 4. Application of PGE(2) and vehicle to scaly hind limbs. (Figure of the same rat). A and B show clearance of scaly lesions on the dorsum of one hind leg of an EFA-deficient rat after two weeks of topical treatment with PGE(2) dissolved in the vehicle. C and D show no change in scaly lesions on the dorsum of the other hind leg of an EFA-deficient rat after two weeks of topical treatment with the vehicle.

These results suggest that a relationship exists between essential fatty acids, PGE(2) and scaly skin lesions. Furthermore, it may be inferred from our present studies that the clearance of the scaly lesions in skin of EFA-deficient rats by PGE(2) may be due in part to its suppression of abnormal sterol ester formation. Further studies are in progress to elucidate this latter view and the mechanisms of action of PGE(2) on the enzymes involved with the synthesis and hydrolysis of sterol esters.

TABLE I. EFFECTS OF PGE(2) ON SCALY LESIONS

Condition	Rats (N)	Treatment	Indiv. Dermal Score*	Avg. Score	Evaluation□
Normal	4	Saline: 0.1 ml, I.P.	0,0,0,0,-,-	0	-
"	4	PGE(2): (1 mg/kg/day) in 0.1 ml saline, I.P.	0,0,0,0,-,-	0	-
"	4	Vehicle+: 0.1 ml, topical	0,0,0,0,-,-	0	-
"	4	PGE(2): in vehicle, 100 microgram/day, topical	0,0,0,0,-,-	0	-
EFA Deficient	4	Saline: 0.1 ml, I.P.	3,3,3,3,-,-	3	Ineffective
"	4	PGE(2): (1 mg/kg/day) in 0.1 ml saline, I.P.	3,3,3,3,-,-	3	"
"	6	Vehicle: 0.1 ml, topical	3,3,3,3,3,3	3	"
"	6	PGE(2): in vehicle, 25 microgram/day, topical	1,1,2,1,2,1	1.5	Slightly effective
"	6	PGE(2): in vehicle, 50 microgram/day, topical	0,1,0,0,1,1	0.5	Effective
"	6	PGE(2): in vehicle, 100 microgram/day, topical	0,0,0,0,0,0	0	Very effective

*Score	Deficiency Symptoms
0	No scaling in the plantar and dorsum of feet
1	Slight scaling in the plantar and dorsum of feet
2	Moderate scaling in the plantar and dorsum of feet
3	Severe scaling in the plantar and dorsum of feet
-	Not evaluated

+The vehicle was a mixture of propylene glycol:ethanol, 3:7 (v/v) used for the topical application of PGE(2).

□The effectiveness of PGE(2) was evaluated after 14 days of treatment. Average score of 1 or less was considered as effective in clearing the scaly lesions.

PROSTAGLANDIN INHIBITION OF IMMEDIATE AND DELAYED HYPERSENSITIVITY *IN VITRO*

L. M. Lichtenstein and C. S. Henney

Johns Hopkins University School of Medicine
The Good Samaritan Hospital
Baltimore, Md.

Prostaglandins have a significant and important inhibitory effect on two pertinent *in vitro* models, one of immediate and the other of delayed hypersensitivity. These inhibiting effects are best viewed in the context of the adenylate cyclase - cyclic AMP system and can be compared to the inhibition demonstrated by other agents which, like certain of the prostaglandins, stimulate adenylate cyclase. The similarities between the effects noted on two quite different immunologic responses are striking and suggest a general inhibitory role for the prostaglandins and other adenylate cyclase stimulators on leukocyte-lymphocyte responsiveness in general; this possibility will be dealt with by Dr. Henry Bourne. In this presentation we shall describe the activities of these inhibitors first on the immediate and then on the delayed system of hypersensitivity and conclude with some comments on the possible therapeutic implications of the effects noted.

IMMEDIATE HYPERSENSITIVITY

The model of immediate allergy involves the antigenically induced release of histamine from the isolated peripheral leukocytes of sensitive human donors (Lichtenstein et al, 1964; Sadan et al, 1969; Lichtenstein, in press). Histamine release has been shown to be dependent upon IgE antibodies "fixed" to the cell surface, and the reaction can be initiated as well with animal antisera to human IgE (Lichtenstein et al, 1970). A defect in this system, especially as it pertains to measurements of adenylate cyclase activity or cyclic AMP levels, is that the IgE antibody is found only on the basophilic leukocytes and the histamine is similarly located only in these cells which represent about 1% of the total (Ishizaka et al, in press). This system is, however, pertinent to the disease state of which it is a model: the sensitivity of a patient's leukocytes to a given allergen correlates in a highly significant manner with the severity of illness noted by that patient on environmental exposure to the allergen (Lichtenstein et al, 1966). Earlier studies, concerned with the mechanism of histamine release following IgE antibody-allergen interaction, indicated that the reaction was non-cytotoxic and suggested that it was similar in many respects to a secretory response (Lichtenstein, 1968).

Following this lead we explored the possible involvement of the cyclic AMP system in the release mechanism: This led to the observation that both the catecholamines and the methylxanthines inhibited the response, and that together they had a synergistic effect (Lichtenstein and Margolis, 1968). This inhibition was rather surprising when viewed in the context of other cyclic AMP modulated secretory systems where these agents enhanced the response (Robison et al, 1971). In clinical terms it was, however, quite reasonable since these drugs are major contributors to the therapy of allergic disorders. Moreover, the concentrations required for <u>in vitro</u> inhibition, 0.2-0.4 mM for the methylxanthines and about 0.01-0.1 microM for the catecholamines were within the pharmacological range (Lichtenstein and Margolis, 1968; Lichtenstein and DeBernardo, in press). Our interpretation of these data as indicating the involvement of cyclic AMP in this process was strengthened by the observation (Fig. 1) that the dibutyryl derivative of cyclic AMP also inhibited histamine release. Since this first demonstration of the activity of these agents, others working with models of immediate allergy utilizing primate and human lung, have confirmed these experiments (Assem and Schild, 1969; Ishizaka et al, 1971).

In exploring the activities of other adenylate cyclase stimulators we observed that PGE(1) and PGE(2) caused inhibition of histamine release in microM concentrations while PGF(1-alpha) had no activity whatsoever, even at much higher concentrations (Fig. 2). We next attempted to divide the histamine release reaction into stages so that we could study the action of these

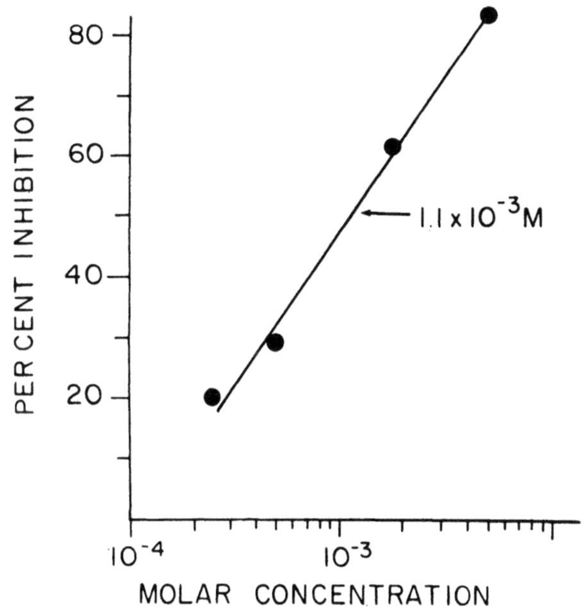

Figure 1. Inhibition of histamine release by dibutyryl-cyclic AMP.

agents on early and late phases of the response. By analogy with other secretory systems it was to be expected that the cyclic AMP involvement would occur shortly after the initiation of the reaction. The separation was accomplished as shown in Fig. 3: a first, antigen dependent, calcium-independent step led to "activated" cells which, when washed free of antigen, released histamine in a second, calcium-dependent step. With the reaction divided in this fashion the catecholamines were found to be active only in the first stage, an observation compatible with their inhibitory activity being mediated through adenylate cyclase. Similarly, the prostaglandins acted only in the early phase of histamine release (Fig. 4). With these agents, however, the inhibition was essentially complete whereas, in most experiments with the catecholamines, maximal inhibition was 40-60%, with higher concentrations of the drugs either causing no further inhibition or actually decreasing the inhibition slightly (Lichtenstein and DeBernardo, in press).

At about the same time we were able to demonstrate that the catecholamines appeared to act through a beta-

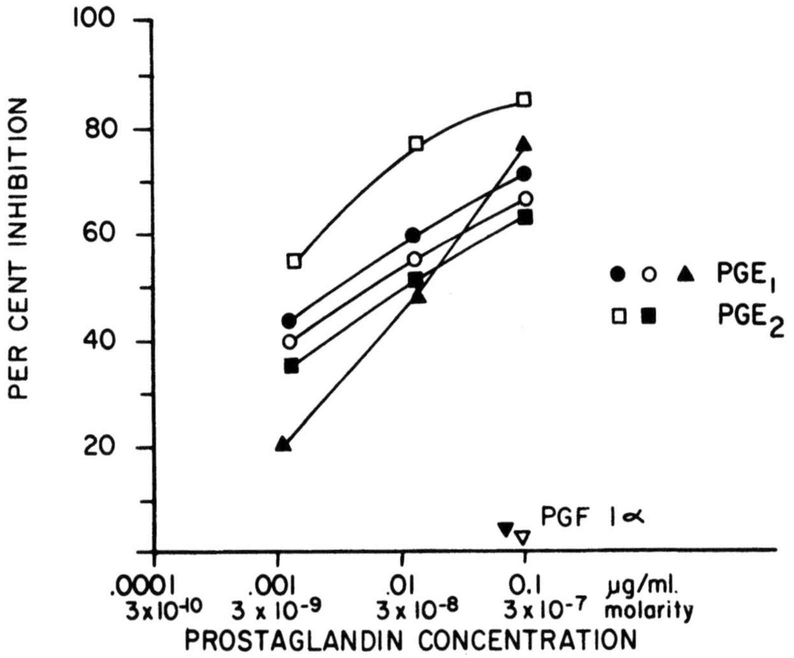

Figure 2. Inhibition of histamine release by the prostaglandins: Dose response curves with PGE(1), PGE(2), and PGF(1-alpha).

receptor, inasmuch as the order of their activity was isoproterenol > epinephrine > norepinephrine > phenylepinephrine (Bourne et al, submitted for publication). Moreover, the action of isoproterenol and epinephrine could be completely or largely abolished by the addition of the beta-blocker, propranolol, (Fig. 5) but was not affected by the alpha-blocker, phentolamine. We found, incidentally, no enhancement of histamine release by the addition of an alpha-receptor stimulating combination such as epinephrine and propranolol to the cells (cf Orange et al, 1971). The prostaglandin inhibition, on the other hand, was shown to operate through a different receptor site than the catecholamines. The experiments shown in Table I have as a positive control the propranolol blocking of isoproterenol inhibition: In the same experiments this beta-blocker had no effect on PGE(1) inhibition.

In every case measurements of cyclic AMP levels in the whole leukocyte preparation showed the changes predicted by the histamine release inhibition data. Thus, the catecholamines stimulated adenylate cyclase activity with the same order of activity (isoproterenol >

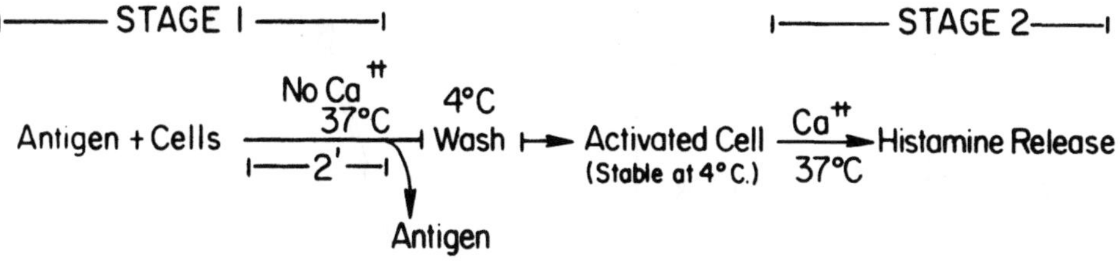

Figure 3. Scheme of separation of histamine release into two stages.

epinephrine..., etc.) as noted for their inhibition of histamine release. This activity on the cyclase was blocked by propranolol. PGE(1) and PGE(2) similarly caused increased intracellular cyclic AMP levels (PGF(1-alpha) did not) and this activity was not interfered with by the beta-blocking agent. These experiments, carried out with Dr. Bourne, are detailed in his presentation at this symposium.

Another cyclase stimulator is perhaps interesting in the context of the reported activity of prostaglandins as mediators of the inflammatory response, rather than as inhibitors (Crunkhorn and Willis, 1971). The mediator which most concerns us, histamine, also can block histamine release. The concentrations required (Fig. 6) are precisely those contained in the cell preparations in use so that the possibility of an in vivo action seems likely. Histamine inhibits in the first stage only, and increases intracellular cyclic AMP levels at appropriate concentrations. Its activity is not blocked by propranolol, nor interestingly enough, by antihistamines in the usual range of concentrations (Bourne et al, 1971). Kinetic experiments make it unlikely that the histamine released from any given cell inhibits further release from that cell (Bourne et al, submitted for publication). Thus, its action is only early in the response and after the release process has started (which in the "whole" reaction is always preceded by a short lag period), its inhibiting ability falls rapidly to zero. It is not unlikely, however, that in vivo release from one mast cell tends to diminish the response of nearby cells thus initiating a multicellular "feedback" type of inhibition. These observations raise the question of whether the prostaglandins also have such a relationship, being released as a result of the allergic reaction as well as inhibiting the phenomenon. This possibility is presently being explored with Dr. Ramwell. However, the human basophil does seem to be different from that of the guinea pig in that we have in no case seen that prostaglandins,

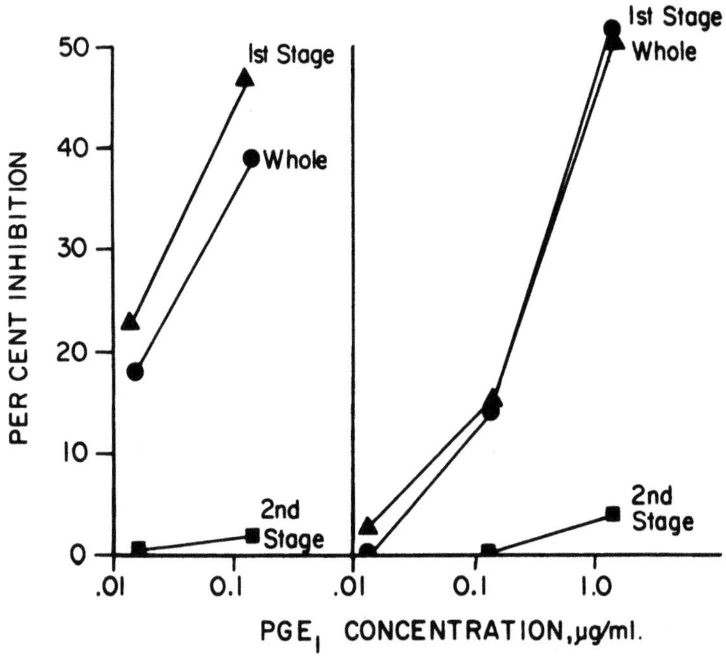

Figure 4. Activity of PGE(1) in first and second stages of histamine release and in the unseparated (whole) reaction.

i.e., PGE(1), PGE(2) or PGF(1-alpha), cause the release of histamine, as is reported to be the case for guinea pig basophils (Sondergaard and Greaves, 1971).

DELAYED HYPERSENSITIVITY

The system used to study cell mediated immunity involves the killing (^{51}Cr release) of mouse mastocytoma cells by splenic lymphocytes obtained from another strain of mice immunized with the target cells 10-11 days previously. This system was originally described by Brunner and, as used in our hands, by Henney (Brunner et al, 1968; Henney and Mayer, in press). *In vitro* models of delayed hypersensitivity have been in use for a shorter period of time than models of the immediate reaction and, in general, their correlations with the *in vivo* or clinical state have been less defined. Our system, however, is a rather direct model, which correlates well with the *in vivo* development of delayed hypersensitivity in animals and recently, using human lymphocytes and mastocytoma cells with tuberculo-proteins affixed to their surface, was shown to correlate well with the human skin response to tuberculin purified protein derivative (PPD)

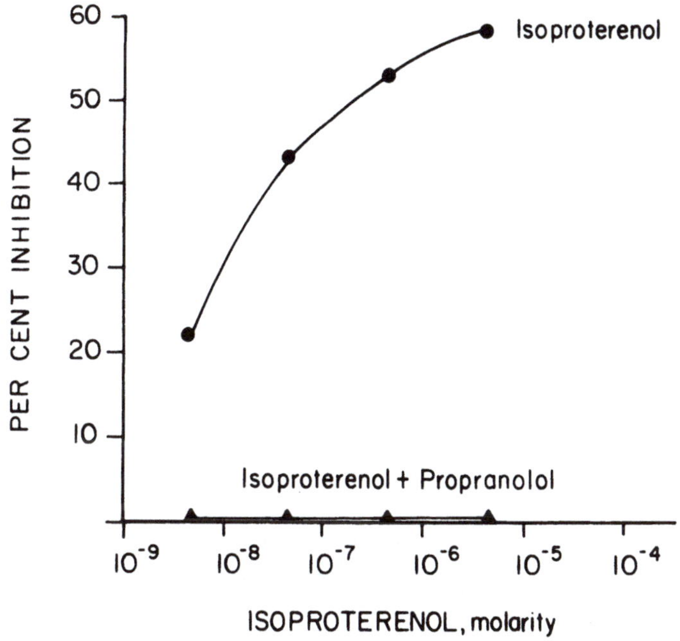

Figure 5. Blocking of isoproterenol inhibition by propranolol.

(Rehn and Henney, in press). The system is highly reproducible; the rate of killing is linear, with clear-cut cell death demonstrable after 1-3 hr of incubation and, depending on the lymphocyte/target cell ratio, maximal killing (90-100%) occurring in 4-9 hr (Henney, in press).

The catecholamines inhibit this reaction. Early in the reaction, with incubation periods of less than one hr, this inhibition appears to be quite profound: during the first 0.5 hr of lymphocyte-mast cell interaction isoproterenol often causes 60-80% inhibition (Fig. 7). At this early period, however, the degree of killing is marginal and therefore the inhibition data are somewhat variable from experiment to experiment. An accurate dose-response curve can be carried out after a 3-6 hr period of killing under which circumstances only 15-25% inhibition by the catecholamines is noted. This inhibition is nonetheless highly significant since duplicate or triplicate samples taken at this time agree within 1-2%. Inhibition can also be obtained with epinephrine. The effect of both catecholamines is regularly noted at a concentration of 0.1 microM, and is completely blocked by propranolol. Theophylline also inhibits this reaction causing a 60-70% or greater decrease in target cell death

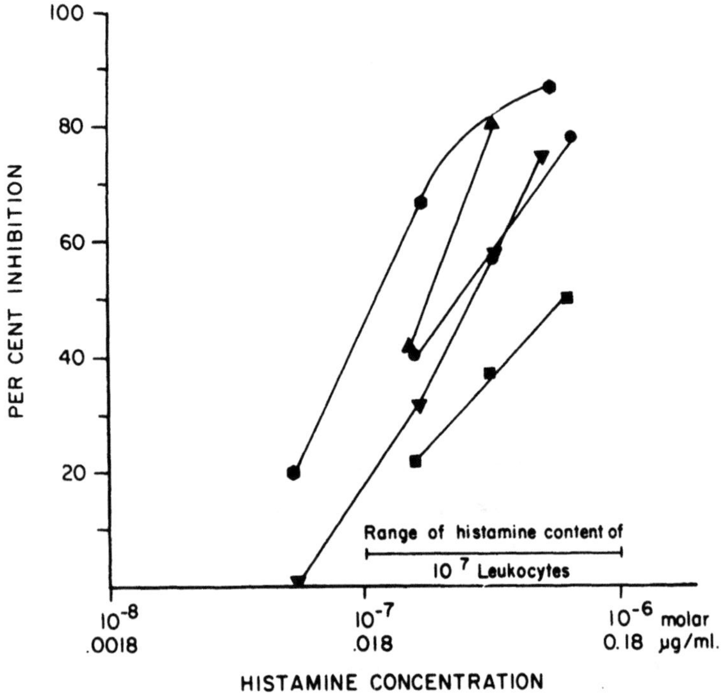

Figure 6. Inhibition of histamine release by histamine.

at 1 mM (Table II). The concentration of theophylline required for 50% inhibition, about 0.5 mM, is almost exactly that which inhibits the immediate reaction to a similar degree. The addition of catecholamines and methylxanthines together produces an effect which is always at least additive and usually somewhat more than additive. Finally, cyclic AMP itself or the dibutyryl derivative both inhibit cell death (50% at 0.1 mM) and the concentrations required to demonstrate this effect are similar to those seen in the immediate response (Table II).

By far the most active inhibitors in this system are the prostaglandins. As in the immediate response, PGE(1) and PGE(2) are active, while PGF(1-alpha) is not (Table II). More importantly, for this prolonged reaction, the inhibition noted with the prostaglandins does not fall with time as is the case with the catecholamines. In the experiment shown in Table III the same set of lymphocytes was exposed to 10 microM PGE(1) and isoproterenol: The typical falloff of isoproterenol inhibition contrasts sharply with the prolonged activity of PGE(1).

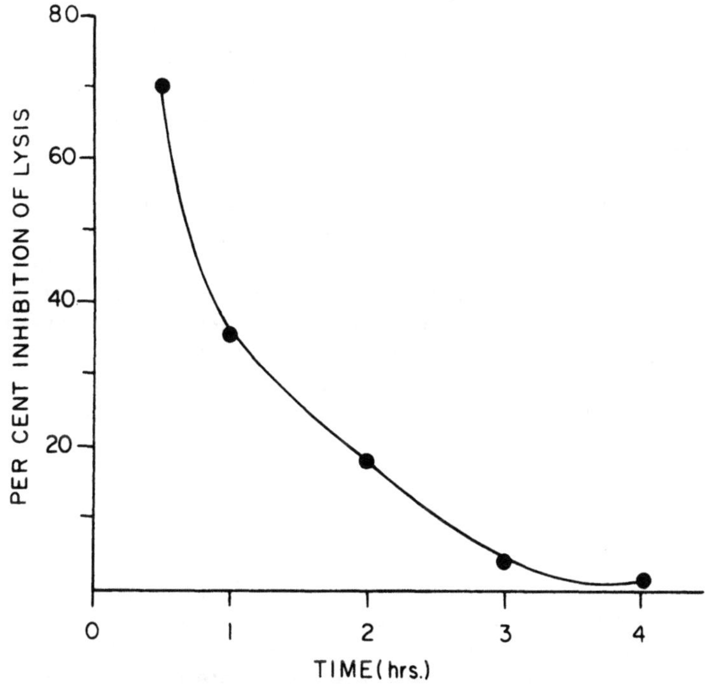

Figure 7. Time course of isoproterenol inhibition of cytolysis. The concentration of isoproterenol is 0.01 mM.

The analogy between the immediate and delayed hypersensitivity systems is maintained with respect to the beta-receptor. In experiments similar to those performed with the histamine release system, propranolcl completely blocks the inhibition caused by the catecholamines but has no effect on the prostaglandins.

Finally, histamine can also block cell-mediated cytolysis. This was of interest because it is at least possible that the increased capillary permeability noted as an early constituent of the characteristic delayed hypersensitivity skin reaction is due to histamine release. Only preliminary experiments have been carried out thus far and the time course of histamine's action is not clear: Inhibition is noted at 10^{-6} to 10^{-5} M histamine after 2 hr of interaction between the lymphocytes and the target cells. The effect is not blocked by propranolol.

Measurements of cellular cyclic AMP levels are, again, very much in accord with the inhibition data. In particular, the stimulation of adenylate cyclase by isoproterenol is transient compared to that caused by the prostaglandins. The detailed data on cyclic AMP levels in this set of experiments will be presented here by Dr. Bourne. It is worth pointing out, however, that these measurements are more relevant than those using the leukocytes, since in this case cell preparations are more than 90% mononuclear.

These experiments on cyclic AMP modulation of delayed hypersensitivity are still very much in progress, though a brief note reports some of them (Henney and Lichtenstein, 1971). It should be pointed out that we have not yet separated the system to show that the effects noted are as a result of inhibition of lymphocyte function rather than protection of the mastocytoma cells. This is, of course, our present interpretation of the data but clear-cut proof is necessary.

CONCLUSIONS

The similarity between the inhibiting abilities of these "cyclic AMP active" drugs on the immediate and delayed system are, of course, striking. A discussion of whether the cyclic AMP system has evolved, in this instance, as a general inhibitor of what might be broadly called the inflammatory response seems in order and Dr. Bourne has taken on that job. I will limit myself here to some general comments on the possible clinical implications of these observations.

With respect to the immediate reaction, there are a number of obvious questions. The catecholamines and methylxanthines clearly are clinically useful in asthma. It is generally supposed that their efficacy in this situation is based on their ability to directly relax smooth muscle, though what the mechanism of such relaxation may be is not clear. It is possible that at least part of this effect is through the inhibition of mediator (histamine, SRS-A, etc.) release. Certainly this seems probable in their anti-allergic activity, but perhaps the action of these drugs on normal smooth muscle is mediated through the inhibition of release of smaller, "physiological" amounts of histamine which maintain the usual tone of muscle. With respect to their efficacy on the <u>in vivo</u> allergic response, inhibition of release has been shown directly. Both Salbutamol, a catecholamine, and theophylline were found to inhibit the wheal and flare response in monkey skin challenged with allergen, while antihistamines, at the levels used, were inactive. On the other hand, these agents did not block the skin response to histamine, which was inhibited by the antihistamines

(Perper, personal communication). These experiments represent a fairly definite confirmation of the hypothesis that the catecholamines and methylxanthines inhibit the allergic response in vivo by stopping mediator release rather than by combating the action of the released mediator on smooth muscle.

The prostaglandins are the most potent inhibitors of allergic histamine release. From a clinical view, however, their action on adenylate cyclase through a receptor different from the beta-receptor is perhaps more interesting. There has been a persistent theory that, in asthma, a basic aspect of the pathogenesis is a "partial beta-blockade" leading to a failure of endogenous epinephrine to induce an adequate compensation to a variety of bronchospastic stimuli (Szentivanyi, 1968). Although there has been little definite evidence to support this concept, Parker has recently shown that the peripheral leukocytes of asthmatics fail to respond to an isoproterenol stimulation with a normal rise in cyclic AMP (Parker, personal communication). The potential for prostaglandins, which bypass the beta-receptor in stimulating adenylate cyclase, is therefore most promising.

Concerning delayed hypersensitivity, no such backlog of information is available and these experiments represent, in fact, the first demonstration that cyclic AMP may have a controlling influence on cell-mediated cytolysis, although it has been shown that blast transformation is inhibited by isoproterenol (Smith et al, 1971). Thus, the clinical potential of these agents is an open question and the possibilities are therefore even more interesting. Certainly, disease processes in which the cell-mediated immunologic response may play a role appear to be chronic, and attempts at suppression are unlikely to be successful with the agents in question which usually have a short half-life or stimulate cyclic AMP transiently. Again, however, the relatively long duration of prostaglandin action makes these agents of special interest. Moreover, new delivery methods and the development of congeners with prolonged activity are surely within our technical ability.

TABLE I

EFFECT OF PROPRANOLOL
ON INHIBITION OF IgE MEDIATED HISTAMINE RELEASE
BY ISOPROTERENOL AND THE PROSTAGLANDINS*

Drug	Conc. (M)	Percent inhibition of release Exp 1	2
Isoproterenol	5×10^{-7}	46	27
+ Propranolol	6×10^{-6}	10	0
PGE(1)	4×10^{-7}	56	55
+ Propranolol	6×10^{-6}	56	59
PGE(1)	4×10^{-7}	63	55
+ Propranolol	6×10^{-6}	69	59

*Inhibition in the first stage of histamine release (see text).

TABLE II

MAST CELL LYSIS BY SENSITIZED LYMPHOCYTES:
INHIBITION BY THE PROSTAGLANDINS,
THEOPHYLLINE AND DIBUTYRYL-CYCLIC AMP

	Concentration, (M)					
	10^{-8}	10^{-6}	10^{-5}	19^{-4}	3×10^{-4}	10^{-3}
Drug	Percent Inhibition of Lysis					
PGE(1)	14.1	29.2	60.5	81.2	----	----
PGE(2)	12.2	31.5	50.2	68.1	----	----
PGF(1-alpha)	0	0	0	5	5	----
Theophylline	----	----	0	8.5	33.6	61.5
D-C AMP	----	----	25.9	46.7	57.1	67.7

TABLE III

TIME COURSE OF ISOPROTERENOL AND PROSTAGLANDIN
INHIBITION OF MAST CELL LYSIS

	Time of Lymphocyte/Mast Cell Incubation, Min				
	30	60	120	180	240
Drug	Percent Inhibition of Lysis				
PGE(1)	64.8	67.2	----	70.5	67.0
Isoproterenol	34.8	29.5	19.1	13.0	9.2

ACKNOWLEDGEMENTS

The work described in this paper is Publication No. 25 from the O'Neill Research Laboratories of The Good Samaritan Hospital. It was supported by grants AI 7290, AI 8290 and AI 10280 from the NIH, National Institute of Allergy and Infectious Diseases.

Dr. Lichtenstein's work was supported by a Research Career Development Award from the NIH, National Institute of Allergy and Infectious Diseases.

REFERENCES

Assem, E. S. K. and Schild, H. O., 1969, Inhibition by sympathomimetic amines of histamine release induced by antigen in passively sensitized human lung, Nature (London), 244:1028.

Bourne, H. R., Lichtenstein, L. M., and Melmon, K. L., (submitted for publication), Pharmacologic control of allergic histamine release *in vitro*: Evidence for an inhibitory role of 3',5'-adenosine monophosphate in human leukocytes.

Bourne, H. R., Melmon, K. L., and Lichtenstein, L. M., 1971, Histamine augments leukocyte cyclic AMP and blocks antigenic histamine release, Science, 173:743.

Brunner, K. T., Mauel, J., Cerottini, J. C., and Chapuis, B., 1968, Quantitative assay of the lytic action of immune lymphoid cells on ^{51}Cr-labelled allogeneic target cells *in vitro*; inhibition by iso-antibody and by drugs, Immunology, 14:181.

Crunkhorn, P., and Willis, A. L., 1971, Cutaneous reactions to intradermal prostaglandins, Brit. J. Pharm. 41:49.

Henney, C. S., (in press), Quantitation of the cell-mediated immune response. I. The number of cytolytically active mouse lymphoid cells induced by immunization with allogenic mastocytoma cells, J. Immun.

Henney, C. S., and Lichtenstein, L. M., 1971, The role of cyclic AMP in the cytolytic activity of lymphocytes, J. Immun. 107:610.

Henney, C. S., and Mayer, M. M., (in press), Specific cytolytic activity of lymphocytes: Effect of antibodies against complement components C2, C3, and C5[1], J. Immunol.

Ishizaka, T., DeBernardo, R., Tomioka, H., Lichtenstein, L. M., and Ishizaka, K., (in press), Identification of basophil granulocytes as a site of allergic histamine release, J. Immun.

Ishizaka, T., Ishizaka, K., Orange, R. P., and Austen, K. F., 1971, The pharmacologic inhibition of the antigen induced release of histamine and slow reacting substance of anaphylaxis (SRS-A) from monkey lung tissues mediated by human IgE, J. Immun. 106:1267.

Lichtenstein, L. M., 1968, Mechanism of allergic histamine release from human leukocytes, In: "International Symposium on the Biochemistry of Acute Allergic Reactions," Edited by E. L. Becker and K. F. Austen, Blackwell Scientific Productions, London.

Lichtenstein, L. M., (in press), The immediate allergic response: in vitro separation of antigen activation, decay and histamine release, J. Immun.

Lichtenstein, L. M., and DeBernardo, R., (in press), The immediate allergic response: In vitro action of cyclic AMP active and other drugs on the two stages of histamine release, J. Immun.

Lichtenstein, L. M., Levy, D. A., and Ishizaka, K., 1970, In vitro reversed anaphylaxis: Characteristics of anti-IgE mediated histamine release, Immunology, 19:831.

Lichtenstein, L. M., and Margolis, S., 1968, Histamine release in vitro: Inhibition by catecholamines and methylxanthines, Science 161:902.

Lichtenstein, L. M., Norman, P. S., Osler, A. G., and Winkenwerder, W. L., 1966, In vitro studies of human ragweed allergy: Changes in cellular and humoral activity associated with specific desensitization, J. Clin. Invest. 45:1126.

Lichtenstein, L. M., and Osler, A. G., 1964, Studies on the mechanisms of hypersensitivity phenomena. IX. Histamine release from human leukocytes by ragweed pollen antigen, J. Exp. Med. 120:507.

Orange, R. P., Kaliner, M. A., and Austen, K. J., 1971, Enhancement of the immunologic release of histamine and SRS-A from human lung, Fed. Proc. 30:653.

Rehn, T. G., and Henney, C. S., (in press), The cytolytic activity of human lymphocytes. I. Cytolytic activity towards PPD coated target cells in tuberculin positive and negative individuals, Clin. Exp. Immun.

Robison, G. A., Butcher, R. W., and Sutherland, E. W., 1971, Cyclic AMP, Academic Press, N. Y.

Sadan, N., Rhyne, M. B., Mellits, D., Goldstein, E. O., Levy, D. A., and Lichtenstein, L. M., 1969, Immunotherapy of pollinosis in children: An investigation of the immunological basis of clinical improvement, N. Eng. J. Med. 280:623.

Smith, J. W., Steiner, A. L., Newberry, W. M., Jr., and Parker, C. W., 1971, Cyclic adenosine 3',5'-monophosphate in human lymphocytes: Alterations after phytohemagglutinin stimulation, J. Clin. Invest. 50:432.

Sondergaard, J., and Greaves, M. W., 1971, Prostaglandin E(1): Effect on human cutaneous vasculature and skin histamine, Brit. J. Derm. 84:424.

Szentivanyi, A., 1968, The beta-adrenergic theory of the atopic abnormality in bronchial asthma, J. Allergy 42:203.

DISCUSSION

KLIGMAN: The problem here is that you are a vitrologist and you must, of necessity, become a vivologist.

LICHTENSTEIN: I would point out to you that in immediate hypersensitivity the catecholamines and theophylline do have some effect on asthma, and we can also block the human skin reaction with these drugs, whereas we cannot block histamine.

KLIGMAN: But there are some very interesting systems in which you can establish relevance even in animals, and therefore, it seems of value to take the prostaglandins into a clinical situation to see if they have any effect on some of the model systems.

LICHTENSTEIN: Immediate hypersensitivity is very complicated. The analogy in rat and guinea pig fail when compared to man. For example, salicylates are very good inhibitors of histamine release in the rat, but they certainly are not in man.

BRAUN: You have observed only inhibition by adenylate cyclase stimulators. In contrast, we have one system in which we can show enhancement of delayed hypersensitivity by adenylate cyclase stimulators, and again this is using poly A:U as the adenylate cyclase stimulator. Therefore, I wonder if what you are saying may be too strong. This is in reference to some data we have collected in mixed lymphocyte reactions using lymphocytes from different mouse strains. We find that in certain strain combinations you get enhancement which you can increase further by adenylate cyclase stimulators. In other systems you get inhibition of mixed lymphocyte reactions.

LICHTENSTEIN: I am aware of this, but I do not know how it fits into the scheme.

WILLIS: In our work with rat skin we showed that prostaglandins seem to act by releasing histamine from the rat mast cells, which suggests that prostaglandins might be responsible for release of histamine. Since salicylate blocked histamine release in rat, this correlates nicely.

THE ROLE OF PROSTAGLANDINS IN MICRO-CIRCULATORY REGULATION AND INFLAMMATION

G. Kaley, E. J. Messina,
and R. Weiner

Department of Physiology
New York Medical College, N. Y.

The mediation of inflammatory changes during tissue injury reactions has been ascribed to a wide variety of biologically active agents, yet the complete description of these factors still awaits definition. Inflammation, irrespective of its cause, sets into motion a series of typical, interwoven events which in its early phases is characterized by vasodilation, adhesion of platelets, increased vascular permeability and emigration of leukocytes into the affected area. Since the recent suggestion that prostaglandins may play an important role in the pathogenesis of inflammation (Kaley and Weiner, 1968) evidence has accrued to indicate that these ubiquitous compounds are not only able to mimic the diverse features of the acute inflammatory reaction (Crunkhorn and Willis, 1971a; Kaley and Weiner, 1971a), but also that they are generated and released from many tissues under conditions when there is a disturbance of normal cell function (Piper and Vane, 1971) and that they are found in inflammatory exudates of different origin (Willis, 1969; Greaves et al, 1971). Additional evidence that prostaglandins may have a paramount role in diverse tissue injury reactions derives from the work of Vane (1971) who offered for consideration the hypothesis that the action of certain anti-inflammatory drugs is based on their ability to interfere with prostaglandin synthesis. Our paper will deal with some of the roles played by prostaglandins in the mediation of the sequelae of injury. The conclusion will be drawn that these highly active agents may very well be involved in the initiation and

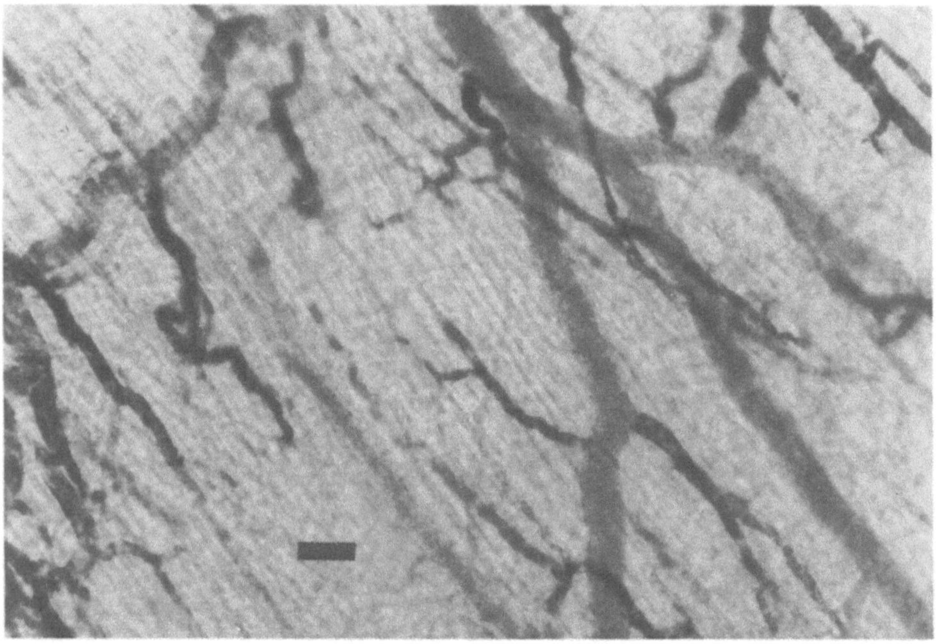

Figure 1. Photomicrograph of vascular leakage evoked by serotonin in rat cremaster muscle. The animal received a local, subcutaneous injection of serotonin (5 micrograms in 0.1 ml) and an intravenous injection of carbon suspension, 1 hr prior to sacrifice. Carbon deposits are exclusively in venules. Scale = 100 microns.

maintenance of many of the essential features of inflammation.

EFFECTS OF PROSTAGLANDINS ON BLOOD FLOW

Following the work of Holmes et al (1963), who showed that PGE(1) reduced the vasopressor effects of catecholamines, angiotensin and vasopressin, we reported that local administration of PGE(1) does reduce vascular responsiveness in the rat mesenteric microcirculation to the above constrictor agents even long after its administration, when the vasodilator effects are no longer manifest (Weiner and Kaley, 1969).

Table I illustrates the effects of local application of PGE(1) and PGA(1) on arterioles of rat cremasteric muscle. _In vivo_ measurements of microvascular diameters

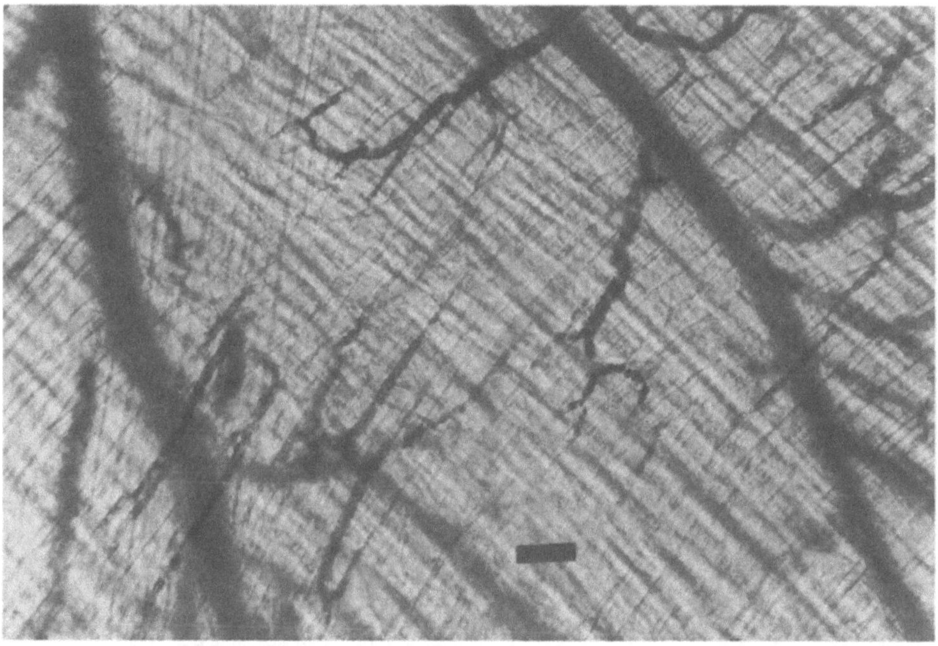

Figure 2. Photomicrograph of vascular leakage evoked by PGE(1) in rat cremaster muscle. The animal received a local, subcutaneous injection of PGE(1) (0.5 micrograms in 0.1 ml) and an intravenous injection of carbon suspension, 1 hr prior to sacrifice. Carbon deposits are exclusively in venules. Scale = 100 microns.

were performed with the aid of a television-microscope recording system as described by Baez (1966). There was a 40% increase in the diameter of arterioles (15 to 35 microns in size) when PGE(1) was added to this skeletal muscle preparation; a change more pronounced than that obtained when similar doses are applied to the mesocecal microcirculation. PGA(1) evoked a 14% increase in arteriolar diameters, a significant change from control, and noteworthy, for in equal doses it does not cause vasodilation in the mesentery.

Table II depicts the effects of PGE(1) on the vasoconstrictor responses of a variety of agents in the rat mesenteric circulation. These experiments were done according to methods previouly described (Weiner and Kaley, 1969). In all of these and the following studies, PGE(1) and the vasoconstrictors were administered topically, in 0.1 ml volumes, and quantitative

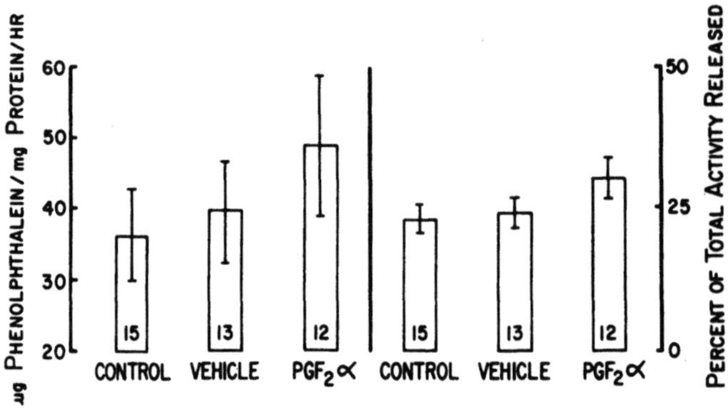

Figure 3. Effect of PGE(2-alpha) (40 microgram/ml) on release of beta-glucuronidase activity from live lysosomal suspensions.

measurements of arteriolar diameters were determined before and 5 min after the administration of PGE(1), at a time when its vasodilator activity had terminated. Vascular responses to all of the vasopressor agents tested were substantially reduced, especially after the higher doses of PGE(1) (10 micrograms) were added to the preparation. PGE(1) also antagonized constrictor responses in the cremasteric microcirculation, as summarized in Table III, albeit, in order to inhibit constrictor responses to a degree similar to those seen in the mesentery, lower doses of pressor agents were utilized. Analogous experiments performed with PGA(1) indicated that it had no effect on vascular responsiveness in the mesentery or cremaster muscle.

This disassociation between the immediate vasodilator effects of PGE(1) and its subsequent effect on vascular reactivity, an observation which has been confirmed and extended by a number of investigators (Viguera and Sunahara, 1969; Hedwall et al, 1971; Kadowitz et al, 1971) renders PGE(1) unique among all other naturally occurring vasodilators and still comprises one of the most puzzling aspects of the vascular actions of prostaglandins. Our findings, when viewed along with those of Hedquist and Brundin (1969), showing that PGE(1) can decrease the rate of release of norepinephrine, support the concept that the vasodilation of inflammation, the sine qua non of the early phase of vascular injury, could be mediated by PGE compounds.

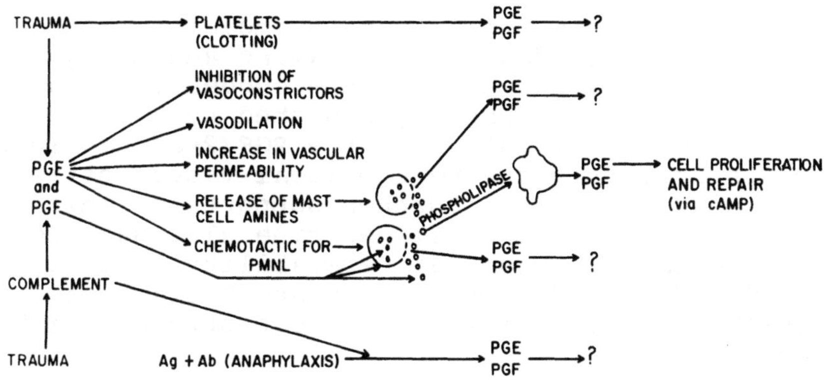

Figure 4. Schematic of interaction of prostaglandins with various substrates in promoting tissue injury reactions.

PROSTAGLANDINS AND VASCULAR PERMEABILITY

Previous reports indicate the effectiveness of PGE(1) in increasing vascular permeability to a degree which rivals that of any of the supposed endogenous mediators of inflammatory reactions (Kaley and Weiner, 1968; Arora et al, 1970). Our present results, represented in Table IV, confirm some of the findings of others (Crunkhorn and Willis, 1971b) in that PGE(2) seems to be as effective an agent in enhancing vascular permeability as PGE(1). In these experiments, and in those to follow, prostaglandins and all other drugs were injected intradermally into the abdominal skin of "blued" rats, in 0.1 ml aliquots. Ten min after the injections animals were sacrificed and the index of permeability change was determined according to the methods previously described (Kaley and Weiner, 1971a). PGA(1) proved much less active than PGE compounds in evoking edema formation (Table IV).

Because of the recent report that in rats PGE compounds bring about the increase in vascular permeability primarily by releasing mast cell amines and that PGF(2-alpha) inhibits the PGE-induced change by interfering with the release of granule-bound histamine (Crunkhorn and Willis, 1971b) we undertook the experiments which are summarized in Table V. In these experiments PGF(2-alpha) or norepinephrine was pre-mixed and injected together with the phlogistic agents into the skin of test animals. While PGF(2-alpha) did reduce vascular permeability responses to PGE(2) and 48/80, it similarly affected, from the point of view of permeability-enhancing properties, equipotent doses of histamine and bradykinin. These results would argue against the suggestion that the effects of PGE(2) are indirectly mediated by mast cell rupture and further intimate that the PGF(2-alpha)-induced inhibition is primarily due to the local vasoconstriction produced by this compound. Further support for this is provided by the fact that norepinephrine also has the capacity to reduce the vascular permeability enhancing effects of PGE(2), 48/80, bradykinin and histamine; effects which are quite similar to those obtained with PGF(2-alpha) (Table V).

Evidence that PGE compounds can mediate vascular permeability changes in tissues other than the skin comes from experiments which show that PGE(1) also produces a marked increase in the permeability of blood vessels of skeletal muscle. By methods previously described (Kaley and Weiner, 1971a), we reexamined the effects of PGE(1) on rat cremaster muscle. On a weight basis, PGE(1) had a more pronounced action than serotonin in enhancing vascular permeability (Fig. 1 and 2).

Generally, the permeability-inducing effects of PGE compounds subside within 20 min, suggesting that the delayed phase of the inflammatory response may not be mediated by these compounds. Yet, this participation is still plausible, were one to assume that prostaglandins are locally synthesized and released in response to trauma and during the development of inflammation (Anggard and Jonsson, 1971). Additional support for the concept that there may be a relationship between prostaglandins and inflammation is provided by recent studies which show that PGE(1) is chemotactic for polymorphonuclear (PMN) leukocytes <u>in vitro</u> (Kaley and Weiner, 1971b), a finding which implicates PGE compounds in the mediation of the cellular phase of the acute inflammatory response, as well.

EFFECT OF PGF(2-ALPHA) ON LYSOSOMES

While PGE(1) has been shown to reduce the release of lysosomal enzymes from phagocytizing leukocytes, via its action on adenylate cyclase (Bourne et al, 1971; Weissmann et al, 1971a), we recently reported that PGF(2-alpha) can exert an opposite effect on lysosomes (Weiner and Kaley, 1971). In these experiments lysosome-rich fractions of rat liver were prepared according to the method of Dingle (1961), and were incubated in the presence of prostaglandins for 30 min at 25°C. After incubation beta-glucuronidase activity in the supernatant, an index of lysosomal fragility, was assessed according to the method of Fishman et al (1948). PGF(2-alpha), as shown in Fig. 3, in a concentration of 40 microgram/ml caused a substantial increase in the release of beta-glucuronidase from lysosomal suspensions, as measured by both the amount of phenolphthalein and the percent of the total enzyme activity present in the supernatant.

Table VI illustrates the effects of various concentrations of PGF(2-alpha) on release of lysosomal beta-glucuronidase. The increase in hydrolase release, in suspensions where PGF(2-alpha) was present in concentrations of 2.8-11.2 x 10^{-5} M was significantly greater than that of controls. Since PGF and PGE compounds are released together during various tissue injury reactions (Piper and Vane, 1969; Greaves et al, 1971; Anggard and Jonsson, 1971) and since PGF compounds do not share with PGE compounds the ability to enhance vascular permeability the above findings, together with those of Weissmann et al (1971b, this Symposium), illustrate how PGF(2-alpha) could still be an important mediator of inflammatory reactions.

The present state of knowledge regarding the multi-faceted interactions of prostaglandins with various substrates in promoting tissue injury reactions is schematically depicted in Fig. 4. In addition to all of the actions of PGE and PGF compounds, which we discussed, there may be a causal relationship between the generation of complement and prostaglandin-like activity in acute inflammation (Giroud and Willoughby, 1970). Prostaglandins have also been suggested as mediators of anaphylaxis (Piper and Vane, 1969). During the clotting process, an inevitable consequence of vascular injury, prostaglandins are also released (Smith and Willis, 1970) and thereby could further contribute to the severity of inflammatory reactions. Equally significant is the release of prostaglandins from mast cells (Anggard and Strandberg, 1971) and the release of "prostaglandin-like" materials from phagocytizing PMN leukocytes (Movat et al, 1971); both of these cell types being importantly involved in the mediation of reactions following immunologic and non-immunologic injury. In addition, phospholipase once liberated from lysosomes of leukocytes (Anderson et al, 1971) and other tissue cells could react with cell membranes to give rise to prostaglandins. The generation of these compounds from so many different sources and diverse pathways may further exacerbate the pattern of response common to all kinds of inflammation. Moreover, during the recovery of inflammation, PGE compounds may participate in promoting cell proliferation, an essential component of the repair of damaged tissues (Franks et al, 1971).

In the final analysis, we can only speculate how prostaglandins affect the reaction of tissues to injury or whether they are the primary and most important mediators of these reactions. Nevertheless, we suggest that they play an essential role in the mediation of the pathogenesis of inflammatory reactions.

TABLE I

VASODILATOR EFFECT OF PGE(1) and PGA(1) IN RAT CREMASTER MUSCLE

	Dose microgram/0.1 ml	No. of Tests	Mean % increase in diameter
PGE(1)	0.1	21	40
PGA(1)	0.1	25	14

TABLE II

INFLUENCE OF PGE(1) ON ARTERIOLAR REACTIVITY TO VASOCONSTRICTOR AGENTS IN RAT MESENTERY

Test Agent	% Decrease in Arteriolar Diameter: Control	After PGE(1) 1.0 microgram	After PGE(1) 10 micrograms
Epinephrine, 50 micrograms	57.7±2.7* (7)	38.0±7.4 (5)	36.2±4.1 (5)
Norepinephrine, 100 micrograms	52.7±2.1 (8)	20.0±8.6 (6)	17.6±6.5 (4)
Angiotensin II, 5 nanograms	50.0±3.9 (8)	23.0±9.4 (5)	8.2±2.8 (5)
Vasopressin, 0.002 I.U.	46.3±2.9 (7)	39.5±8.7 (7)	22.8±6.1 (5)

*Values are means ± standard errors.
Number of observations are given in parentheses ( ).

TABLE III

INFLUENCE OF PGE(1) ON ARTERIOLAR REACTIVITY TO VASOCONSTRICTOR AGENTS IN RAT CREMASTER MUSCLE

Test Agent	% Decrease in Arteriolar Diameter Control	After PGE(1) 1.0 microgram	After PGE(1) 10 micrograms
Epinephrine, 1 microgram	48.4±2.8* (7)	26.1±6.7 (4)	26.3±9.7 (4)
Norepinephrine, 10 nanograms	46.4±2.0 (6)	35.2±3.1 (4)	17.8±5.3 (4)
Angiotensin II, 0.05 nanograms	51.3±5.1 (6)	33.8±9.5 (4)	17.2±4.7 (4)
Vasopressin, 0.0002 I.U.	31.5±2.8 (6)	23.9±2.8 (4)	10.0±2.5 (4)

*Values are means ± standard errors.
Numbers of observations are in parentheses ( ).

TABLE IV

PERMEABILITY-ENHANCING EFFECTS OF PROSTAGLANDINS IN RAT SKIN*

	No. of Rats	Average Responses 100 nanograms	500 nanograms
PGE(1)	25	2.5	3.5
PGE(2)	18	2.7	3.4
PGA(1)	14	1.1	1.5

*Each response was graded from 0 to 4+; responses for each agent were averaged.

TABLE V

EFFECTS OF PGF(2-ALPHA) ON
VASCULAR PERMEABILITY RESPONSES

	Micro-grams	No. of Rats	PGF(2-alpha) 0.5 micrograms	No. of Rats	Norepinephrine 0.5 micrograms
PGE(2)	0.1	16	25	10	30
	0.5	12	70	7	70
48/80	0.025	10	20	8	28
	0.4	12	80	8	75
Brady-kinin	0.1	7	35	6	30
	1	6	72	6	66
Hista-mine	0.3	9	30	10	30
	3	10	60	10	50

*Vascular permeability responses expressed as percentage of control responses.

TABLE VI

PROSTAGLANDIN F(2-ALPHA)-INDUCED
RELEASE OF BETA GLUCURONIDASE
FROM THE LARGE GRANULE FRACTION OF RAT LIVER

Agent microgram/ml		Beta-Glucuronidase*	2P+
Control		100	
PGF(2-alpha)	1	104±4	<0.2
PGF(2-alpha)	10	113±5	<0.05
PGF(2-alpha)	40	125±4	<0.005

*Results are given as percent ± S.E. of control.
+Significance of paired samples, using 2-tailed "t" test.

ACKNOWLEDGEMENTS

The authors are grateful to Dr. John E. Pike of Upjohn for the supply of prostaglandins. The excellent technical assistance of Mr. Asefa Gebrewold is appreciated. These studies were supported (in part) by N.I.H. Grant HE-12342.

REFERENCES

Anderson, A. J., Brocklehurst, W. E., and Willis, A. L., 1971, Evidence for the role of lysosomes in the formation of prostaglandins during carrageenin induced inflammation in the rat, Pharmacol. Res. Comm. 3:13.

Anggard, E., and Jonsson, C. E., 1971, Efflux of prostaglandins in lymph from scalded tissues, Acta Physiol. Scand. 81:440.

Anggard, E., and Strandberg, K., 1971, Efflux of prostaglandin E(2) from cat paws perfused with compound 48/80, Acta Physiol. Scand. 82:333.

Arora, S., Lahiri, P. K., and Sanyal, R. K., 1970, The role of prostaglandin E(1) in inflammatory process in the rat, Int. Arch. Allergy 39:186.

Baez, S., 1966, Recording of microvascular dimensions with an image-splitter television microscope, J. Appl. Physiol. 21:299.

Bourne, H. R., Lehrer, R., Cline, M. J., and Melmon, K. L., 1971, Cyclic 3',5'-adenosine monophosphate in the human leukocyte: Synthesis, degradation, and its effect on neutrophil activity, J. Clin. Invest. (in press).

Crunkhorn, P., and Willis, A. L., 1971a, Cutaneous reactions to intradermal prostaglandins, Brit. J. Pharmacol. 41:49.

Crunkhorn, P., and Willis, A. L., 1971b, Interaction between prostaglandins E and F given intradermally in the rat, Brit. J. Pharmacol. 41:507.

Dingle, J. T., 1961, Studies on the mode of action of excess vitamin A. III. Release of a bound protease by the action of vitamin A, Biochem. J. 179:509.

Fishman, W. H., Springer, B., and Brunetti, R., 1948, Application of an improved glucuronidase assay method to the study of human blood beta-glucuronidase, J. Biol. Chem. 173:449.

Franks, D. J., McManus, J. P., and Whitfield, J. F., 1971, The effect of prostaglandins on cyclic AMP production and cell proliferation in thymic lymphocytes, Biochem. Biophys. Res. Comm. 44:1177.

Giroud, J. P., and Willoughby, D. A., 1970, The interrelations of complement and a prostaglandin-like substance in acute inflammation, J. Pathol. 101:241.

Greaves, M. W., Sondergaard, J., and McDonald-Gibson, W., 1971, Recovery of prostaglandins in human cutaneous inflammation, Brit. Med. J. 1:258.

Hedqvist, P., and Brundin, J., 1969, Inhibition of prostaglandin E(1) of noradrenaline release and of effector response to nerve stimulation in the cat spleen, Life Sci. 8:389.

Hedwall, P. R., Abdel-Sayed, W. A., Schmid, P. G., Mark, A. L., and Abboud, F. M., 1971, Vascular responses to prostaglandin E(1) in gracilis muscle and hindpaw of the dog, Am. J. Physiol. 221:42.

Holmes, S. W., Horton, E. W., and Main, I. H. M., 1963, The effect of prostaglandin E(1) on responses of smooth muscle to catecholamines, angiotensin and vasopressin, Brit. J. Pharmacol. 21:182.

Kadowitz, P. J., Sweet, C. S., and Brody, M. J., 1971, Differential effects of prostaglandins E(1), E(2), F(1-alpha), and F(2-alpha) on adrenergic vasoconstriction in the dog paw, J. Pharmacol. Exp. Therap. 177:641.

Kaley, G., and Weiner, R., 1968, Microcirculatory studies with prostaglandin E(1), in Prostaglandin Symp. Worcester Found. Exp. Biol., (P. W. Ramwell and J. E. Shaw, eds.), pp. 321-328, J. Wiley, New York.

Kaley, G., and Weiner, R., 1971a, Prostaglandin E(1): A potential mediator of the inflammatory response, Ann. N. Y. Acad. Sci. 180:338.

Kaley, G., and Weiner, R., 1971b, Prostaglandins and Inflammation: The effect of prostaglandin E(1) on migration of leukocytes, Nature, (in press).

Movat, H. Z., Macmorine, D. R. L., and Takeuchi, Y., 1971, The role of PMN-leukocyte lysosomes in tissue injury, inflammation and hypersensitivity. VIII. Mode of action and properties of vascular permeability factors released by PMN-leukocytes during "in vitro" phagocytosis, Int. Arch. Allergy 40:218.

Piper, P. J., and Vane, J. R., 1969, Release of additional factors in anaphylaxis by anti-inflammatory drugs, Nature 223:29.

Piper, P., and Vane, J. R., 1971, The release of prostaglandins from lung and other tissues, Ann. N. Y. Acad. Sci. 180:363.

Smith, J. B., and Willis, A. L., 1970, Formation and release of prostaglandins by platelets in response to thrombin, Brit. J. Pharmacol. 40:545.

Vane, J. R., 1971, Inhibition of prostaglandin synthesis as a mechanism of action for aspirin-like drugs, Nature New Biology 231:232.

Viguera, M. G., and Sunahara, F. A., 1969, Microcirculatory effects of prostaglandins, Can. J. Physiol. Pharmacol. 47:627.

Weiner, R., and Kaley, G., 1969, Influence of prostaglandin E(1) on the terminal vascular bed, Am. J. Physiol. 217:563.

Weiner, R., and Kaley, G., 1971, Lysosomal fragility induced by prostaglandin F(2-alpha), Nature (in press).

Weissmann, G., Dukor, P., and Zurier, R. B., 1971a, Effect of cyclic AMP on release of lysosomal enzymes from phagocytes, Nature New Biology 231:131.

Weissmann, G., et al, 1971b, (this Symposium).

Willis, A. L., 1969, Parallel assay of prostaglandin-like activity in rat inflammatory exudate by means of cascade superfusion, J. Pharmacol. 21:126.

DISCUSSION

KADOWITZ: Your data with PGE(1) and PGA(1) in the vasculature of the rat agree with our results in the regional circulation of the dog, that is to say, that PGE(1) is about 100 times more potent than PGA(1) in its ability to attenuate vasoconstrictor responses. You have shown that the vasodilator and blocking actions of PGE(1) are separable in that blockade persisted at a time when vasodilation was no longer observed. We agree that this blockade of vasoconstrictor responses is probably not due to a physiological antagonism. We have arrived at this very same conclusion by comparing the effect of other vasodilators such as histamine and glyceryl trinitrate with prostaglandin. Infusions of histamine and glyceryl trinitrate in concentrations that produce the same amount of vasodilation as the prostaglandins were found to be without effect on vasoconstrictor responses to norepinephrine and angiotensin.

I would like to address myself to a statement by Dr. Kligman that corticosteroids are vasoconstrictors. We have been interested in the vascular actions of steroids for a long time now and have failed to observe any vasoconstriction in the regional circulation of the cat or dog when a wide range of doses of hydrocortisone was injected intravenously or intra-arterially into the hindpaw, gracilis muscle, kidney, intestine or forelimb. In fact we have given as much as 150 mg/kg IV and have observed no significant alteration in arterial blood pressure or vascular resistance in any of several vascular beds. The results of some of these studies are shown in Fig. 1, 2, and 3. In Fig. 1, it can be seen that injection of hydrocortisone (10 mg/kg IV) produces no change in arterial pressure or hindquarters perfusion pressure, and in 20 such experiments perfusion pressure was 110 ± 5 mm Hg before and 111 ± 4 mm Hg after administration of the steroid. Fig. 2 shows that when this amount of hydrocortisone is injected intravenously into the dog, there is no change in hindpaw perfusion pressure; hydrosortisone does not modify vasoconstrictor responses to intra-arterial norepinephrine or sympathetic nerve stimulation, but it enhances the vasoconstrictor response to intra-arterial epinephrine. This figure is typical of results obtained in six such experiments. Fig. 3 shows that hydrocortisone is without effect on venomotor tone in the superficial veins of the forelimb or hindpaw, and that it does not alter venomotor responsiveness to norepinephrine.

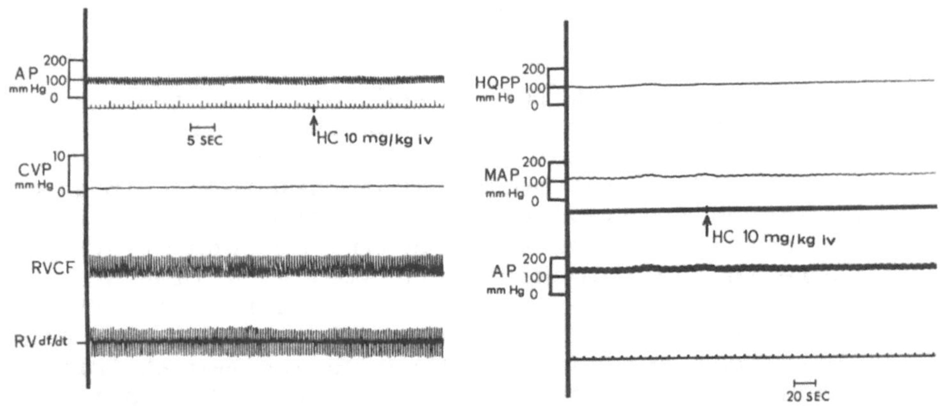

Figure 1. Records from two experiments showing that hydrocortisone, 10 mg/kg IV, is without effect on the arterial pressure (AP) or hindquarters perfusion pressure (HQPP) or central venous pressure (CVP), right ventricular contractile force (RVCF) or its first derivative (RV df/dt) in the cat.

These results demonstrate that hydrocortisone does not constrict the hindpaw vascular bed, which is about 90% skin, or the hindquarters vasculature, which is about 40% skin, and that it does not modify venomotor tone in the veins that carry the venous outflow from the cutaneous vasculature. These results do show that hydrocortisone has the ability to selectively enhance the vasoconstrictor effect of epinephrine in the skin vessels, and suggest that if any epinephrine is present in the skin its effect may be increased by hydrocortisone. You have made a statement about the effect of steroids on the microcirculation; I wonder if you could say a little more about this.

KALEY: There are a few reports in the literature which show that some of the steroids, especially the glucocorticoids, have a tendency to increase the responsiveness of blood vessels to vasoconstrictor agents (Altura, B. M., 1966, Am. J. Physiol. 211:(No. 6):1393).

KLIGMAN: When you put corticosteroid on topically, the skin blanches: This could represent several mechanisms but obviously indicates vasoconstriction. Dr. Kaley, I am concerned about some of the inconsistencies of the reports with the prostaglandins. For example, Dr. L. Juhlin injected PGE(1) and PGE(2) into the skin and did not observe the presence of weals. He got only erythema which lasted about 24 hr. These are simple systems which can be

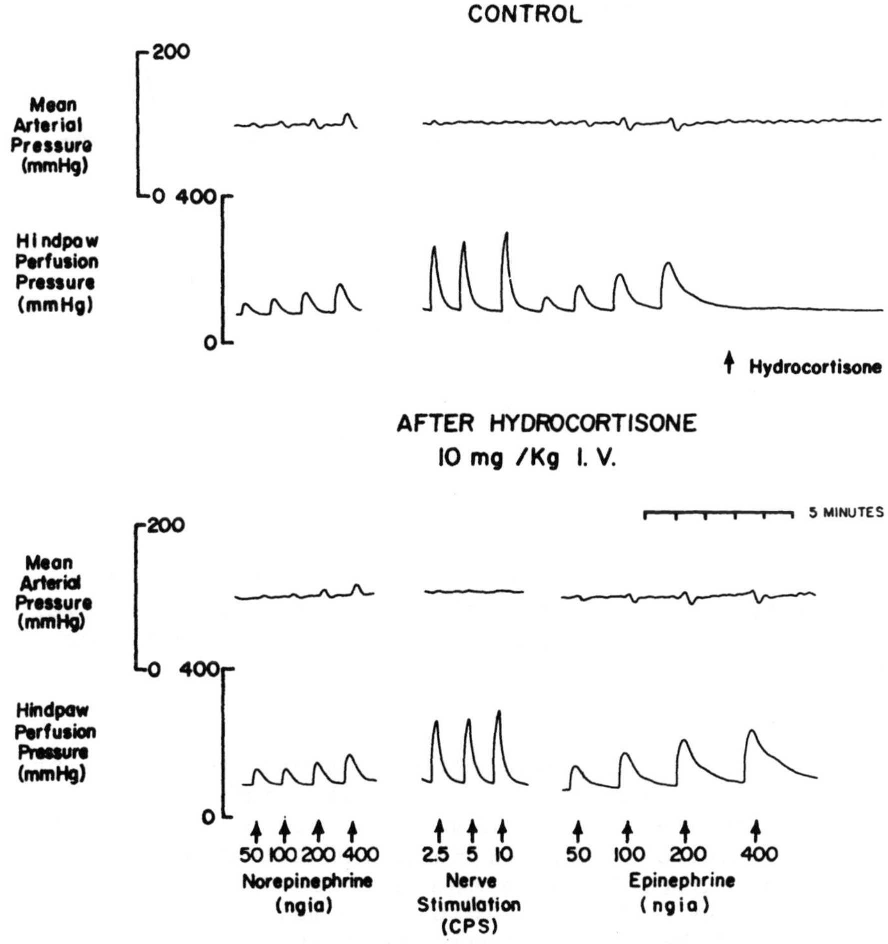

Figure 2. Record from an experiment showing that hydrocortosone, 10 mg/kg IV, is without effect on mean arterial pressure or hindpaw perfusion pressure in the dog, and that it does modify vasoconstrictor responses to norepinephrine or nerve stimulation in the hindpaw, but that it does enhance the response to intra-arterial epinephrine.

used to test these responses, so I cannot understand why observations differ. Juhlin, furthermore, is in the holy land, Sweden!

KALEY: It may be that depending upon the animal species or even strain differences, or differences between various cell populations, responses to prostaglandins may vary quite substantially.

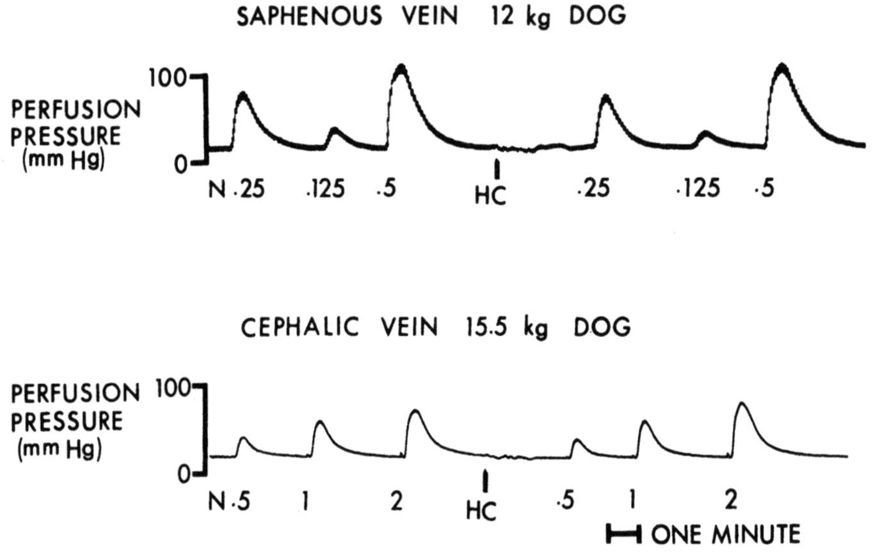

Figure 3. Records from two experiments showing that hydrocortisone, 10 mg, directly into the saphenous vein or cephalic vein does not alter basal venomoter tone or responsiveness to norepinephrine.

SUNAHARA: As Dr. Kaley has already mentioned, some of the work we have done is with the systemic application of PGE(1), and the results we have seen are very similar, i.e., inhibition of topically-applied catecholamine-induced vasoconstriction in the mesocecal microcirculation. This inhibition remains from 60 to 90 min. I would like to add that after intravenous administrations of PGE(2), PGF(2-alpha) or arachidonic acid, the effects of topically-applied noradrenaline are quite different from those of PGE(1). After arachidonic acid is infused there is a marked potentiation of catecholamine reactions. We are at a loss as to how to explain this. It might also be noted that the previous dosing with indomethacin and aspirin does not seem to alter the above effect (i.e., topically-applied noradrenaline continues to be potentiated).

NAKANO: What do you think is the underlying mechanism for the sustained PGE(1) effects?

KALEY: This is a very important question. My favorite hypothesis is that stimulation of cyclic AMP synthesis is involved.

NAKANO: Zimmerman has shown that when you give theophylline in an animal pretreated with PGE(1), the action is much more potent.

KUEHL: I think your statement that levels of prostaglandin are very important in terms of their responses, is a very critical one indeed. We utilize the cyclic AMP response as an assay system for prostaglandins. In this system we obtain a bell-shaped dose response curve; with PGE(1) at higher levels (20 microgram/ml) the response begins to diminish and thus added prostaglandin becomes inhibitory rather than stimulatory.

KALEY: In this regard it is also interesting that minute amounts of prostaglandins, possibly via activation of adenylate cyclase, can inhibit the release of mediators from sensitized cells during immunologically-induced inflammatory reactions, whereas somewhat larger doses of prostaglandins can reproduce non-immunologically induced tissue injury reactions.

KADOWITZ: We have been able to demonstrate profound effects on the vasculature with prostaglandins at concentrations as low as 10^{-10} M. I was wondering whether these concentrations might have any effect on cyclic AMP in smooth muscle.

KALEY: I do not know. We have not measured cyclic AMP in vascular smooth muscle.

LICHTENSTEIN: There is a great deal of difference, as I have mentioned, in rats and man. And with respect to the drugs that turn off histamine, the ones that work in humans and primates do not work in rats.

FRIEDBERG: Dr. Willis has found that there are observable mast cell degranulations 30 min after exposure of the mesentery to prostaglandins. And you have found just the opposite. May I suggest the possibility that mast cell degranulation occurs secondary to the edema which is formed. In Willis' system he has injected the prostaglandins intraperitoneally into the rat, and you have used an *in vitro* system.

KALEY: In our experiments PGE(1) did not cause mast-cell degranulation. We are saying that PGE compounds enhance vascular permeability directly; this does not, however, exclude the possibility that in certain strains of rats, degranulation of mast cells may partially contribute to the PGE-induced increase in vascular permeability.

PRO-INFLAMMATORY EFFECTS OF CERTAIN PROSTAGLANDINS

E. M. Glenn, B. J. Bowman, and N. A. Rohloff

The Upjohn Company
Kalamazoo, Mich.

INTRODUCTION

Several groups of workers have suggested that certain prostaglandins, especially PGE(1), elicit vascular permeability responses following their local inoculation into the skin of man and experimental animals (Willis, 1969; Arora et al, 1970; Giroud and Willoughby, 1970; Kaley and Weiner, 1971; Crunkhorn and Willis, 1971; Sondergard and Greaves, 1971). Prostaglandins of various types have been isolated from inflammatory lesions of both species (Willis, 1969; Greaves et al, 1971). In consequence, certain prostaglandins are listed among the numerous other chemical mediators of the inflammatory process. Prostaglandins are released by almost all kinds of disturbing influences on the microenvironment of cells (Piper and Vane, 1971). Moreover, it has been shown that the synthesis and release of certain prostaglandins is inhibited by aspirin and indomethacin in the isolated spleen, sensitized lungs and isolated platelets (Smith and Willis, 1971; Vane, 1971; Ferriera et al, 1971).

The following study is an extension of published work in the area and involves comparisons of the pro-inflammatory actions of certain prostaglandins in a number of animal species and a variety of experimental circumstances. The work to be reported here also contains original data concerning the influence of orally-administered non-steroidal anti-inflammatory drugs on the serum concentrations of PGF(2-alpha) in rats.

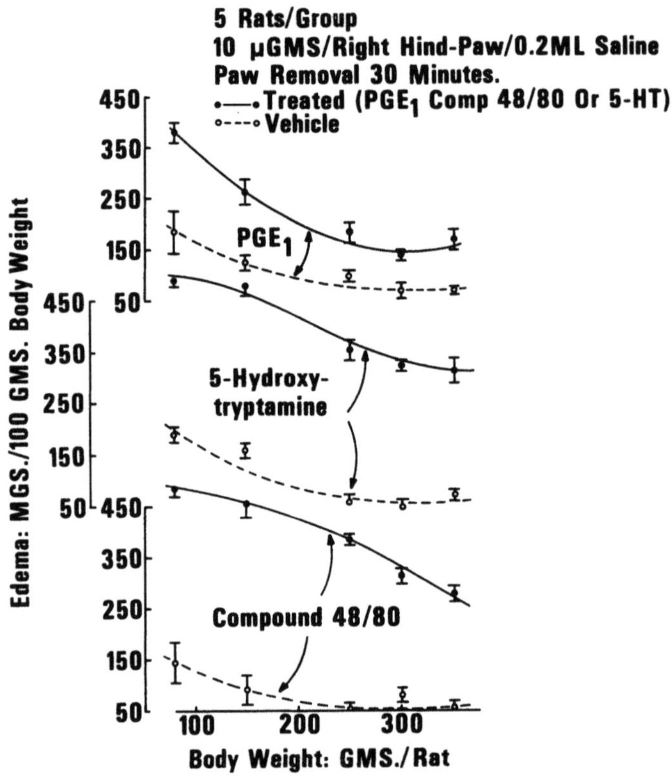

Figure 1. Effect of PGE(1), serotonin, and compound 48/80 on acute inflammation in rats of varying weights.

METHODS

Prostaglandins and other mediators in 0.1-0.2 ml saline are inoculated locally into the right hindpaws of 16-hr fasted Carworth male rats of the Sprague-Dawley strain, weighing 225-250 g. Paws are inoculated with vehicle simultaneously in separate groups of animals. Both inoculated and uninoculated paws are removed with a Harvard guillotine from chloroformed-killed animals at varying times. Weight differences of injected and uninjected paws are expressed as "mg of edema per 100 g of final body weight" (mg/100 g fbw). Five to ten animals are used per group. Other well-known "mediators" of inflammation are injected simultaneously as a further check on the methodology.

Granuloma pouches are made by standardized techniques either by the intrapouch injection of 0.5 ml of d-alpha-tocopherol or croton oil (Glenn et al, 1963). Local

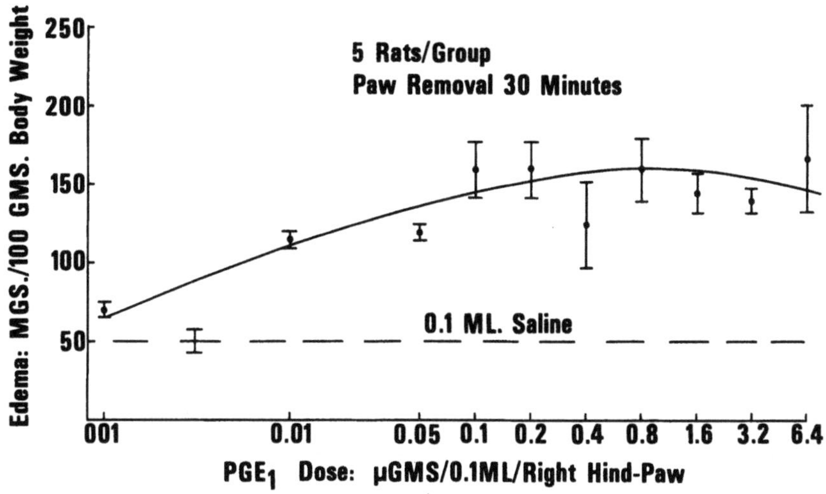

Figure 2. Acute inflammation in the rat induced by PGE(1).

injections of probable mediators are made in 0.1 ml saline into the lumen of pouches twice each day for 7.5 days and, on the eighth day, animals are killed and volumes of exudate measured in a graduate cylinder. Ten animals per group are used and the results are expressed as the mean with the standard error of the mean.

PGF(2-alpha) serum concentrations are determined by radioimmunoassay (Kirton and Cornette, 1971) in 16-hr fasted rats at varying times after the oral administration of drugs. Fasting does not alter the serum concentrations of PGF(2-alpha) in rats. The various prostaglandins are obtained from Dr. John Pike of The Upjohn Company and other agents were purchased from various commercial sources.

RESULTS

Compound 48/80, serotonin, and PGE(1) elicit inflammatory edema in the hindpaws of rats of varying weights. PGE(1) is less effective quantitatively than the others (Fig. 1). The response of PGE(1) is dose-related, 0.01 micrograms per paw eliciting maximal effects (Fig. 2). Sodium Gyclopal-induced anesthesia does not affect the multidose quantitative aspects of inflammatory edema occurring in response to histamine, serotonin, compound 48/80, or PGE(1) in the rat. When compared to other prostaglandins [PGE(1), PGE(2) and PGA(2)], PGF(2-alpha) does not elicit inflammatory edema in the rat paw (Fig. 3A). Similar results are obtained in male hamsters, (Fig.

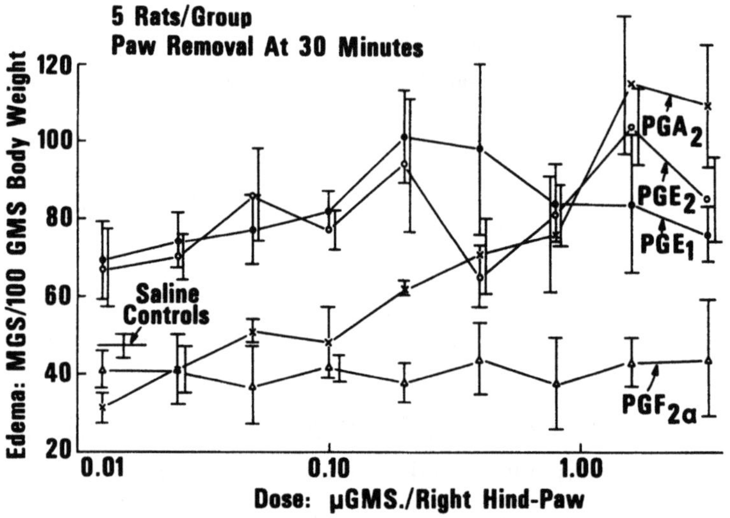

Figure 3A. Acute inflammation induced with various prostaglandins in rats.

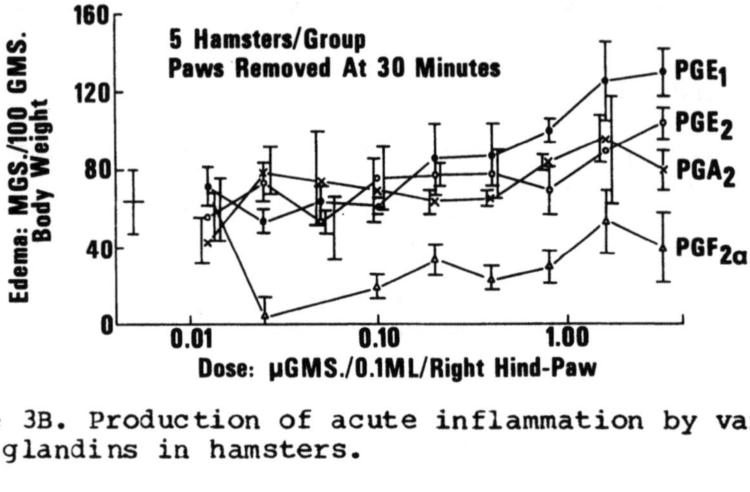

Figure 3B. Production of acute inflammation by various prostaglandins in hamsters.

3B). The time course for the production of inflammatory edema in response to serotonin, compound 48/80, and PGE(1) is the same for all compounds. Inflammation occurs within a short time, reaches a maximum in 20-30 min and rapidly subsides to lower levels within 120 min (Fig. 4). The inflammatory edema persists for at least 6 hr in all cases.

When examined in male rats, mice, hamsters, guinea pigs and gerbils, histamine alone elicits significant hindpaw edema in the guinea pig. PGE(1) is as effective as histamine, serotonin, and compound 48/80 in mice and gerbils. It is more effective than the two latter agents in hamsters (Fig. 5). PGE(1) causes hemorrhagic reactions in the paw of male hamsters and erythematcus reactions in rat paws.

When compared to bradykinin triacetate at low dosages, PGE(1) is more effective on a weight basis (Fig. 6). When the dosages of both drugs are increased, the response to PGE(1) disappears; that to bradykinin continues to increase (Fig. 7). When varying amounts of PGE(1) are inoculated into the hindpaws of rats (alone or simultaneously with varying amounts of histamine, serotonin, or carrageenan) it elicits an additive effect in all instances and not a synergistic one. At 100 times the concentration of PGE(1), locally-inoculated caffeine and isoproterenol (100-500 microgram/paw) fail to elicit inflammatory edema. The combination of PGE(1) (1 microgram/paw) with either agent alone (50 microgram/paw) or all of them together(1 + 50 + 50) fails to display inflammatory edema greater than that caused by PGE(1) alone.

When inoculated locally into on-going granulomatous reactions of the rat, PGE(1) elicits greater accumulation of fluid in the pouch (Fig. 8). Medrol (methylprednisolone), a steroidal anti-inflammatory drug, produces the expected inhibition.

In another experiment, illustrated also in Fig. 8, serotonin also produces greater accumulation of fluid into the pouch. PGE(2) and PGF(2-alpha) are less effective, if at all. The increased accumulation of pouch fluid in response to PGE(1) is dose-related in both d-alpha-tocopherol and croton oil-induced granuloma pouches (Fig. 9). Similar quantitative results are obtained with serotonin (5-50 microgram/pouch/day) in both types of granulomatous reaction. Neither 5-HT nor PGE(1) are effective at higher doses when given directly into air pouches alone which are devoid of irritants like croton oil and d-alpha-tocopherol. Arachidonic acid, a percursor of certain prostaglandins, fails to cause greater accumulation of croton oil-induced granuloma pouch fluid when given at 500 microgram/pouch/day.

When PGE(1) is injected at maximal pro-inflammatory concentrations (1.0 microgram/paw) every 30 min, there is no evidence that the hindpaw of the rat fails to respond to repeated injections. The quantitative aspects of edema

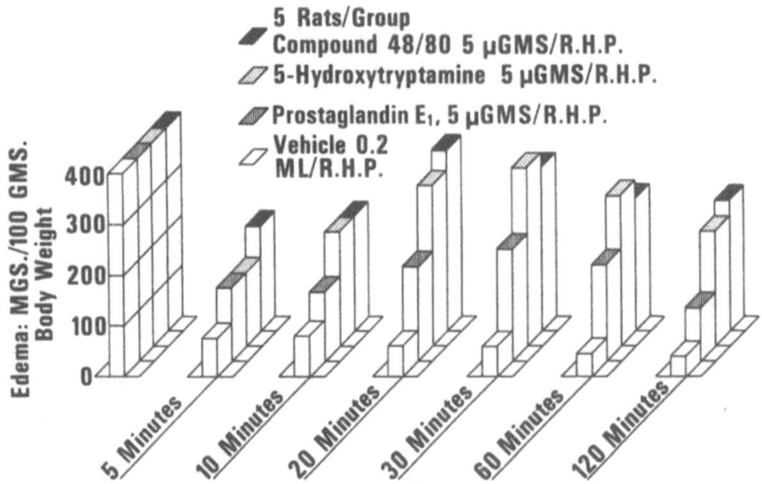

Figure 4. Time-course of inflammation when induced by various phlogistins.

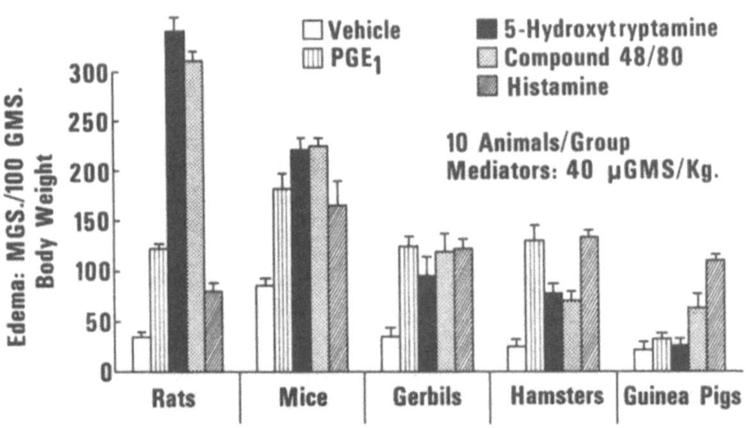

Figure 5. Species differences in response to various phlogistins.

formation are additive at all the different times (30 min intervals for 6 hr). Similar results are obtained in rats when the local inoculations are made once every 24 hr (5 microgram/paw) for at least one week, followed by paw removal within 30 min at 24 hr intervals and after another local inoculation on the day of paw removal (0.2 microgram/paw). Similar negative results are obtained with serotonin and compound 48/80 pre-treatment.

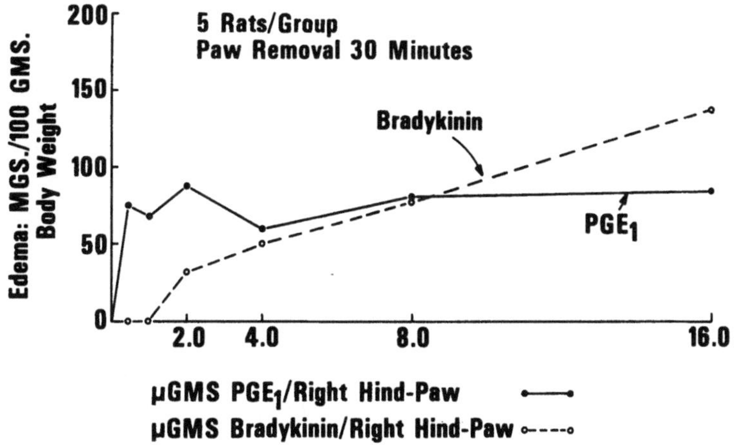

Figure 6. Acute inflammation to low dosages of PGE(1) and bradykinin.

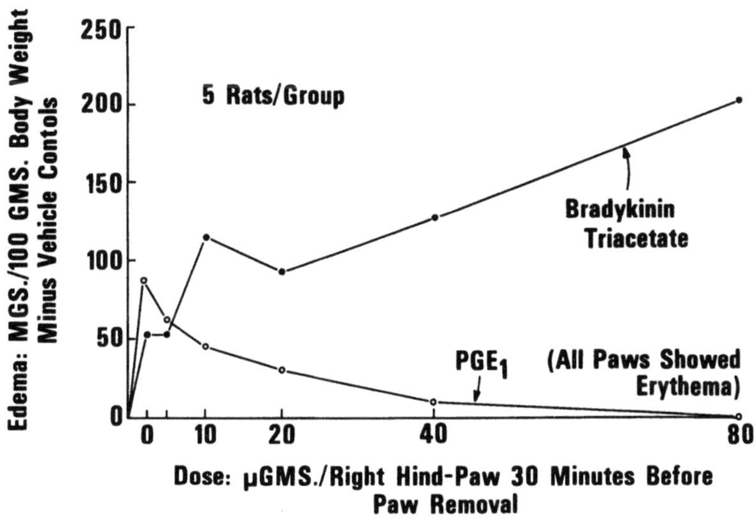

Figure 7. Acute inflammation to high dosages of PGE(1) and bradykinin.

In rats, the serum concentrations of PGF(2-alpha) are increased with age (Fig. 10), especially in rats at least 8 months of age, weighing 650+ grams (Fig. 11). The elevated serum concentrations of PGF(2-alpha) are decreased significantly within 3 hr by the oral administration of indomethacin, flufenamic acid, phenylbutazone, and aspirin (Fig. 11).

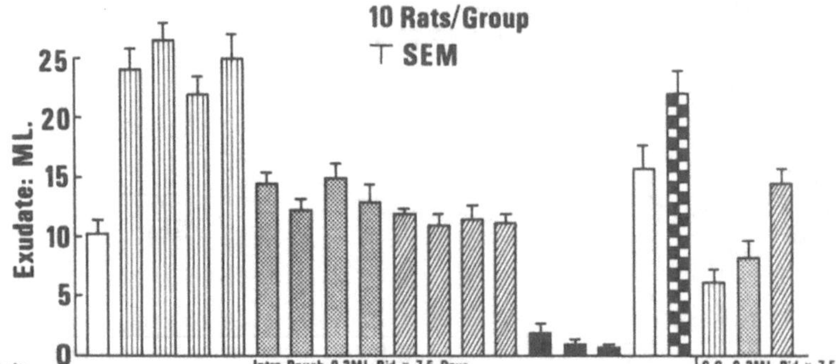

Injection Route: Intra-Pouch 0.2ML Bid × 7.5 Days | S.C. 0.2ML Bid × 7.5

Drug	Saline	PGE$_1$ (U-10136)				PGE$_2$ (U-12062)				PGF$_{2\alpha}$Salt (U-14583 E)				Medrol (U-7532)			Serotonin		PGE$_1$	PGE$_2$	PGF$_{2\alpha}$
Dose: μGMS/INJ.	0	5	10	20	40	5	10	20	40	6.5	13	26	52	12.5	25	50	0	50	40	40	52
Body Wt. Gain: GMS.	45	38	18	5	23	31	22	24	26	25	22	26	29	21	17	10	55	49	9	23	29

*Tromethamine Salt

Figure 8. Effect of locally injected prostaglandins and serotonin on granuloma pouch exudation.

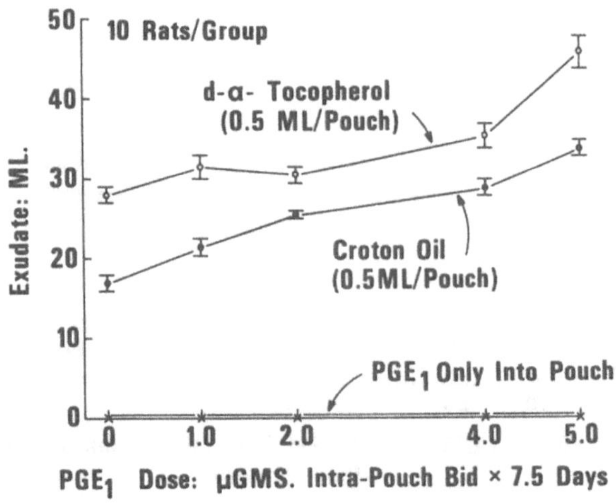

Figure 9. Effect of PGE(1) on exudation in granuloma pouches.

In inflammatory reactions caused by the local inoculation of thrombin (100 units into both hindpaws), the serum concentrations of PGF(2-alpha) are somewhat decreased and they return to control levels after the oral administration of high dosages of Medrol (Fig. 12). Lower dosages appear to cause a slight depression of the serum concentrations. These results differ somewhat from those obtained with indomethacin where the PGF(2-alpha) serum concentrations are decreased at all dosages in thrombin-induced inflammatory reactions of the hindpaws (Fig. 13).

DISCUSSION AND SUMMARY

When inoculated locally, PGE(1), PGE(2) and PGA(2) produce measurable inflammatory edema in the hindpaws of rats. PGF(2-alpha) is pro-inflammatory in rats, mice, gerbils, hamsters, but not in guinea pigs. Histamine alone, but not serotonin, PGE(1) and compound 48/80, elicits inflammatory edema of the hindpaw in the guinea pig. PGE(1), while being less effective on a weight basis than serotonin and compound 48/80 in rats, especially at higher locally-inoculated dosages, e.g., those above 1.0 microgram/paw, is either equally effective or more effective in gerbils, mice and hamsters. PGE(1) causes erythematous reactions in rat paws and hemorrhages in hamster paws.

PGE(1), compound 48/80 and serotonin are equally effective as phlogistins in anesthetized and conscious rats. The time-response to all the different mediators alluded to here is the same, the peak response occurs within 20-30 min and subsides slowly thereafter. The inflammatory response to PGE(1) plateaus at 0.01 microgram/paw and, at much higher concentrations (40-80 micrograms/paw), PGE(1) fails entirely to elicit a pro-inflammatory effect. This response differs from the other mediators which produce far greater amounts of total edema than does PGE(1) at high locally-inoculated doses.

When PGE(1) is inoculated into the hindpaw, along with carrageenan, serotonin or compound 48/80, it fails to elicit a synergistic acute pro-inflammatory effect. The results are additive. When inoculated locally on a chronic basis into croton oil or d-alpha-tocopherol-induced granuloma pouches, twice each day for a week, PGE(1) produces greater amounts of exudation. These pro-inflammatory effects are dose-related. Similar dose-related results are obtained with serotonin. These two agents do not elicit granulomatous reactions with fluid accumulation when inoculated locally at high dosages (50-100 micrograms b.i.d., for 7.5 days) into air pouches containing no irritant.

All of these various data, taken together, indicate that certain prostaglandins, PGE(1), PGE(2) and PGA(2), are capable of causing acute inflammatory reactions and,

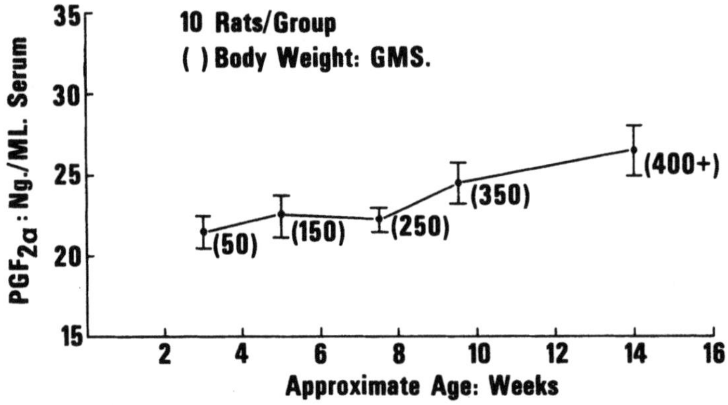

Figure 10. PGF(2-alpha) serum concentrations in rats of varying ages.

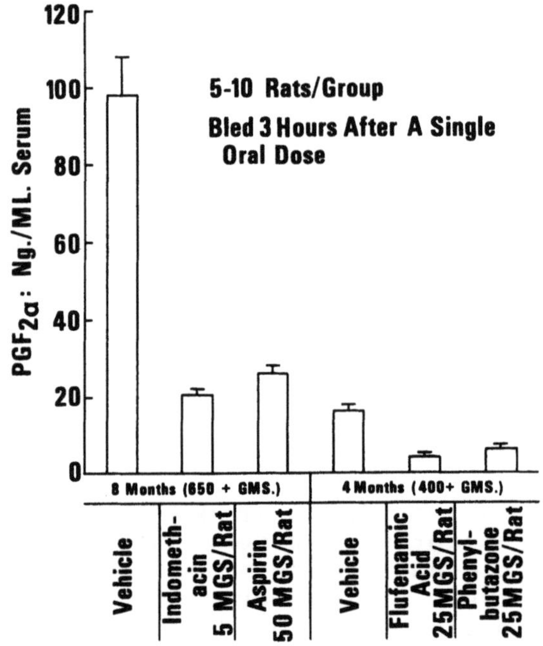

Figure 11. Effect of orally-administered drugs on PGF(2-alpha) serum concentrations in rats.

when inoculated locally into areas where chronic inflammation is occurring already, PGE(1), especially, is capable of causing greater exudative reactions. Certain prostaglandins may play a "permissive" or "supportive" role in the maintenance of chronic inflammatory edema. They are not unique in this regard. Serotonin acts similarly.

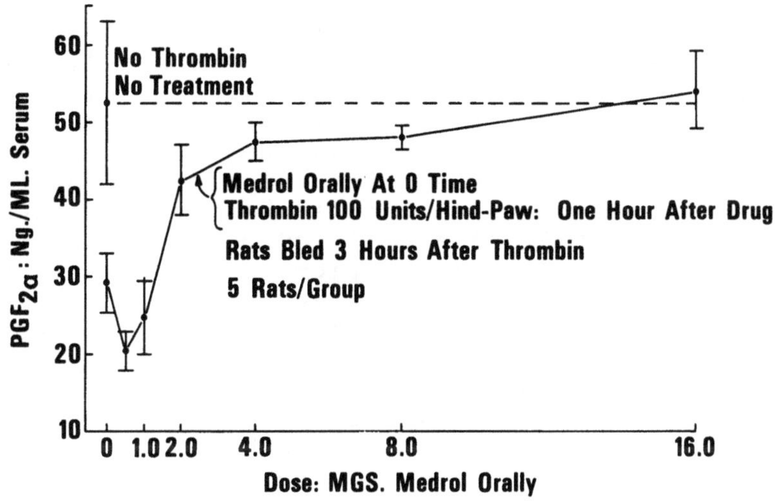

Figure 13. Effect of indomethacin on PGF(2-alpha) serum concentrations in rats.

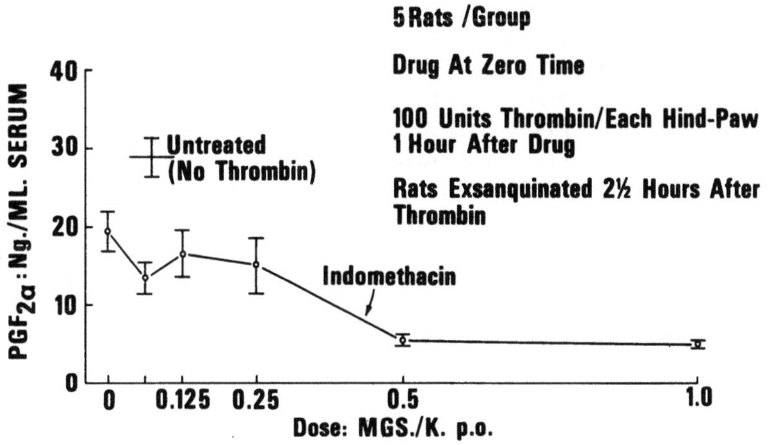

Figure 12. Effect of Medrol on serum concentrations of PGF(2-alpha).

A further extension of these studies is concerned with decreases in PGF(2-alpha) serum concentrations within 3 hr after the oral administration of the non-steroidal anti-inflammatory drugs, indomethacin, aspirin, phenylbutazone and flufenamic acid. Medrol, a steroidal anti-inflammatory drug, produces slight decreases in PGF(2-alpha) serum concentrations at low dosages and elevates them at higher dosages. Indomethacin, on the other hand, decreases the serum concentrations of PGF(2-alpha) at all dosages examined here.

Many "mediators" of inflammation exist. Additional mediators of the tissue response to injury are likely to be discovered in the future. It is unlikely that a specific inhibitor of a single mediator now known to exist, including certain prostaglandins, will inhibit inflammation significantly and for very long. The final "control" of acute and chronic inflammatory processes will depend upon a multitude of "specific" mediator inhibitors. On the other hand, and as attractive as the data appear here and in the literature (Smith and Willis, 1971; Vane, 1971; Ferriera et al, 1971), it is unlikely that the anti-inflammatory effects of the non-steroidal anti-inflammatory drugs, and especially the steroidal anti-inflammatory drugs, can be explained entirely upon their ability to decrease prostaglandin synthesis and/or release in the whole organism. Possibly, the inhibition of synthesis release of prostaglandins by some of these drugs is a reflection of more fundamental aspects of their action on cell membranes (Glenn et al, 1971; Juby and Hudyma, 1970; Seeman, 1970).

The isolated pro-inflammatory effects of certain prostaglandins, along with decreases in PGF(2-alpha) serum concentrations after giving several non-steroidal anti-inflammatory drugs, does not prove that the "obvious" interrelationships are either direct or that the one is necessarily related to the other. The observations do imply that, along with their other tissue-sparing actions, some of the non-steroidal anti-inflammatory drugs may affect the release and synthesis of certain prostaglandins *in vivo*. Whether this explains their major action or whether it is even one of crucial pharmacological importance, remains to be decided.

REFERENCES

Arora, S., Lahiri, P. K., and Sanyal, R. K., 1970, Int. Arch. Allergy 39:186.

Collier, H. O. J., 1971, Nature 232:17.

Crunkhorn, P., and Willis, A. L., 1971, Br. J. Pharmacol. 41:49.

Ferriera, S. H., Moncada, S., and Vane, J. R., 1971, Nature New Biology 231:237.

Giroud, J. P., and Willoughby, D. A., 1970, J. Pathol. 101:241.

Glenn, E. M., Bowman, B. J., Lyster, S. C., and Rohloff, N. A., 1971, Proc. Soc. Exptl. Biol. Med. 137:

Glenn, E. M., Miller, W. L., and Schlagel, C. A., 1963, in "Recent Progress in Hormone Research" (Gregory Pincus, ed.), p. 107.

Greaves, M. W., Sondergard, J., and McDonald-Gibson, W., 1971, Brit. Med. J., 2:258.

Juby, P. F., and Hudyma, T. W., 1970, in "Annual Reports in Medicinal Chemistry" (Cornelius K. Cain, ed). p. 182, Academic Press, N. Y.

Kaley, F., and Weiner, R., 1971, Ann. N. Y. Acad. Sci. 180:338.

Kirton, K. L., and Cornette, 1971, (in preparation).

Piper, P., and Vane, J., 1971, Ann. N. Y. Acad. Sci. 180:363.

Seeman, Phillip, 1970, in "Permeability and Function of Biological Membranes" (Bolis, L., and Katchalsky, A., eds), p. 40, North Holland Publishing Co.

Smith, J. B., and Willis, A. L., 1971, Nature New Biology 231:235.

Sondergard, J., and Greaves, M., 1971, Brit. J. Dermatol. 84:424.

Vane, J. R., 1971, Nature New Biology 231:232.

Willis, A. L., 1969, in "Prostaglandins Peptides and Amines" (P. Mantegazza and E. W. Horton, Eds.), p. 31, Academic Press, N. Y.

DISCUSSION

<u>KLIGMAN</u>: I think that when one is confronted with material that is so pluripotential, so varied in what it does, one begins to wonder whether a model exists which is <u>not</u> affected by prostaglandin. Is there a biological system that is unreactive? I think perhaps we are concentrating too much on inflammation. This material can have many other effects, e.g., suppression of feather development in chick embryos.

<u>GLENN</u>: Perhaps we are looking at the wrong thing insofar as prostaglandin action is concerned. In fact, looking at the prostaglandins in this area solely, may deter us from looking at more basic and more important physiologic and metabolic actions of these substances.

STUDIES ON THE MODE OF ACTION OF NON-STEROIDAL ANTI-INFLAMMATORY AGENTS

E. A. Ham, V. J. Cirillo, M. Zanetti,
T. Y. Shen, and F. A. Kuehl, Jr.

Merck Institute for Therapeutic Research
Rahway, N. J.

INTRODUCTION

The recent proposal by Vane and his associates at the Royal College of Surgeons in London (Vane, 1971; Smith and Willis, 1971; Ferreira et al, 1971) that the anti-inflammatory activity of indomethacin may be related to its ability to inhibit prostaglandin biosynthesis prompted the present studies. Our purpose was to compare the ability of a substantial number of established non-steroidal anti-inflammatory agents and related compounds to inhibit prostaglandin biosynthesis with their anti-inflammatory activity *in vivo*, as defined by the carrageenin foot edema assay. This latter assay (Winter et al, 1963), following its initial application to the development of indomethacin analogs, has been widely employed in the search for other anti-inflammatory agents. These compounds, consisting chiefly of the aryl acid type, as exemplified by substituted indole acetic acids, phenylacetic acids and anthranilic acids, have been shown to possess a wide spectrum of biological activities which include inhibition of mucopolysaccharide biosynthesis, suppression of proteolysis and other less well-defined functions involving membrane structures, such as prevention of platelet aggregation, inhibition of leukocyte migration (Bertelli and Houck, 1969) and erythrocyte stabilization (Gorog and Kovacs, 1970). The multiplicity of effects of the non-steroidal anti-inflammatory agents and the doses required to elicit them suggest that these properties are not associated with their primary action as anti-inflammatory agents.

However, the hypothesis of Vane and his colleagues is attractive in light of recent studies implicating the prostaglandins as causative agents in the inflammatory process (Willis, 1970; Greaves et al, 1971), and it also satisfies the requirement that the in vitro effect be related to the in vivo levels of these drugs required for therapeutic efficacy.

METHODS

Frozen sheep seminal vesicles were thawed, denuded of fat and connective tissue, minced and homogenized in 0.125 M Na-EDTA buffer (pH 8.3) - 1 part tissue/2 parts buffer - using a Polytron model PT 10 ST homogenizer at a pulse-frequency of 7300 c.p.s. for 30 sec. The homogenate was centrifuged for 10 min at 1200 x g and the supernatant phase was decanted through cheesecloth. This coarse filtrate was the source of the prostaglandin synthetase enzyme system. The enzyme preparation was separated into 1.0 ml aliquots and stored at -20°C. When stored in this manner, thawing only one tube for each assay, the frozen material retained full enzymatic activity for at least three months. For experimental purposes the 1.0 ml aliquots were diluted 1:10 with 0.125 M Na-EDTA buffer (pH 8.3).

The reaction mixture, modified from Wallach and Daniels (1971), consisted of 500 micrograms bovine serum albumin (Pentex, fraction V), 5×10^{-4} M hydroquinone, 2×10^{-3} M reduced glutathione, 3×10^{-5} M arachidonic acid, 0.05 microcuries [1-^{14}C]-arachidonic acid, the test drug in 0.01 ml methanol and 0.5 ml (1.2 mg protein) of the enzyme preparation in a final volume of 1.0 ml of Na-EDTA buffer (pH 8.3). The incubations were carried out for 30 min at 37°C in a Dubnoff shaker. Routinely 40% of the labelled arachidonic acid was converted to PGE(2).

Each reaction was terminated by the addition of 0.4 ml of citric acid (0.3 M), and 5 micrograms of PGE(2), PGF(2-alpha) and arachidonic acid were added. This was followed by extraction with ethyl acetate (2 x 2 ml). The combined organic phase was then backwashed with 1 ml of H2O, and the ethyl acetate reduced to dryness under nitrogen. The residue was dissolved in benzene, applied to a 1 x 20 cm lane on a silica gel G plate (Analtech) and subjected to chromatography in Andersen's (1969) FVI solvent system: ethyl acetate/acetone/acetic acid (90:10:1). After exposure to I2 vapor to visualize the prostaglandins, the PGE(2) zone (Rf=0.40) was scraped from the TLC plate and added to a toluene/ethanol (70:30) scintillation solution containing 4 g 2,5-diphenyloxazole and 50 mg p-bis[2-(4-methyl-5-phenyloxazolyl)] benzene per l. The radioactivity in each zone was determined in a Packard Tri-Carb liquid scintillation spectrometer.

The microsomal fraction used for the kinetic studies was obtained from the undiluted homogenate. Centrifugation at 10,000 x g for 20 min yielded a supernatant phase from which a microsomal pellet was isolated by subsequent centrifugation at 105,000 x g for 60 min. The incubation mixture consisted of 1 x 10^{-3} M hydroquinone, 2 x 10^{-3} M reduced glutathione, 3 to 30 x 10^{-5} M arachidonic acid, 0.1 microcuries [1-^{14}C]-arachidonic acid, 2 to 6 micrograms fluoroindomethacin in 0.01 ml methanol and 0.2 mg protein equivalent of microsomal material in a final volume of 1.0 ml of Na-EDTA buffer (pH 8.3). The incubations were run for 2 to 5 min at 37°C in a Dubnoff shaker.

RESULTS AND DISCUSSION

As shown in Table I, there is good agreement between the *in vivo* and *in vitro* activities for individual members within a given group of structurally related compounds (i.e., indomethacin analogs, arylacetic acids, pyrazolones and the salicylates). Such agreement does not occur for the fenamates. When comparing the anti-inflammatory drugs from all groups a reasonable correlation is seen between their ability to inhibit prostaglandin synthetase and their *in vivo* activity: indomethacin > ibuprofen > azapropazone > metiazinic acid; fenamates > phenylbutazone > oxyphenylbutazone > aspirin > aminopyrine. Of particular significance is the finding that the high degree of stereospecificity for two pairs of enantiomers is in agreement with their inhibitory effect upon prostaglandin synthetase. To our knowledge, this represents the first demonstration of such a correlation between an *in vitro* enzymatic effect and the *in vivo* activity of these anti-inflammatory agents. The activity in the prostaglandin synthetase system is not restricted to acidic anti-inflammatory agents. Indoxole and thiabendazole, for example, effect inhibition in proportion to their *in vivo* activity.

The specificity of the *in vitro* assay is demonstrated by the lack of inhibition by three non-anti-inflammatory drugs: the uricosuric agent probenecid, the slow-acting anti-rheumatic drug chloroquine and the anti-asthmatic agent chromoglycate.

Several kinetic studies, utilizing Lineweaver-Burk plots, showed that fluoroindomethacin was a competitive inhibitor (Ki = 8 x 10^{-6} M) of arachidonic acid (Km = 2 x 10^{-4} M). During these experiments two observations, as yet unexplained, were noted: Higher levels of fluoroindomethacin were required for inhibition when the microsomal system without BSA replaced the homogenate system with BSA, and below 5 x 10^{-5} M arachidonic acid the inhibitor showed more activity than would be expected from Michaelis kinetics.

Absolute correlation between in vivo and in vitro data is not found in every instance nor should it be expected, since the challenge of distribution and metabolism are absent from the in vitro assay described. Also, prostaglandin formation may represent only one component of a sequential series resulting in inflammation. Hence, inhibition or antagonism at other loci might be expected to contribute to anti-inflammatory activity.

As stated by Collier (1971), difficulties exist in accommodating the Vane hypothesis with a primary action of prostaglandins solely upon the inflammatory process. The widespread occurrence of the prostaglandins and their role as regulatory agents now appear well-established. Indeed, it has been proposed that prostaglandins play an essential intermediate role in the actions of luteinizing hormone (LH) to stimulate cyclic AMP and progesterone formation in the mouse ovary (Kuehl et al, 1970). The fact remains, however, that relatively high levels of fluoroindomethacin (100 microgram/ml) have no inhibitory effect upon LH-induced cyclic AMP formation (Kuehl et al, 1971) or progesterone biosynthesis (Ham et al, 1971) in ovarian tissue. The nature of this apparent specificity is the subject of current studies.

In summary, we believe that the data presented here support the concept of Vane and his associates that the biological activity of indomethacin and other non-steroidal anti-inflammatory agents is related to their ability to inhibit prostaglandin biosynthesis involved in the inflammatory process. It is not possible, however, to exclude actions at other loci involved in the inflammatory response.

TABLE I

COMPARISON OF THE IN VITRO AND IN VIVO ACTIVITIES OF NON-STEROIDAL ANTI-INFLAMMATORY AGENTS

Group	Compound	Prostaglandin Synthetase ID(50) (microgram/ml)	Foot Edema* ED(50) (mg/kg)	Approximate† Daily Clinical Dose (mg)
Indomethacin Analogs	Fluoroindomethacin (a)	0.24	1.1	-
	Indomethacin	0.16	2.4	75-100
	Compound I (b)	0.16	4	200**
	Compound II (c)	0.70	4	200-400**
	Compound III (d)	0.80	30	1000-1500**
	(-) isomer of Cpd. III (d)	13.5	>50	-
Arylacetic Acids	Compound IV (e)	0.05	0.2	-
	(-) isomer of Cpd. IV (e)	>0.36	>2	-
	Ibuprofen	0.31	12	600-800
	Metiazinic Acid	0.60	25	1000-1500
	Fenclozic Acid	0.70	20	300-400**
	Naproxen	1.4	3	-
Fenamates	Flufenamic Acid	0.70	25	600
	Mefenamic Acid	0.50	75	500-1000
	Niflumic Acid	0.34	20	500-1000
Salicylates	Flufenisal	11	25	600**
	Desacetylflufenisal	0.80	25	-
	Aspirin	15	75	2500

TABLE I (CONTINUED)

Group	Compound	Prostaglandin Synthetase ID(50) (microgram/ml)	Foot Edema* ED(50) (mg/kg)	Approximate+ Daily Clinical Dose (mg)
Pyrazolones	Azapropazone	0.30	35	600-1200
	Phenylbutazone	3.8	60	300-400
	Oxyphenylbutazone	16	60	300-400
	Aminopyrine	21	-	300-400
Non-Acidic	Indoxole	0.37	100	-
	Thiabendazole	3.3	150	-
Miscellaneous	Probenecid	Inactive	Inactive	-
	Chromoglycate	Inactive	Inactive	-
	Chloroquine	Inactive	Inactive	-

(a) 1-(p-fluorobenzoyl)-5-methoxy-2-methylindole-3-acetic acid
(b) 1-(p-chlorobenzoyl)-5-dimethylamino-2-methyl-indole-3-acetic acid
(c) <u>cis</u>-1-(p-chlorobenzylidenyl)-5-methoxy-2-methyl-indene-3-acetic acid
(d) (+)-5-methoxy-2-methyl-1-(p-methylthiobenzyl)-indole-3-alpha-propionic acid
(e) (+)-3-chloro-4-cyclohexyl-alpha-methylphenyl acetic acid

*We thank C. A. Winter, E. A. Risley, and G. W. Nuss for permission to use their foot edema data.

**Experimental clinical dosage. Not a currently availble drug.

+cf. Shen (1972).

REFERENCES

Andersen, N. H., 1969, Preparative thin-layer and column chromatography of prostaglandins, J. Lipid Res. 10:316.

Bertelli, A., and Houck, J. C., (eds.), 1969, Inflammation Biochemistry and Drug Interaction. Williams and Wilkins Co., Baltimore.

Collier, H. O. J., 1971, Prostaglandins and aspirin, Nature 232:17.

Ferreira, S. H., Moncada, S., and Vane, J. R., 1971, Indomethacin and aspirin abolish prostaglandin release from the spleen, Nature New Biology 231:237.

Gorog, P., and Kovacs, I. B., 1970, The inhibitory effect of non-steroidal anti-inflammatory agents on aggregation of red cells *in vitro*, J. Pharm. Pharmac. 22:86.

Greaves, M. W., Sondergaard, J., and McDonald-Gibson, W., 1971, Recovery of prostaglandins in human cutaneous inflammation, Brit. Med. J. 2:258.

Ham, E. A., Cirillo, V. J., and Kuehl, F. A., 1971, (unpublished observations).

Kuehl, F. A., Humes, J. L., Tarnoff, J., Cirillo, V. J., and Ham, E. A., 1970, Prostaglandin receptor site: Evidence for an essential role in the action of luteinizing hormone, Science 169:883.

Kuehl, F. A., Humes, J. L., Cirillo, V. J., and Ham, E. A., 1971, Cyclic AMP and prostaglandins in hormone action, in "Advances in Cyclic Nucleotide Research, Vol. 1: International Conference on the Physiology and Pharmacology of Cyclic AMP," (P. Greengard, R. Paoletti, and G. A. Robison, eds.), Raven Press, New York, (in press).

Shen, T. Y., 1972, Perspectives in Nonsteroidal Anti-inflammatory Agents, Angewandte Chemie, (in press).

Smith, J. B., and Willis, A. L., 1971, Aspirin selectively inhibits prostaglandin production in human platelets, Nature New Biology 231:235.

Vane, J. R., 1971, Inhibition of prostaglandin synthesis as a mechanism of action for aspirin-like drugs, Nature New Biology 231:232.

Wallach, D. P., and Daniels, E. G., 1971, Properties of a novel preparation of prostaglandin synthetase from sheep seminal vesicles, Biochim. Biophys. Acta 231:445.

Willis, A. L., 1970, Identification of prostaglandin E(2) in rat inflammatory exudate, Pharmacol. Res. Comm. 2:297.

Winter, C. A., Risley, E. A., and Nuss, G. W., 1963, Anti-inflammatory and antipyretic activities of indomethacin, 1-(p-chlorobenzoyl)-5-methoxy-2-methylindole-3-acetic acid, J. Pharmacol. Exp. Ther. 141:369.

CONTRACTION OF ISOLATED CUTANEOUS VASCULAR SMOOTH MUSCLE AND ITS RESPONSE TO PROSTAGLANDINS

R. K. Winkelmann, W. M. Sams, Jr., and M. E. Goldyne

Dept. of Dermatology, Mayo Clinic
Rochester, Minn.

INTRODUCTION

The development of a technique for the preparation of continuous helical strips of smooth muscle from small arterioles showed that cutaneous vascular smooth muscles could be studied directly and that their responses to natural and pharmacological compounds are consistent (Bohr et al, 1961). The pharmacological responses are dependent on the species and organs selected for vessel study. The size of the vessel and even the region of the organ are important (Bohr, 1965). Such studies have provided the basis for a receptor theory of small coronary vessel response to catecholamine (Bohr, 1967). Similar studies of human cutaneous vessels have indicated the presence of alpha-receptors and are important in understanding the physiology and pathology of the skin (Winkelmann et al, 1970). Our purpose is to summarize the data obtained by this method and to show the effect of prostaglandins on isolated human cutaneous vascular strips.

METHODS

The samples of skin studied were obtained at surgery from the paw, ear, leg and flank of the dog, from the ear of the rabbit and from the breast of the human. Vessels that were 300 to 500 microns in outer diameter were dissected from the dermis and subcutaneous tissue. Helical strips were cut as outlined by Bohr et al (1961) and by Sams and Winkelmann (1967) and were mounted in a bath of physiological salt solution at 37°C that was equilibrated with O2 and CO2 (95:5 v/v). The composition (mM/l) of the physiological salt solution was: NaCl, 119.0; KCl, 4.7; KH2PO4, 1.18; MgSO4, 1.17; NaHCO3, 14.9; glucose, 5.5; sucrose, 50.0; CaCl2, 1.6; and edetate calcium-disodium (Versenate), 0.026. One end of each strip was attached with 6-0 silk suture to a fixed point, and the other end was attached to the movable arm of a force displacement transducer (Grass Model FT03C) (Fig. 1). Contractions were recorded with a polygraph (Grass Model 5) at 0.25 mm/sec. The muscle strip was equilibrated for 2 hr with a tension of 100 mg, during which time it was washed with physiological salt solution every 15 min. The concentration of drugs utilized is reported as the final concentration in the bath in g/ml. The duration of exposure to test substances was 1 min if the response was immediate, or up to 5 min if the response was slow. Dose-response curves were constructed by comparison to the response to 100 mM of potassium chloride.

RESULTS

Spontaneous Activity

Vascular smooth muscle has a spontaneous tone, as found by Uchida and Bohr (1969) in studying small vessels from the rat, rabbit, dog and monkey. However, the mesenteric vessels studied by them was without a spontaneous tone. Larger vessels had less tone than small ones. Most isolated vascular smooth muscle does not have a spontaneous phasic contractile activity, but Johansson and Bohr (1966) demonstrated that skin vessel strips from the paw of the dog have spontaneous activity. Passive stretch and potassium stimulation augmented the spontaneous activity, and high concentration of calcium reduced it. Our studies of dog paw vessels confirm Johansson and Bohr's findings, but the absence of spontaneous activity in skin vessel strips from other regions of the dog such as the ear, back and leg indicate that this spontaneous activity is related to the region and not to the species. An exception may be human skin, for which we have observed spontaneous activity in most of the vessel strips from breast and extremity skin (Winkelmann et al, 1970). The spontaneous activity of the human vessels is augmented by potassium chloride and at times by catecholamines.

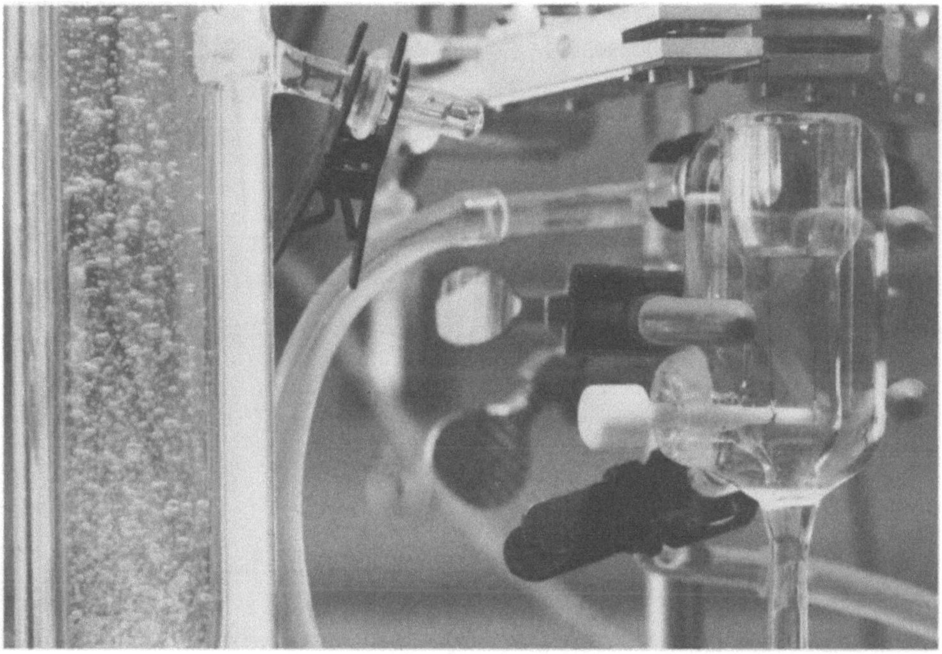

Figure 1. Apparatus for study of isolated cutaneous vascular smooth muscle strips, showing isolated constant temp chamber and strips attached to transducers by 6-0 silk thread. Oxygenated physiological solution reservoir at constant temp is at the left.

Physiological Responses

Blood vessel strips from the skin show a contractile response to the catecholamines, norepinephrine and epinephrine. They respond similarly to isoproterenol at an increase in concentration of 1,000 or more. No vaso-relaxing effect of isoproterenol was observed in skin vessels of the human, dog or rabbit. The use of an alpha-blocking agent, phentolamine (Regitine), completely blocked the response to catecholamine. Beta-blockade with propranolol had no effect. Such studies indicate that skin vessels have alpha-receptors. Studies by Bohr (1967) demonstrated alpha- and beta-receptors in skeletal muscle vessels and alpha- and beta-receptors in coronary vessels, except in very small coronary vessels, in which beta-activity predominated. No beta-receptor activity in skin vessels demonstrated even when continuous contraction with potassium chloride was utilized.

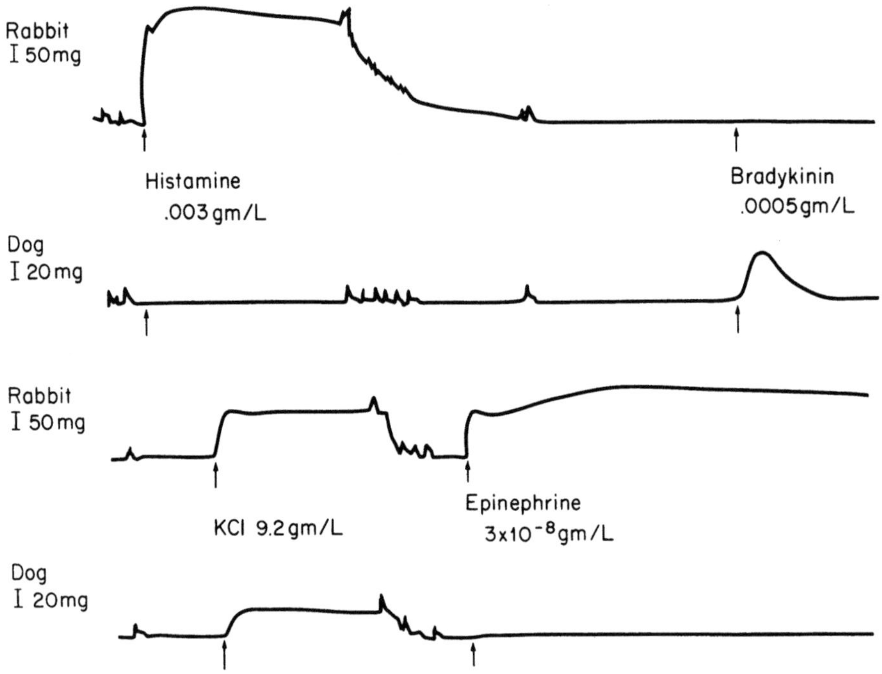

Figure 2. Ear skin vessel strips of the rabbit and dog in the same chamber, demonstrating response of rabbit tissue to histamine and of dog tissue to bradykinin. (From Winkelmann, R. K., Sams, W. M. Jr., and Bohr, D. F., 1969, Effect of nicotinate ester, acetylcholine, and other vasodilating agents on cutaneous and mesenteric vascular smooth muscle, Circ. Res. 25:687. By permission of the American Heart Association.)

Additional receptors that can be demonstrated in human skin vessel strips include those for serotonin and angiotensin II. The human skin vessel strips are responsive to ATP also but not to cyclic AMP. Caffeine (2×10^{-4} g/ml) depresses the response of cutaneous vascular smooth muscle to catecholamine, decreasing the strength of contraction and even at times increasing the threshold of catecholamine response. Theophylline did not have a specific effect. Preparations of vessel strips from the human breast skin do not respond to histamine, bradykinin, acetylcholine or methocholine in physiological or pharmacological concentrations (up to 10^{-3} g/ml).

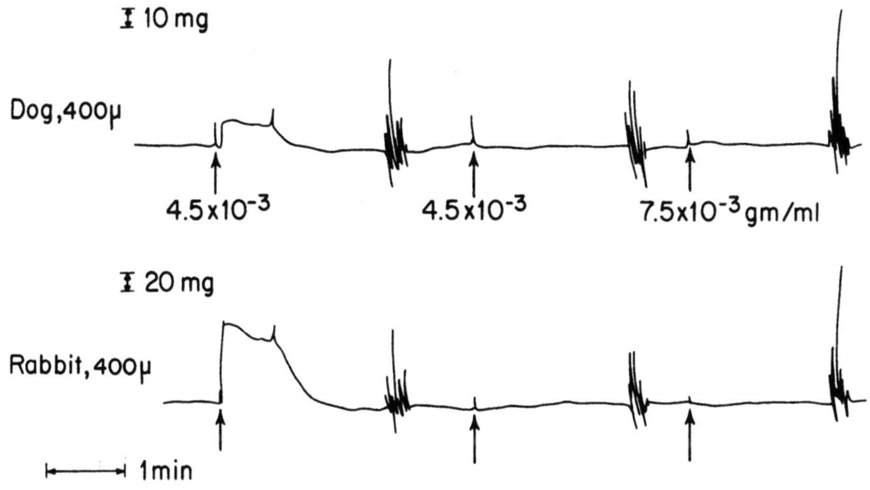

Figure 3. Direct effect of papaverine on ear skin vessel strips from the dog and rabbit. Papaverine causes relaxation of epinephrine contractile response (arrow) and continues to suppress the response subsequently (two arrows). Papaverine was added once at a concentration of 6×10^{-4} g/ml at the plateau of contraction.

Regional and Species Variations

Individuality of vascular smooth-muscle responses has been demonstrated for vessels of cerebral, pulmonary, coronary, mesentery, renal or skin origin (Bohr et al, 1961; Sundt and Winkelmann, unpublished). Pulmonary vessels respond to methacholine and to acetylcholine. Mesentery, renal and cutaneous vessels of the dog and rabbit have the same range of response. Species individuality also exists (Winkelmann et al, 1969). Thus, the vessels from all organs of the rabbit respond to histamine; the vessel strips of the dog and human fail to respond (Fig. 2). The vessel strips of the dog give a single tachyphylactic response to bradykinin; the vessel strips of the rabbit and human do not respond. At present, we have identified catecholamine, serotonin and angiotensin as general stimuli for isolated vascular smooth muscle; special stimuli for vascular tissue from special species or organs include histamine, bradykinin and acetylcholine.

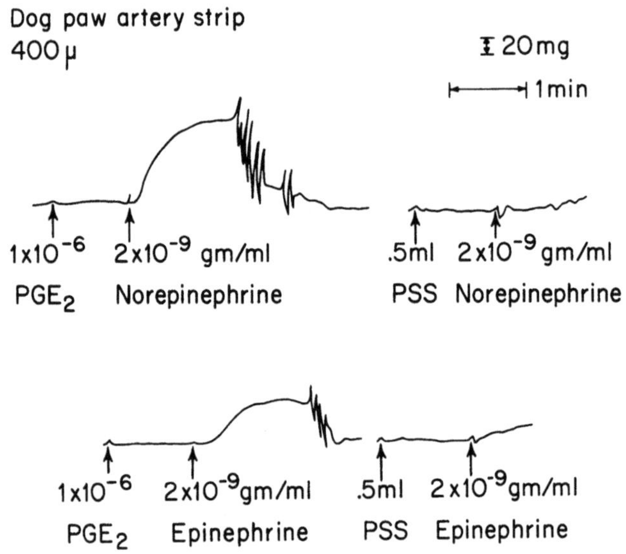

Figure 4. Potentiation of norepinephrine and epinephrine by subthreshold concentration of PGE(2) in a dog paw skin vessel strip.

Physical Factors

The nature of the contractile response is modified by the physical environment. Studies by Sams and Winkelmann (1969) indicated that the optimal temp for contraction of skin vascular smooth muscle is between 27°C and 30°C, a temp that may reflect the special requirements of the skin circulation. The depressed contraction at low temp was further depressed by low concentrations of calcium. Similarly, ultraviolet light changes the responses of the vascular smooth muscle to catecholamines (Sams and Winkelmann, 1969). The vessel strips immediately lose their contractile response to catecholamine after stimulus with ultraviolet light.

Pharmacological Agents

Pharmacological agents applied to vascular smooth muscle may produce alteration in response to a known stimulus or may cause a response themselves. Similarly, such agents may alter the spontaneous tone or the rhythmic activity. Finally, such agents may block a physiological or pharmacological response.

The corticosteroids applied to cutaneous vascular smooth muscle of the dog and rabbit may augment contraction to catecholamine or potassium chloride and, as Sams and Winkelmann (1967) showed, an increase in concentration may change the response to a pronounced relaxation. The corticosteroids may be roughly ranked in potency by the concentration necessary to produce relaxation effect, hydrocortisone being most effective, followed by methylprednisolone and then by dexamethasone. Human vessels relaxed to large corticosteroid doses, but no added contraction occurred at any strength.

Vasodilating agents can be classified according to their activity on isolated cutaneous vascular smooth muscle (Winkelmann et al, 1969). The type I vasodilator agent is one that is often found in the body, as histamine or acetylcholine, and one that causes either no reaction or a contraction of the isolated cutaneous vascular smooth muscle strip. Some species or organ variation in response to type of vasodilation should be expected. The type II vasodilator agent has a relaxing effect on contracted smooth muscle or prevents the contraction if administered first. Examples are nicotinic acid, papaverine (Fig. 3) and procaine. Their action is the same in all species and on vessel strips from all organs.

Prostaglandins

Cutaneous vascular smooth muscle from the paw of the dog responds to PGE(2) with contraction, usually at a concentration of 10^{-5} g/ml. The threshold is normally above 10^{-6} g/ml. Concentrations higher than 10^{-4} g/ml are not water soluble. Vessel strips from the skin of the dog ear or back show a significantly smaller number of responses than do vessel strips from paw skin. Only 18% of the ear vessel strips responded to any dose of PGE(2), whereas half of the vessel strips from the back skin and 70% of the vessel strips from the paw skin responded. In addition, only 10% of 40 human breast skin strips contracted at any concentration, and only one of eight human finger vessel strips responded. The alpha-blocking drug phentolamine and the beta-blocking drug propranolol did not change the contractile response to PGE(2). These responses of skin vessel strips from the dog and human are interpreted to indicate variability in species sensitivity in response to PGE(2). Regional differences also are

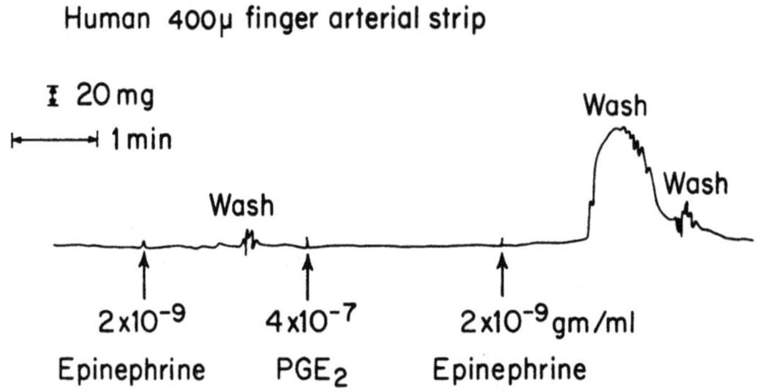

Figure 5. Potentiation of subthreshold epinephrine by subthreshold PGE(2) in a skin vessel strip of the human finger.

indicated because the dog vessel responses vary so greatly according to the area sampled.

The use of a subthreshold dose of PGE(2) potentiated the response of the vessel strip to catecholamine. Subthreshold amounts of epinephrine or norepinephrine with prostaglandin repeatedly produced dramatic contractions concurrently (Fig. 4). Either the prostaglandin or the epinephrine could be added first. Potentiation could be readily demonstrated 90% of the time in the skin from both the paw and ear of the dog, despite the difference in response to prostaglandin alone. Similarity in response to catecholamine-prostaglandin potentiation was also observed in the finger and breast skin of the human and was demonstrated in half of the specimens from each source (Fig. 5). The potentiation expressed itself in increase in the strength of contraction and also in decrease in the threshold of catecholamine required for a contraction. Dose-response curves often demonstrated a catecholamine threshold change of 10^{-2} g/ml or more (Fig. 6).

Study of the vessels from the dog paw revealed that low concentration of calcium inhibits the responses to prostaglandin and the prostaglandin-catecholamine potentiation. As the concentration of calcium is increased to normal isotonic values and then to twice normal values, the contraction to prostaglandin increases and the response to combined catecholamine-prostaglandin increases.

Studies with PGA(2) and PGA(2-alpha) have demonstrated a similar but less pronounced response of the skin vessel strips to PGE(2) (Fig. 7). Catecholamine

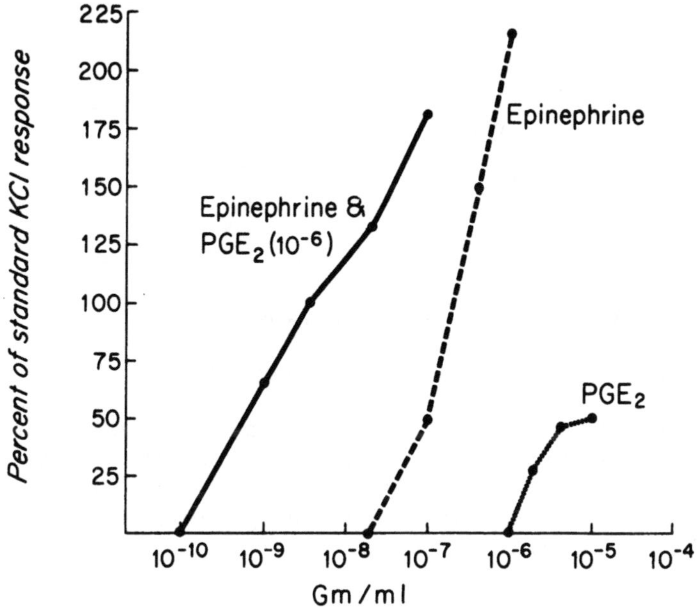

Figure 6. Dose-response curves demonstrating increase in force of contraction and shift to lower threshold produced in a dog paw vessel strip by subthreshold PGE(2).

potentiation with PGA(2) and PGF(2-alpha) occurs, but to a smaller degree than with PGE(2). The threshold values are unchanged or only changed slightly.

COMMENTS

Prostaglandins cause vasodilation when injected into the vascular tree. Similarly, the subcutaneous injection of prostaglandin causes a delayed, chronic vasodilation. The events in vivo are a complex interaction of nerve, ion and mediator response, and both Starr and West (1966) and De la Lande and Rand (1965) have demonstrated that nerve stimulation can change vasoconstrictor action into vasodilation. This may explain why prostaglandin causes isolated vessel strips to contract whereas the same agent causes vasodilation in the intact animal. However, vasoconstriction has been observed by direct action of prostaglandin on vessels in vivo. Stovall and Jackson (1967) and Anggard (1969) showed that vasoconstriction of nasal mucosa occurred in man and dog whether the prostaglandin was injected into the carotid artery or whether it was applied locally.

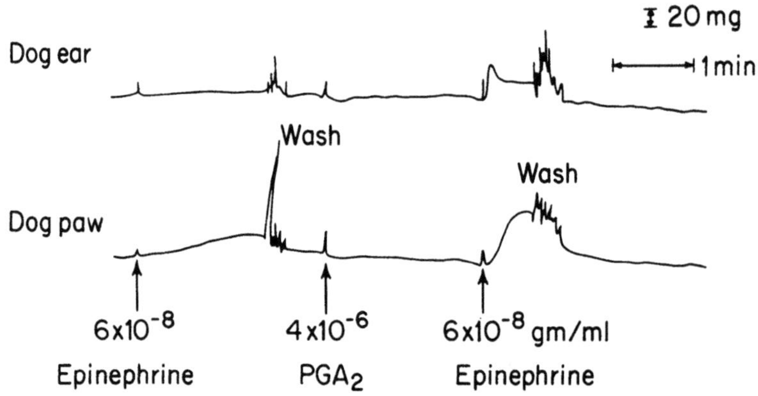

Figure 7. Potentiation of epinephrine produced in paw and ear skin vessel strips of the dog by PGA(2). Note that unresponsiveness of ear tissue to PGA(2) and to epinephrine alone does not preclude potentiation of epinephrine by PGA(2).

Strong and Bohr (1967) found a dual effect of prostaglandin when it was applied directly to the renal, mesenteric, or skeletal muscle artery strips of the dog. With dilute solutions (10^{-9} to 10^{-7} g/ml), relaxation occurred, but at higher concentrations contraction occurred. Their studies of rabbit aortic and coronary strips showed contraction only at any concentration. Our studies of cutaneous arteries of the human and dog did not show relaxation at any of the concentrations studied. We believe that relaxation or vasodilation _in vivo_ must have another and more complex explanation than the one of a direct effect on the vascular smooth muscle.

Prostaglandins activate the adenylate cyclase system, with increase in cellular cyclic AMP. This change has been measured as an effect of prostaglandin in the adrenal, thyroid and pituitary glands and in the corpus luteum, lung, spleen, diaphragm, kidney and bone (Ramwell and Shaw, 1970). Such a general response of adenylate cyclase to prostaglandins suggests that this could be the specific mechanism for its action. However, Ramwell and Shaw (1970) have postulated an effect of prostaglandin on the ion transport within epithelial membranes, the effect being based on prostaglandin's capacity to displace calcium ions. This is supported by studies with the erythrocytes and the uterus of the rat in which the adenylate cyclase activity and the prostaglandin effect can be separated. Our studies support this latter concept because increases in exogenous cyclic AMP or in endogenous cyclic AMP produced by caffeine or theophylline do not enhance contraction of vessel strips or the response to prostaglandin. Our studies also indicate a calcium dependence for prostaglandin response of cutaneous vascular smooth muscle as, indeed, for catecholamine and potassium chloride also. Therefore, an obvious relationship between the adenylate cyclase system and the prostaglandin effect on the response of cutaneous blood vessel smooth muscle cannot be supported. The prostaglandin effect could be related to ion movements, particularly calcium, which would explain most easily its potentiation of catecholamine responses, despite absence of cyclic AMP effect.

SUMMARY

Skin blood vessel strips from the human, dog, and rabbit give contractile responses to catecholamines, serotonin, and angiotensin II. Alpha-receptors for catecholamine can be demonstrated. Species and organ variations occur in the response of small vessel tissue to bradykinin (dog), histamine (rabbit) and acetylcholine (lung). A prostaglandin-contractile activity could be demonstrated on cutaneous vessels. This activity was most pronounced in paw vessels of the dog and rarely occurred in human skin. Potentiation of catecholamine contractions by PGE(2), PGA(2), and PGF(2-alpha) could be demonstrated. The prostaglandin receptor was not blocked by alpha- or beta-blocking agents. The prostaglandin response was related to concentration of calcium and was unrelated to exogenous or endogenous cyclic AMP.

REFERENCES

Anggard, A., 1969, The effect of prostaglandins on nasal airway resistance in man, Ann. Otol. Rhonol. Laryngol. 78:657.

Bohr, D. F., 1967, Adrenergic receptors in coronary arteries, Ann. N. Y. Acad. Sci. 139:799.

Bohr, D. F., 1965, Individualities among vascular smooth muscles in electrolytes and cardiovascular diseases. Vol. I. Edited by E. Bajusz, S. Basel, and A. G. Karger, pp 342-355.

Bohr, D. F., Goulet, P. L., and Taquini, A. C., Jr., 1961, Direct tension recording from smooth muscle of resistance vessels from various organs, Angiology, 12:478.

De la Lande, I. S., and Rand, M. J., 1965, A simple isolated nerve-blood vessel preparation, Aust. J. Exp. Biol. Med. Sci. 43:639.

Johansson, B., and Bohr, D. F., 1966, Rhythmic activity in smooth muscle from small subcutaneous arteries, Am. J. Physiol. 210:801.

Ramwell, P. W., and Shaw, J. E., 1970, Biological significance of the prostaglandins, Recent Prog. Horm. Res. 26:139.

Sams, W. M. Jr., and Winkelmann, R. K., 1967, Effect of corticosteroids on isolated vascular smooth muscle, J. Invest. Dermatol. 49:519.

Sams, W. M. Jr., and Winkelmann, R. K., 1969, The effect of ultraviolet light on isolated cutaneous blood vessels, J. Invest. Dermatol. 53:79.

Sams, W. M. Jr., and Winkelmann, R. K., 1969, Temperature effects on isolated resistance vessels of skin and mesentery, Am. J. Physiol. 216:112.

Starr, M. S., and West, G. B., 1966, The effect of bradykinin and anti-inflammatory agents on isolated arteries, J. Pharm Pharmacol. 18:838.

Stovall, R., and Jackson, R. T., 1967, Prostaglandins and nasal blood flow, Ann. Otol. Rhinol. Laryngol. 76:1051.

Strong, C. G., and Bohr, D. F., 1967, Effects of prostaglandins E(1), E(2), A(1) and F(1-alpha) on isolated vascular smooth muscle, Am. J. Physiol. 213:725.

Sundt, T. M., and Winkelmann, R. K., Responses of cerebral and other vascular tissue to pharmacological stimuli, (unpublished data).

Uchida, E., and Bohr, D. F., 1969, Myogenic tone in isolated perfused vessels: occurrence among vascular beds and along vascular trees, Circ. Res. 25:549.

Winkelmann, R. K., Sams, W. M. Jr., and Bohr, D. F., 1969, Effect of nicotinate ester, acetylcholine, and other vasodilating agents on cutaneous and mesenteric vascular smooth muscle, Circ. Res. 25:687.

Winkelmann, R. K., Sams, W. M. Jr., and King, J. H., 1970, Human cutaneous vascular smooth muscle responses to catecholamines, histamine, serotonin, bradykinin, angiotensin, and prostaglandins (abstract), J. Clin. Invest. 49:103a.

DISCUSSION

AMER: If you don't think cyclic AMP is involved in your system, how do you explain the effects of papaverine?

WINKELMANN: I have no way of explaining the response to papaverine except to say that it is apparent that these type-2 vasodilators, nicotinate, procaine, and papaverine, actually affect the vascular contractions so definitely that they will prevent production of contractions. Under these circumstances the vasodilators substances actually turn off the contraction capacity totally. Perhaps this is related to adenylate cyclase, but I know of no specific evidence on isolated tissue for this.

AMER: Is the weight of the tissue you are dealing with great enough to determine the levels of cyclic AMP?

WINKELMANN: The amount of tissue we are using is less than a milligram, so we cannot determine cyclic AMP levels in it. We have to approach the problem by determining the exogenous and endogenous effects of increased cyclic AMP concentrations. To date, no increase in contractions have been induced by such an increase.

GOTH: Did I understand you to say that human blood vessels do not have a receptor for histamine?

WINKELMANN: That's what I said.

GOTH: How do you explain the localized area of redness when you inject histamine interdermally?

WINKELMANN: I don't think that necessarily reflects the response of a vascular smooth muscle. As I indicated, all of the studies by Majno and the others in the last 5 or 10 years have indicated that this is a venular permeability phenomena, and I don't think the axon flare part of it is related to an arteriolar change. There is no reason to believe that what we have seen here is inconsistent with the clinical picture.

BUCKLES: Where did you get the specimens?

WINKELMANN: We get surgical specimens. The patients are either having orthopedic surgery or having breast amputations at the present time. The dogs are exsanguinated. The human vessels are placed in physiological saline in the refrigerator at 4°C overnight and the saline solution is changed to avoid anesthetic effects where these responses all disappear, with the exception of ATP and potassium chloride with ether anesthesia. We do have receptors that we can change with anesthesia.

BUCKLES: Do we know if vascular smooth muscles produce prostaglandin?

WINKELMANN: Not so far as I know.

RAMWELL: Clarke in our laboratories showed that rabbit aorta will synthesize prostaglandins from arachidonic acid.

WINKELMANN: It would be very interesting to see, and we are working on a radioimmunoassay for prostaglandin right now. It would also be interesting to see if this tissue can convert tritiated arachidonic acid to prostaglandin.

BUCKLES: Are your samples maintained at the normal oxygen tensions?

WINKELMANN: Yes. The reservoir is saturated with oxygen and the solution is moved in during the course of the experiment. Oxygen is maintained at approximate normal tension.

BUCKLES: Recently there has been quite a bit of data to show that so much oxygen is lost by diffusion into the walls of large arteries that in fact the average oxygen tension in arterials is down to the order of 60 mm Hg. This changes the responsiveness quite a bit. Have you tested this at low oxygen?

WINKELMANN: No, we haven't. Dr. Bohr has done this and came to 37°C as a reasonable circulating fluid temp. The more significant thing actually might be temp of contraction. One of the things we have found when we studied the ambient temp, is that the optimum temp of the contraction was 27-30°C in the skin vessels. That's quite logical and that's much closer to what you might anticipate for surface vessels.

STRONG: Detar and Bohr (1968, Am. J. Physiol. 214:241) looked at the oxygen requirements of vascular smooth muscle. 100 mm Hg was the threshold below which catecholamine-induced contractions began to decrease linearly with decrease in O2 tension. We have found that the oxygen requirements for prostaglandin-induced contractions of isolated rabbit aorta smooth muscle was greater than that for catecholamine-induced contractions.

KLIGMAN: I have a technical question because I am concerned about the question of why histamine does not cause contraction or relaxation. You have a strip of muscle and it is attached at both ends, but in its normal situation, a muscle doesn't contract this way.

WINKELMANN: That's why we cut the vessels in a helical pattern, so that you have a continuous muscle strip. Whenever you find an instance where acetylcholine will not make the tissue respond, and yet you inject it into the skin and you get a reddening, the question is, "What happened?". The answer is that in living tissue we have a tonus produced by catecholamines and the normal mediators that are already present, we have the tonus produced by the nerve, ion concentration, pH and oxygenation. All of these things affect the system, and acetylcholine can produce its effect through many of these systems. What we've done is isolate the vascular smooth muscle machinery and found that it did not have a surface receptor. The effect of a material like histamine has to be explained at another level. We know that the venules respond, and it seems to me that this should be enough.

BLOCH: As an anatomist I would like to make a slight emendation. You are not dealing only with smooth muscle in the aorta, but also with other tissues, that can contract elastic fibers. What you can say is that a strip of blood vessel is contracting.

INTERACTIONS OF PROSTAGLANDIN E(1) AND

CATECHOLAMINES IN ISOLATED VASCULAR SMOOTH MUSCLE

C. G. Strong and J. T. Chandler

Mayo Clinic and Mayo Foundation
Rochester, Minn.

INTRODUCTION

Prostaglandin E compounds are potent vasodepressors in situ, decreasing blood pressure and vascular resistance. Von Euler (1936) showed that prostaglandin lowers blood pressure and induces vasodilatation in the perfused hindlimb of the cat. Lee et al (1965) concluded that the vasodepressor action of PGE(1) was attributable to a direct relaxation effect on peripheral arteriolar beds. Weiner and Kaley (1969) have shown that PGE(1) has a direct transient vasodilator effect on muscular microvessels and another independent effect whereby it antagonizes the vasoconstrictor action of angiotensin, vasopressin, epinephrine, and norepinephrine. Hedwall et al (1970) also demonstrated that the direct venodilator effect of PGE(1) could be dissociated from its inhibition of venoconstrictor responses to nerve stimulation, norepinephrine, and angiotensin.

Strong and Bohr (1967) found that whereas PGE(1) in low concentrations caused relaxation of small isolated strips (200 to 1000 microns outer diameter [o.d.]) of canine skeletal muscle, mesentery, and renal arteries, it paradoxically caused contraction of isolated rabbit aortic strips. Khairallah et al (1967) also showed that PGE(1) caused direct contraction of isolated rabbit aortic strips and that the subcontractile concentration of PGE(1) (10^{-9} g/ml) potentiated the contractile effects of angiotensin, serotonin, and vasopressin but not those of norepinephrine.

The present study was done to evaluate factors modifying the direct effects of PGE(1) on isolated rabbit aortic and mesenteric arterial smooth muscle, and those modifying the indirect effects of PGE(1) on catecholamine-induced contractions of these differing vascular smooth muscle models. For comparison, the influence of PGE(1) on vascular tone induced by potassium, calcium, and barium was investigated.

METHODS

New Zealand white rabbits were killed by neck stroke. Helical strips (2 mm x 10 mm) were cut from their aortas. Strips of smooth muscle from small mesenteric arteries (200 to 600 microns o.d.) were prepared under a dissecting microscope; the artery segment was slipped onto a fine stainless steel rod, the rod was mounted in a Petri dish containing physiological salt solution, and a helical strip approximately 5 mm long and 200 microns wide was cut. Fine (6-0) waxed silk thread was tied to both ends of the strips for mounting in 3 ml muscle baths filled with physiological salt solution (PSS) of the following composition (mM/l): NaCl, 119; KCl, 4.7; KH_2PO_4, 1.18; $MgSO_4$, 1.17; $NaHCO_3$, 14.9; dextrose, 5.5; sucrose, 50; $CaCl_2$, 1.6; and calcium disodium versenate, 0.026. The versenate was added to chelate possible traces of heavy metals that catalize the auto-oxidation of catecholamines (Bohr and Johansson, 1966).

The PSS was aerated with gas mixtures of O_2:CO_2 (95:5 v/v), or N_2:CO_2 (25:5 V/V). Constant bath temp was adjustable over the range of 27°C to 47°C. The bath was exposed to light of constant intensity.

One end of each strip was tied to a metal rod connected to an isometric tension transducer (Grass force-displacement transducer FT03), and the other end was tied within the bath to a fixed point that could be adjusted to change the passive tension of the strip. Each strip was stretched to approximately 150% of its resting length. The preparations were allowed to equilibrate in the bath for 1 to 2 hr. The strips were given repeated "standard stimulation" by increasing the bath concentration of KCl to 50 mM and, after 5 min, replacement of the bath fluid with PSS. When the "standard response" of a strip had thus been determined, it was used thereafter as a point of comparison for subsequent responses by that strip. The isometric tension of the strips was recorded by an ink-writing oscillograph (Grass 4-channel polygraph).

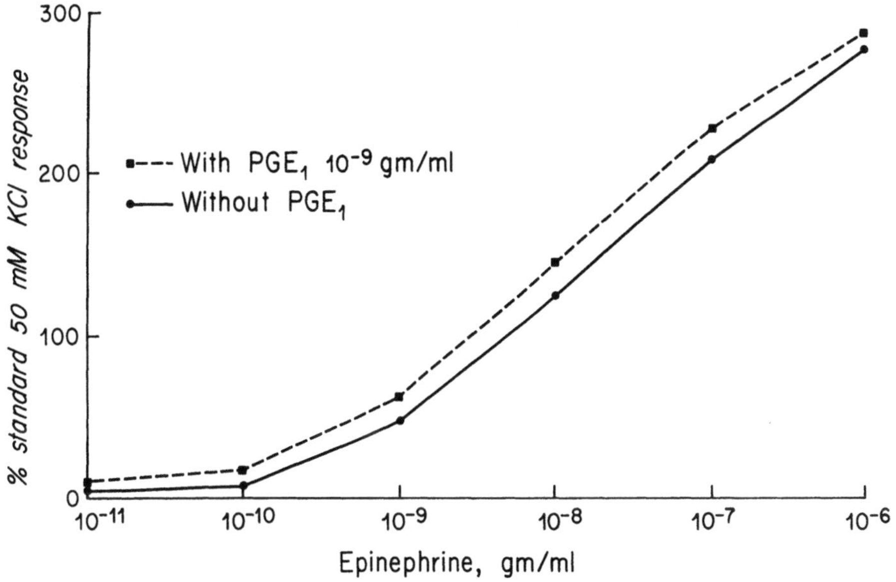

N (pairs)	4	12	28	22	19	10
$\bar{d}$	6.70	10.08	15.47	20.00	19.74	30.00
SE$\bar{d}$	3.72	2.85	2.55	5.40	8.30	25.33
p	0.16	0.02	0.001	0.001	0.03	0.28

Figure 1. Potentiation of epinephrine-induced contractions of isolated rabbit aortic smooth muscle by PGE(1) (10^{-9} g/ml). Mean responses are displayed on the graph. The mean differences for the paired observations (d), the standard errors of these mean differences (SEd), and the P value derived from the two-sided t test (Dixon and Massey, 1969) are tabulated for each concentration of epinephrine. The subcontractile concentration of PGE(1) (10^{-9} g/ml) significantly potentiated the contractile responses to epinephrine (10^{-10}, 10^{-9}, 10^{-8}, and 10^{-7} g/ml). (From Chandler, J. T., and Strong, C. G., The actions of prostaglandin E(1) on isolated rabbit aorta, submitted for publication.)

The potassium-depolarizing solution used had the following composition (mM/l): K2SO4, 100; KH2PO4, 1.18; MgSO4, 1.17; NaHCO3, 14.9; and dextrose, 5.5.

Mesenteric artery strips bathed with depolarizing solution containing 100 mM potassium but no calcium contracted initially and then relaxed spontaneously. The strips were given active tension with a bath concentration of calcium or barium (0.75 mM) and tested for relaxation by PGE(1).

The following drugs were used: epinephrine (Adrenalin), norepinephrine (Levophed), and phentolamine hydrochloride (Regitine). The crystalline PGE(1) was dissolved in the bathing salt solution at 10^{-4} g/ml and used immediately or frozen and stored. All concentrations refer to final bath concentrations in the muscle chamber and, except for electrolytes, are expressed as g/ml.

RESULTS

Effect of PGE(1) on Catecholamine-Induced Contractions of Isolated Smooth Muscle

Contractions by Epinephrine. The subcontractile PGE(1) concentration of 10^{-9} g/ml caused statistically significant potentiation of the contractile responses of aortic strips to epinephrine (10^{-10}, 10^{-9}, 10^{-8}, and 10^{-7} g/ml). Since observations were paired, in that each strip was measured with epinephrine and epinephrine plus PGE(1) (10^{-9} g/ml) at each concentration of epinephrine, analysis was done using the method of paired observations and a two-sided t test (Dixon and Massey, 1969). The mean responses, mean differences for the pairs, and standard error of these mean differences are shown in Fig. 1.

In contrast, PGE(1) caused statistically significant inhibition of contractile responses of 20 mesenteric artery (200 to 600 microns o.d.) strips to epinephrine (10^{-8}, 3×10^{-8}, and 10^{-7} g/ml) (Fig. 2).

Contractions by Norepinephrine. The subcontractile PGE(1) concentration (10^{-9} g/ml) caused statistically significant potentiation of contractile responses of rabbit aortic strips to norepinephrine (10^{-10}, 10^{-9}, and 10^{-8} g/ml) (Fig. 3). Responses to norepinephrine (10^{-11}, 10^{-7}, and 10^{-6} g/ml) were not potentiated.

In contrast, PGE(1) (10^{-9} g/ml) caused statistically significant inhibition of contractile responses of mesenteric artery strips ($n = 20$) to norepinephrine (10^{-8} to 3×10^{-7} g/ml) (Fig. 4).

Modifying Factors. Cooling the muscle bath from 37°C to 27°C decreased the contractile responses of aortic smooth muscle to epinephrine and norepinephrine by 20% to 50% but caused both relative and absolute increases in the potentiation produced by PGE(1) (10^{-9} g/ml) ($n = 20$) (Fig. 5).

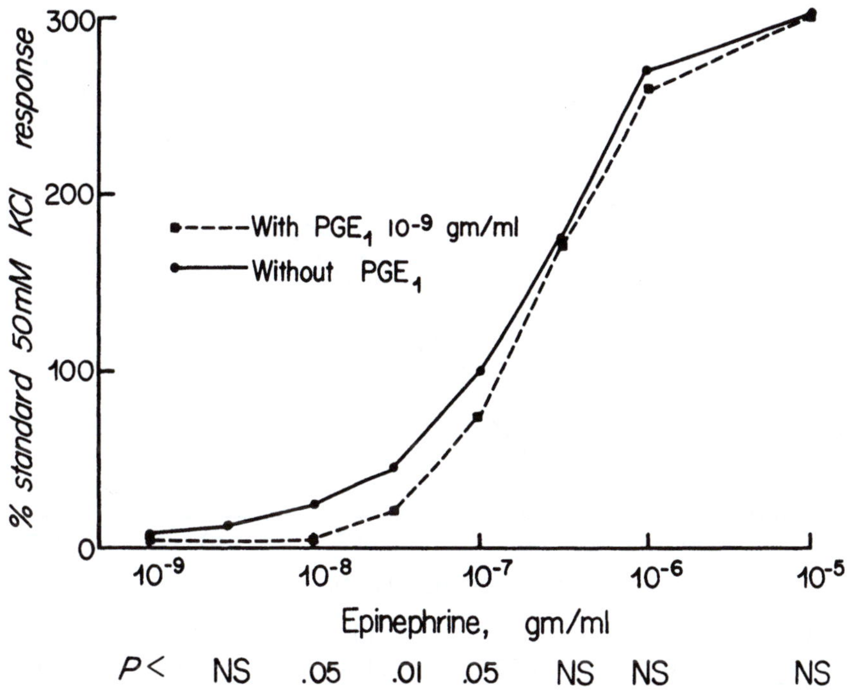

Figure 2. Inhibition of epinephrine-induced contractions of 20 isolated strips of rabbit mesenteric artery smooth muscle by PGE(1) 10^{-9} g/ml. Mean responses are displayed. The P values derived from the two-sided t test are tablulated for each concentration of epinephrine. PGE(1) (10^{-9} g/ml) significantly inhibited the contractile responses to epinephrine (10^{-8}, 3 x 10^{-8}, and 10^{-7} g/ml).

At 37°C, PGE(1) (10^{-9} g/ml) did not cause potentiation of contractions induced by norepinephrine, (10^{-6} g/ml). However, on cooling to 27°C, the tension induced by norepinephrine (10^{-6} g/ml) was reduced to a mean of 45% of that at 37°C, and at this lower temp PGE(1) (10^{-9} g/ml) potentiated these norepinephrine-induced contractions by an average of 20% (n = 8).

Similarly, when the contractile responses of the strips to norepinephrine (10^{-7} g/ml) were reduced 75% by partial alpha-adrenergic blockade with phentolamine (10^{-13} g/ml), the potentiation by the subcontractile concentration of PGE(1) (10^{-9} g/ml) was increased by 100% (n = 8). However, when the response to norepinephrine (10^{-7} g/ml) was completely blocked by phentolamine (10^{-5} g/ml), PGE(1) (10^{-9} g/ml) did not cause potentiation or contraction.

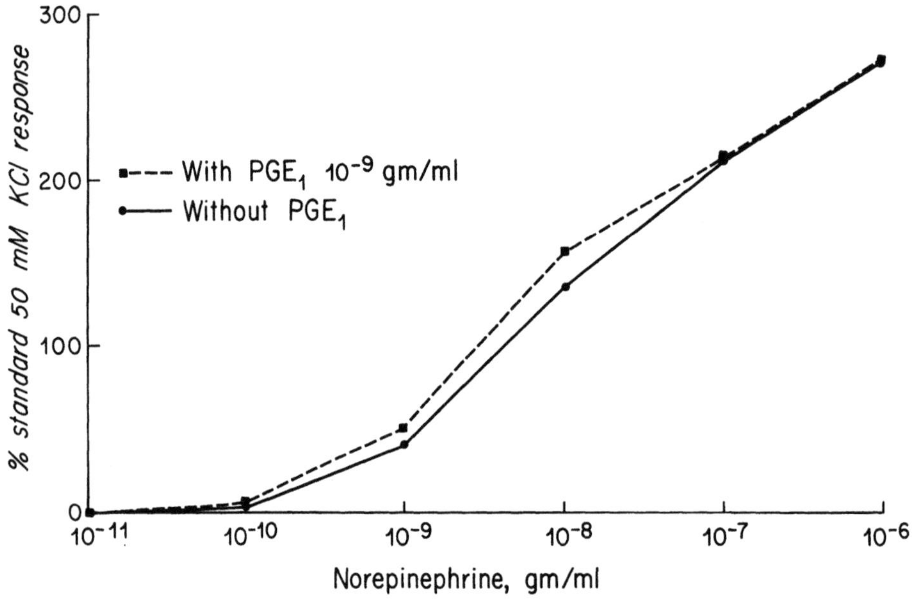

N (pairs)	12	20	10	12	5
$\bar{d}$	3.50	10.95	21.50	1.417	1.60
SE$\bar{d}$	1.38	1.61	7.23	2.811	1.60
p	0.04	0.001	0.02	0.62	0.38

Figure 3. Potentiation of norepinephrine-induced contractions of isolated strips of rabbit aortic smooth muscle by PGE(1) 10^{-9} g/ml. This subcontractile concentration of PGE(1) caused statistically significant potentiation of contractile responses to norepinephrine (10^{-10}, 10^{-9}, and 10^{-8} g/ml). (From Chandler, J. T., and Strong, C. G., The actions of PGE(1) on isolated rabbit aorta, submitted for publication.)

Effect of PGE(1) on Potassium-Induced Contractions of Aortic Smooth Muscle

Near maximal contractions induced by the addition of 50 mM KCl to the bath were not potentiated by 10^{-9} g/ml of PGE(1). As the level of the initial contraction was lessened by decreasing the concentration of KCl to 20 mM, 10^{-9} g/ml of PGE(1) caused increasingly greater potentiation. The potentiated responses never exceeded the response, which was maximal for potassium-induced contractions and which was elicited by potassium concentrations greater than 50 mM.

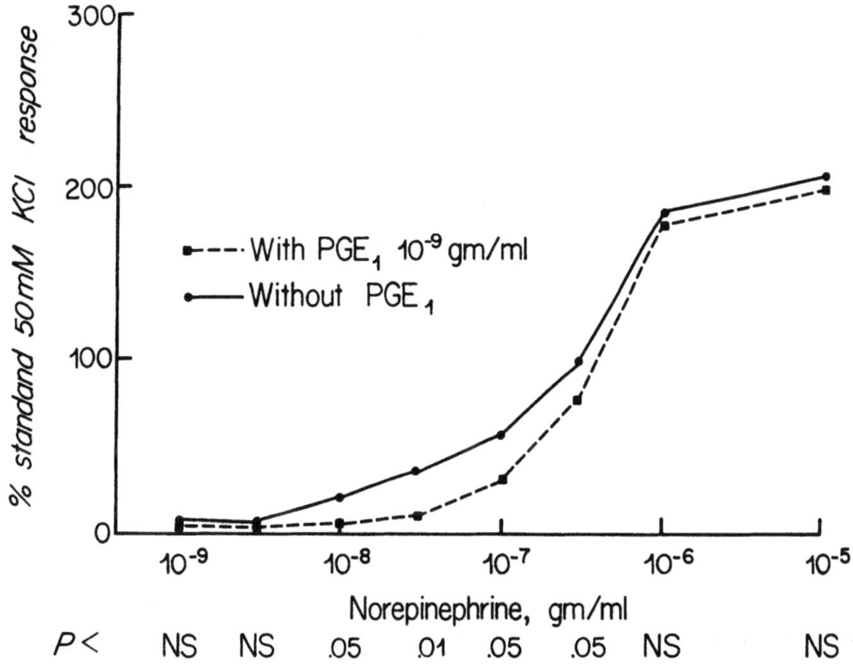

Figure 4. Inhibition of norepinephrine-induced contractions of 20 isolated strips of rabbit mesenteric artery smooth muscle; PGE(1) (10^{-9} g/ml) significantly inhibited the contractile responses to norepinephrine (10^{-8}, 3 x 10^{-8}, 10^{-7}, and 3 x 10^{-7} g/ml.)

In eight experiments PGE(1) (10^{-9} g/ml) did not cause potentiation of contractions induced by KCl (50 mM) at 37°C. When the aortic strips were cooled to 27°C the contractile responses to KCl (50 mM) were reduced by a mean of 35% compared with those at 37°C. These lesser 27°C contractions were potentiated by PGE(1) (10^{-9} g/ml), an average of 30%.

Relaxation of Isolated Mesenteric Arterial Smooth Muscle by PGE(1)

The model proposed (Strong and Bohr, 1967) for study of relaxation of smooth muscle from small arteries was initially evaluated using strips given active tension with KCl (23 mM). Our study has demonstrated that when the strips are given active tension with norepinephrine (10^{-8}, 10^{-7}, and 10^{-6} g/ml), relaxation in response to PGE(1) is progressive and dose related. Fig. 6 shows the responses of 35 mesenteric artery strips to several concentrations of PGE(1). In the strips where tension is induced by addition of KCl (23 mM) to the bathing PSS, the response was biphasic, PGE(1) (10^{-9} and 10^{-8} g/ml) causing significant relaxation and PGE(1) (10^{-6} and 10^{-5} g/ml) causing contraction. In contrast, when the strips were given active tension with epinephrine or norepinephrine, PGE(1) caused progressively increased relaxation with increasing concentrations over the range 10^{-9} to 10^{-5} g/ml.

Decreasing the oxygen tension of the PSS from 400 mm Hg to 100 mm Hg (by aeration with (95:5, N_2:CO_2 v/v) had no effect on the contractions with norepinephrine or on the relaxation by PGE(1) (10^{-9} to 10^{-5} g/ml) (n = 8).

When 20 rabbit mesenteric strips were bathed with depolarizing solution containing K_2SO_4 (100 mM) and no calcium, the strips contracted and then spontaneously relaxed. Addition to the bath of Ca^{++} or Ba^{++} (0.75 mM) caused the strips to contract. When thus contracted, the strips were relaxed in a progressive and dose-dependent manner by PGE(1) (10^{-9} to 10^{-5} g/ml).

DISCUSSION

The potentiating, subcontractile range of PGE(1) concentrations was narrow, usually 1 to 3 x 10^{-9} g/ml. Direct contraction of aortic smooth muscle began at PGE(1) 10^{-8} g/ml. This differs from the threshold contracting concentration of PGE(1) 10^{-10} g/ml reported by Strong and Bohr (1967), but in their study the aortic smooth muscle strips had been given active tension with angiotensin (10^{-8} g/ml) or KCl (22-26 mM). As we view our present results, the lower PGE(1) concentrations reported by Strong and Bohr (1967) would be considered to be potentiating the contraction induced by angiotensin or KCl rather than having a direct contractile effect.

Khairallah et al (1967) found that the threshold concentration for direct contraction of aortic smooth muscle by PGE(1) was 3 to 5 x 10^{-9} g/ml, and that lower concentrations of PGE(1) (10^{-10} and 10^{-9} g/ml) had a potentiating effect on contractions by angiotensin, serotonin, and bradykinin. Our results differ from theirs in that they reported that PGE(1) (10^{-9} g/ml) failed to potentiate contraction of four aortic strips by norepinephrine (10^{-8} g/ml). Our demonstration of potentiation of norepinephrine-induced contractions by PGE(1) 10^{-9} g/ml differs from their observations for

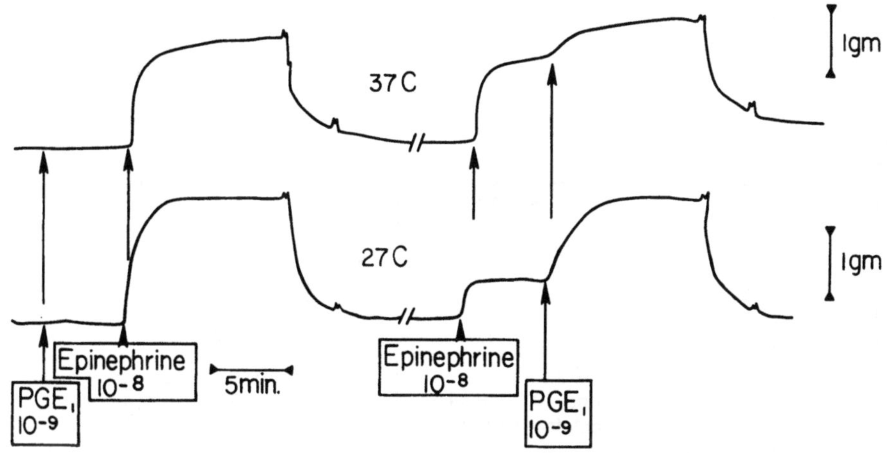

Figure 5. Effect of temp on potentiation of epinephrine-induced contractions by PGE(1) (10^{-9} g/ml). At both 37°C and 27°C, PGE(1) (10^{-9} g/ml) is a subcontractile concentration. Compared with responses at 37°C, the response of the same rabbit aortic strip to epinephrine (10^{-8} g/ml) is reduced at 27°C, but the potentiation is increased. (From Chandler, J. T., and Strong, C. G., The actions of prostaglandin E(1) on isolated rabbit aorta, submitted for publication.)

reasons that are not clear, since their reported drug concentrations, ionic conditions, temp, and method of vessel strip preparation resemble those reported herein.

Temp reduction from 37°C to 27°C reduces the direct contractile responses of aortic smooth muscle to PGE(1) as it does those in response to catecholamines. In contrast, the same temp reduction increases the potentiation by PGE(1) of contractions induced by catecholamines and potassium. This effect may not be a direct effect of temp since any factor that reduced the initial catecholamine or potassium-induced contraction produced this result. Parallel effects were observed with temp lowering, decreased concentration of agonist, and partial alpha-adrenergic blockade. In general, the lower the initial agonist-induced tension, the greater the potentiation by prostaglandin, both in absolute magnitude and relative to the initial contraction. Potentiation by PGE(1) was limited by the magnitude of the maximal contraction that could be elicited by the particular agonist. At maximal contraction by potassium, PGE(1) (10^{-9} g/ml) produced no

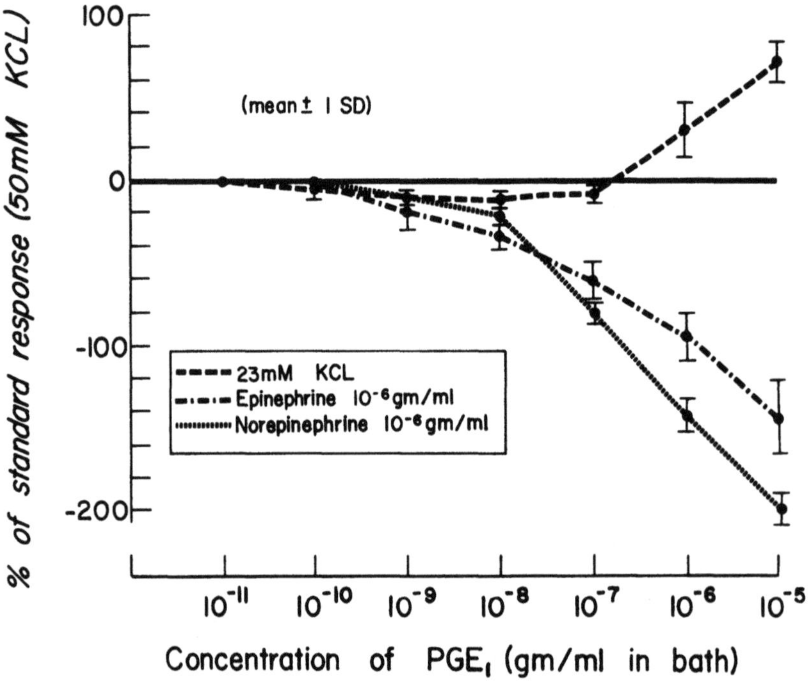

Figure 6. Responses of 35 mesenteric artery strips (200 to 600 microns o.d.) to PGE(1). Means and standard deviations are displayed on the graph. In strips given active tension by addition of KCl (23 mM) to the bathing PSS, concentrations of PGE(1) (10^{-9} and 10^{-8} g/ml) cause significant relaxation and concentrations of PGE(1) (10^{-6} and 10^{-5} g/ml) cause contraction. In contrast, when the strips are given active base-line tension with epinephrine or norepinephrine, PGE(1) causes progressive relaxation with increasing concentrations over the range 10^{-9} to 10^{-5} g/ml.

potentiation. As the potassium concentration was reduced, potentiation by PGE(1) (10^{-9} g/ml) appeared and increased, but the potentiated response was never greater than the maximal response to the agonist (potassium) alone. That this was not limited by the strength of the contractile mechanism was shown by the much greater (approximately double) response of the same strips to large concentrations of catecholamines.

Inhibition by PGE(1) (10^{-9} g/ml) of contraction of isolated mesenteric arterial smooth muscle by epinephrine (10^{-8} to 10^{-7} g/ml) and by norepinephrine (10^{-8} to 3×10^{-7} g/ml) parallels the _in vivo_ observations (Weiner and Kaley, 1969; Hedwall et al, 1970) of inhibition of vascular responsiveness to these catecholamines, nerve stimulation, and angiotensin.

Direct dose-dependent PGE(1)-induced relaxation of isolated mesenteric arterial smooth muscle given active base-line tension with epinephrine or norepinephrine differed from the biphasic response to PGE(1) when base-line tension was induced with KCl (23 mM). Furthermore, relaxation by PGE(1) was preserved when the strips were bathed in depolarizing solution containing KCl (100mM) and then given base-line tension with low concentrations of calcium or barium. These observations indicate a site of PGE(1) action that is intracellular and close to the site where barium (Bohr, 1964) may substitute for calcium in inactivating the troponin-tropomyosin ATPase inhibitory system (Ebashi and Ebashi, 1964; Hartshorn et al, 1969). The precise mechanism of action of PGE(1)-induced relaxation of vascular smooth muscle remains to be elucidated by techniques other than those employed in this study.

Relaxation of mesenteric vascular smooth muscle and inhibition of vasoconstrictor responsiveness by PGE(1) are physiologically more important than PGE(1)-induced contraction and potentiation of catecholamine-induced aortic smooth muscle contraction, since most in vivo studies have shown only a vasodepressor effect (Carlson and Oro, 1966; Weeks, 1969), reflecting dominance of resistance vessels over large arteries in those studies and the individuality of vascular smooth muscle from various sites and various orders of branching of the arterial tree (Bohr, 1967).

ACKNOWLEDGEMENTS

The work described in this paper was supported in part by Research Grant HE-12584 from the National Heart and Lung Institute, P.H.S.

The crystalline PGE(1) used in the studies was a gift of Dr. J. E. Pike, Department of Chemistry, the Upjohn Company, Kalamazoo, Mich.

REFERENCES

Bohr, D. F., 1964, Contraction of vascular smooth muscle, Can. Med. Assoc. J. 90:174.

Bohr, D. F., 1967, Individuality in smooth muscle control, in "Ureteral Reflux in Children," (J. F. Glenn, ed.), pp. 57-66, National Academy of Sciences - National Research Council, Washington, D. C.

Bohr, D. F., and Johansson, B., 1966, Contraction of vascular smooth muscle in response to plasma: Comparison with response to known vasoactive agents, Circ. Res. 19:593.

Carlson, L. A., and Oro, L., 1966, Effect of prostaglandin E(1) on blood pressure and heart rate in the dog: Prostaglandin and related factors 48, Acta Physiol. Scand 67:89.

Dixon, W. J., and Massey, F. J., Jr., 1969, "Introduction to Statistical Analysis," 3rd edition, McGraw-Hill Book Company, Inc., New York.

Ebashi, S., and Ebashi, F., 1964, A new protein component participating in the superprecipitation of myosin B, J. Biochem. (Tokyo) 55:604.

Hartshorne, D. J., Theiner, M., and Mueller, H., 1969, Studies on troponin, Biochim. Biophys. Acta 175:320.

Hedwall, P. R., Abdel-Sayed, W. A., Schmid, P. G., and Abboud, F. M., 1970, Inhibition of venoconstrictor responses by prostaglandin E(1), Proc. Soc. Exp. Biol. Med. 135:757.

Khairallah, P. A., Page, I. H., and Turker, R. K., 1967, Some properties of prostaglandin E(1) action on muscle, Arch. Int. Pharmacodyn. Ther. 169:328.

Lee, J. B., Covino, B. G., Takman, B. H., and Smith, E. R., 1965, Renomedullary vasodepressor substance, medullin: Isolation, chemical characterization and physiological properties, Circ. Res. 17:57.

Strong, C. G., and Bohr, D. F., 1967, Effects of prostaglandins E(1), E(2), A(1), and F(1-alpha) on isolated vascular smooth muscle, Amer. J. Physiol. 213:725.

Von Euler, U. S., 1936, On the specific vaso-dilating and plain muscle stimulating substances from accessory genital glands in man and certain animals (prostaglandin and vesiglandin), J. Physiol. (London) 88:213.

Weeks, J. R., 1969, The prostaglandins: biologically active lipids with implications in circulatory physiology, Circ. Res. 24:(Suppl. 1):123.

Weiner, R., and Kaley, G., 1969, Influence of prostaglandin E(1) on the terminal vascular bed, Amer. J. Physiol. 217:563.

DISCUSSION

BUCKLES: Have you tried any divalent cations, other than barium, that might act similarly? Is there any predeliction in your mind for barium?

STRONG: We chose barium because it has been suggested that diazoxide relaxes vascular smooth muscle by an action that opposes the intracellular effects of both ionic calcium and barium. We sought a parallel action of PGE(1).

KALEY: I wondered if you have tried PGE(2), and whether, in view of the findings of Khairallah et al (Khairallah, P. A., Page, I. H., and Turker, R. K., 1967, Arch. Int. Pharmacodyn. 169: (No. 2): 328), you evaluated the effects of PGE(1) on the angiotensin-induced contraction of aortic smooth muscle?

STRONG: No, we didn't look at the angiotensin in the aortic smooth muscle strip. We have looked at PGE(2), but not so extensively.

NAKANO: In reference to this vascular smooth muscle tone you have created with KCl, do you have anything to say about its relation to *in vivo* prostaglandin?

STRONG: When dealing with isolated vascular smooth muscles, we often see behavior that we would not see *in vivo*. For example, in 1967, we showed that prostaglandins caused contraction of coronary arterial smooth muscle, whereas *in vivo* this is not true. With respect to the model, I think it is fair to give the strip tone, because *in situ* the vessels have tone. Furthermore, unless the muscle strips are given active tension, it would not be possible to demonstrate relaxation *in vitro*.

KADOWITZ: If Dr. Winkelmann were to induce tonus in his cutaneous vessels with a background amount of phenylephrine, he'd see a beautiful relaxation to acetylcholine and histamine. The response to acetylcholine can be blocked by very small amounts of atropine, and the response to histamine can be blocked by very small amounts of antihistamines.

WINKELMANN: We have done this in the isolated system, and it does not reproduce the response that Dr. Kadowitz indicated. The same thing is true, of course, with Dr. Strong's work. We have done exactly what Dr. Strong has done, but in the skin vessels. And this is why we think they are important; the relaxation does not occur whether the tone is given with potassium chloride or with epinephrine or with any other previous tone-giving substance. We do not get the relaxation with any PGE(2) concentration from 10^{-9} to 10^{-5} M, so we think that this is an important difference. It may relate to the difference in the prostaglandin we have each studied; it may be an important regional or organ vascular difference.

IBERALL: Regarding the problem of the arterial pulse, as you make the transition down to 20 micron diameter vessels, you are now beginning to deal with the question in which both hydrodynamics and transport are intermingled, and in which we think the catecholamines are very strongly implicated. So the transport system that you are dealing with is very much mechanically influenced. Are you dealing with artifacts here, or can you shift your results to resemble anything that would approach the real physiological domain?

STRONG: There is no question that the passive stretch that you give the strips influences their responsiveness, and every effort is made to give them the optimal passive stretch. Optimal passive tension occurs with an approximate 40 to 50% increase in length.

IBERALL: Do you attempt to do anything to match the existing tethering that you have in various size tubes? Tethering is an important aspect of determining the pulse characteristic of the vascular bed. The existing tethering would give a bias to the system at various sizes of tubes *in vivo*. Thus, there are stresses on the tubes to start with. Do you try to match this kind of characteristic to assess the kind of dynamic response you will get in larger or smaller tubes?

STRONG: No, we don't.

CELLULAR ASPECTS OF PROSTAGLANDIN SYNTHESIS AND TESTICULAR FUNCTION

L. C. Ellis, J. M. Johnson, and J. L. Hargrove

Dept. of Zoology and Div. of Biochemistry
Utah State University
Logan, Utah

INTRODUCTION

It is somewhat surprising that testicular prostaglandin synthesis and the relationship of prostaglandins to testicular function have been neglected for so long though evidence for the existence of this class of hormones was first associated with human seminal fluid (Kurzrok and Lieb, 1930). Investigations in our laboratory on prostaglandin synthesis and their actions on the testis were initiated after preliminary studies showed that lipid peroxidation had a direct relationship to androgen synthesis (Ellis and Baptista, 1969) and possibly served some endocrine function in the testis. Reports by other workers (Samuelsson et al, 1966; Euler and Eliasson, 1967) that lipid peroxidation was involved in prostaglandin synthesis in other organs, added impetus to our studies.

Prostaglandins have now been isolated from rat testes (Carpenter and Wiseman, 1970) and the enzymes for inactivation of prostaglandins (e.g., 15-hydroxy prostaglandin dehydrogenase and prostaglandin-delta^{13}-reductase) have been observed in swine (Anggard et al, 1971) and rat (Nakano et al, 1971) testes. We report here on: the presence of prostaglandin synthetase in rat testes, the pituitary and adrenocorticol control of prostaglandin synthetase activity, and the relationship of androgen synthesis to lipid peroxidation and prostaglandin synthesis. We also present information about the action of prostaglandins on contractions of the rabbit testicular

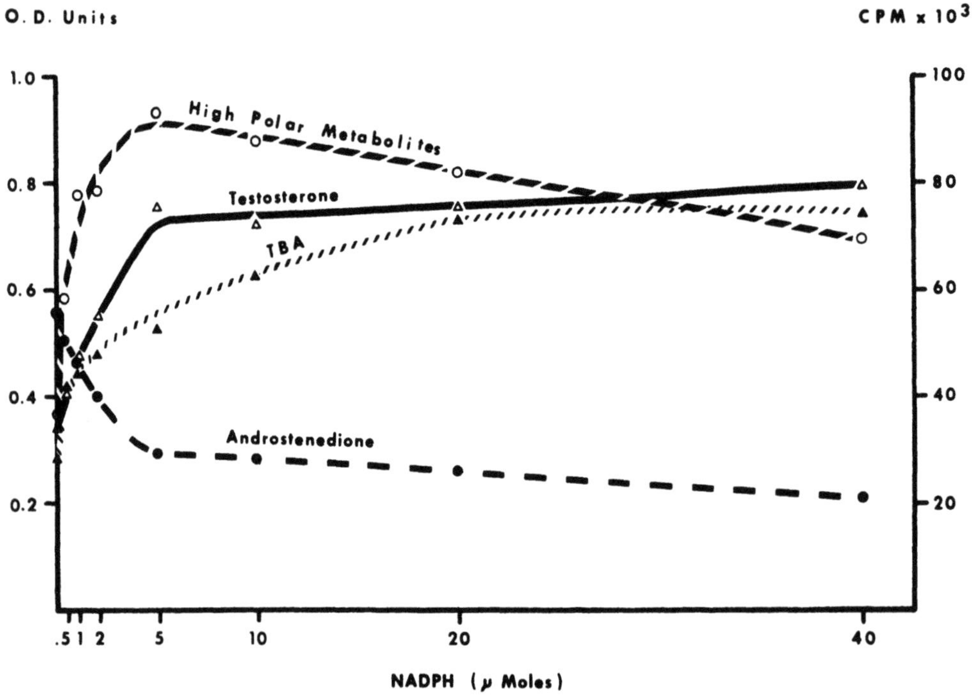

Figure 1. Lipid peroxidation and the biotransformation of androstenedione into testosterone and high-polar intermediates by rat testicular tissue _in vitro_. The incubation media consisted of 0.05 microcuries of androstenedione-4-^{14}C (58.0 millicuries/mM), 5 ml of a balanced phosphate buffer (pH 7.4), 20 microliters propylene glycol, 0.3 g of homogenized tissue, and increasing amounts of NADPH. The samples were incubated in air for 1 hr at 37.5°C.

capsule, and cellular implications of lipid peroxidation and prostaglandin synthesis as they relate to aging.

MATERIALS AND METHODS

Androgen synthesis was measured by using radioactively labeled steroid precursors (pregnenolone-7-alpha-^{3}H, progesterone-4-^{14}C, and androstenedione-4-^{14}C - New England Nuclear Corp) that were rigorously purified by paper chromatography (Ellis and Berliner, 1963, 1965) prior to use. The appropriate steroid precursors were incubated with either minced or homogenized rat testicular tissue as described in the legends and footnotes of the figures and tables. The resulting steroid metabolites were extracted, purified and quantified, as described in detail elsewhere (Ellis and Berliner, 1963, 1965).

Lipid peroxidation was measured by the TRA procedure (2-thiobarbituric acid assay) of Kitabachi (1967) with minor modifications. Minced rat testicular tissue was incubated with arachidonic acid, norepinephrine, and reduced glutathione (Nakano et al, 1971) for measurement of prostaglandin synthetase activity. The effects of prostaglandins on contractions of the smooth muscle were determined with a transducer (Statham) and a polygraph (Gilson) at 35°C in oxygenated Tyrode's solution (Hargrove et al, 1971; Johnson et al, 1971).

RESULTS

Lipid Peroxidation and Steroidogenesis

Adding NADPH to incubations of testicular tissue markedly increased lipid peroxidation and also enhanced the biotransformation of androstenedione into testosterone and other high-polar intermediates (Fig. 1). NADH similarly increased lipid peroxidation, but to a lesser extent than did NADPH (0.420 O.D. units for 4 mM NADPH vs 0.264 O.D. units for 4 mM NADH per 0.5 ml of incubation media). Similarly, 20 microliters of linoleic acid increased lipid peroxidation from 0.261 to 0.469 O.D. units per 0.5 ml of incubation media.

Addition of low concentrations of ascorbic acid to the incubation mixture (Fig. 2) increased lipid peroxidation, but higher concentrations markedly inhibited this phenomenon. Ascorbic acid at higher concentrations also inhibited the conversion of androstenedione into testosterone. Hydrogen peroxide at low concentrations increased lipid peroxidation, but decreased it when added at higher concentrations (Fig. 3); the conversion of androstenedione into testosterone was diminished by 1-5 microliters of 30% H_2O_2. The appearance of high-polar intermediates was increased when 1-100 microliters of H_2O_2 were added to the incubations.

D-alpha-tochopherol concentrates markedly reduced lipid peroxidation, but had only transient effects on the conversion of androstenedione into testosterone and high-polar intermediates. Melatonin at low concentrations reduced the conversion of pregnenolone and progesterone into testosterone, but increased the production of lipid peroxides.

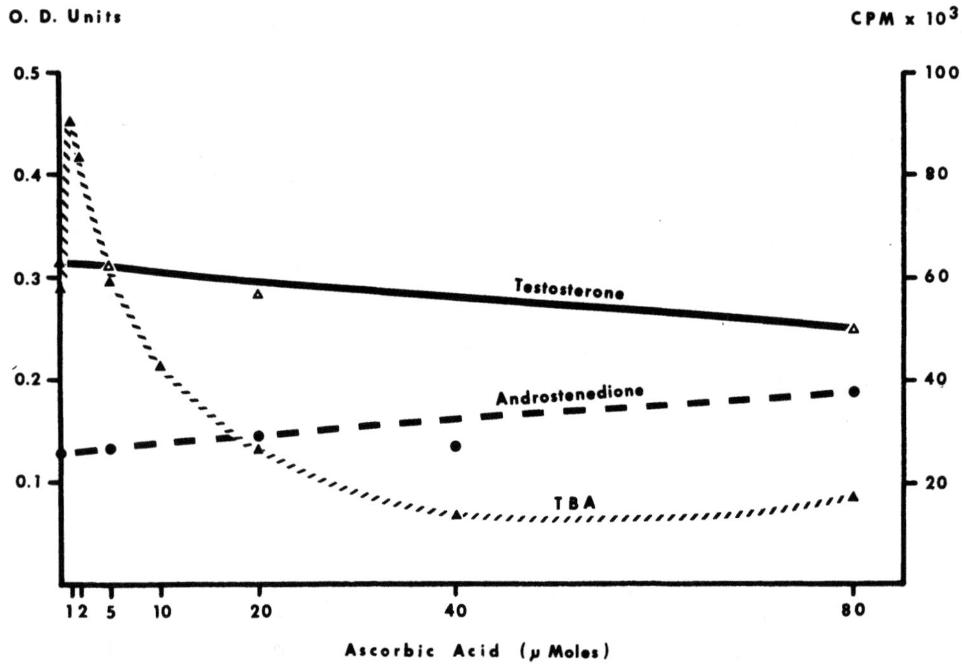

Figure 2. Lipid peroxidation and the biotransformation of androstenedione into testosterone by rat testicular tissue *in vitro*. The incubation mixture consisted of 0.05 microcuries of androstenedione-4-^{14}C (58.0 millicuries/mM), 5 ml of a balanced phosphate buffer (pH 7.4), 20 microliters of propylene glycol, 0.3 g of homogenized tissue, and increasing amounts of ascorbic acid. The samples were incubated in air for one hr at 37.5°C. (Ellis and Baptista, 1969).

Hypophysectomy (Table I) decreased testicular weights and lipid peroxidation, but when a combination of growth hormone and ACTH were injected into hypophysectomized animals, there was an increase in both testicular weight and lipid peroxidation. When minced testicular tissue and teased-tubular preparations were evaluated for androgen synthesis from pregnenolone and progesterone, and for lipid peroxidation (Table II), both phenomena had the same distribution with respect to the seminiferous tubules and the parenchyma of the testis.

In Table III are reported the effects of hypophysectomy and adrenalectomy on prostaglandin synthetase activity in rat testicular tissue. 5 weeks after hypophysectomy there was a reduction in prostaglandins synthesized; this effect was not seen at 3

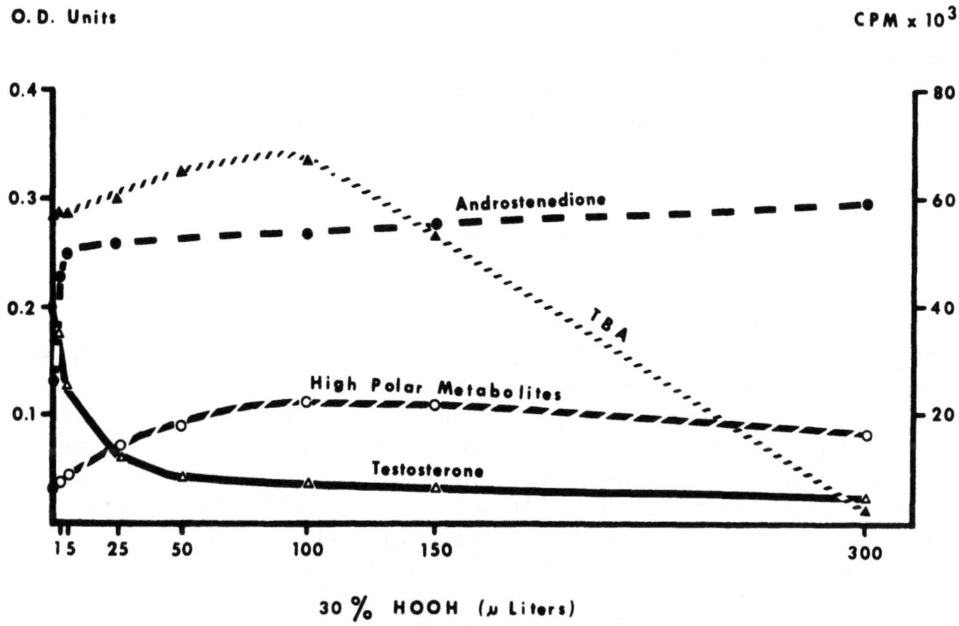

Figure 3. Lipid peroxidation and the biotransformation of androstenedione into testosterone and high-polar steroid intermediated by rat testicular tissue *in vitro*. The incubation mixture consisted of 0.05 microcuries of androstenedione-4-^{14}C (58.0 millicuries/mM), 5 ml of a balanced phosphate buffer (pH 7.4), 20 microliters propylene glycol, 0.3 g of homogenized tissue, and from 1 to 300 microliters of 30% hydrogen peroxide. The samples were incubated in air for one hr at 37.5°C. (Ellis and Baptista, 1969).

weeks past hypophysectomy. 10 days following adrenalectomy there was a 75% reduction in testicular prostaglandin synthetase activity.

In an aging study, we determined that a positive relationship existed with respect to testicular development (size), androgen synthesis, and lipid peroxidation. A marked increase in lipid peroxidation was observed between 28 and 34 days of age. This increase was associated with a decrease in testosterone synthesis but not androstenedione synthesis. The testosterone:androstenedione ratio changed from less than one prior to 20 days of age to greater than one after 24 days of age. Incubation of rat testicular tissue with PGA(1) (Fig. 4), resulted in an accumulation of progesterone and 17-alpha-hydroxyprogesterone and a decrease in androstenedione, testosterone and total androgens. Lipid peroxidation was increased by PGA(1), while PGE(1) and PGF(1-alpha) promoted an accumulation of 17-alpha-hydroxyprogesterone, and PGE(1) decreased testosterone production. Both PGE(1) and PGF(1-alpha) diminished lipid peroxidation.

Effect of Prostaglandins on Rabbit Testicular Capsules

PGE(1) decreases the rate and overall tone of autorhythmic smooth muscle preparations. Moreover, this response was dose dependent with contractions effaced with higher concentrations of the compound. The solvent, when added to the preparation in concentrations greater than those used in this study, had only minor effects on the rate of contractions. PGE(1) also decreased rate and amplitude of the contractions without altering tone. In dose-dependent reactions, PGF(1-alpha) increased tone, the rate of contractions, and the rhythmicity of spontaneous contractions. PGF(1-alpha) also increased the tone of the preparation in the absence of contractions and initiated rhythmic contractions in these quiescent preparations.

DISCUSSION

The following observations suggest that steroidogenesis, lipid peroxidation and prostaglandin synthesis in the testis are related: (a) both lipid peroxidation and steroid biotransformations were NADPH dependent; (b) these phenomena closely followed one another during the early aging process; (c) both had the same relative distribution in the testis with respect to the seminiferous tubules and the parenchyma; (d) prostaglandin synthesis is known to involve lipid peroxidation (Samuelsson et al, 1966; Euler and Eliasson, 1967); (e) prostaglandin synthetase activity is diminished following hypophysectomy as were lipid peroxidation and androgen synthesis; (f) prostaglandins have been isolated from rat testis (Carpenter and Wiseman, 1970); (g) rat testes can metabolize prostaglandins (Anggard et al, 1971; Nakano et al, 1971). Of possible significance is the observation (Fig. 4) that, of the prostaglandins tested, only PGA(1) appreciably modified lipid peroxidation or androgen synthesis. These effects were probably due more to end-product inhibition than to physiological effects since large quantities of the prostaglandins were required to observe an effect.

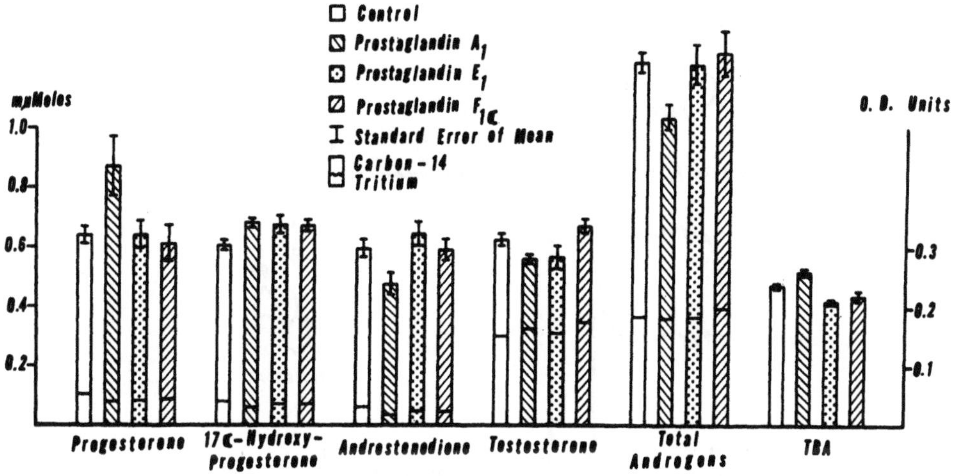

Figure 4. The effects of prostaglandins A(1), E(1), and F(1-alpha) on lipid peroxidation and the biotransformation of pregnenolone and progesterone into 17-alpha-hydroxyprogesterone, androstenedione, testosterone, and total androgens by rat testicular minced preparations <u>in vitro</u>. The incubation mixture consisted of 0.5 microcuries pregnenolone-7-alpha-^{3}H (2.275 nanoM), progesterone-4-^{14}C (2.275 nanoM), 5 ml of a balanced phosphate buffer (pH 7.4), 20 mocriliters propylene glycol, 0.3 g minced tissue and 1 mg of the respective prostaglandin. The samples were incubated in air for one hr 15 37.5°C. The prostaglandins used in this study were generously provided by Dr. John E. Pike of the Upjohn Co. (From Ellis and Baptista, 1969).

Contrary to these points, however, ascorbic acid and D-alpha-tochopherol had pronounced effects on lipid peroxidation but only minor effects on steroid biotransformations. H2O2 and melatonin, on the other hand, obviously increased lipid peroxidation and inhibited the biotransformation of steroids. An increase in high-polar steroid intermediates was noted with additions of H2O2, but not melatonin (unpublished observations).

Also of possible importance is the observation that antioxidants may alter androgen synthesis (unpublished observations), and prostaglandin synthetase activity (Nugteren, 1970). Moreover, the type of prostaglandin synthesized (E or F series) appears to be determined by the amount and nature of the cofactors (antoxidants) (Van Dorp, 1967). In this respect, melatonin and serotonin have been observed to have a marked effect on androgen

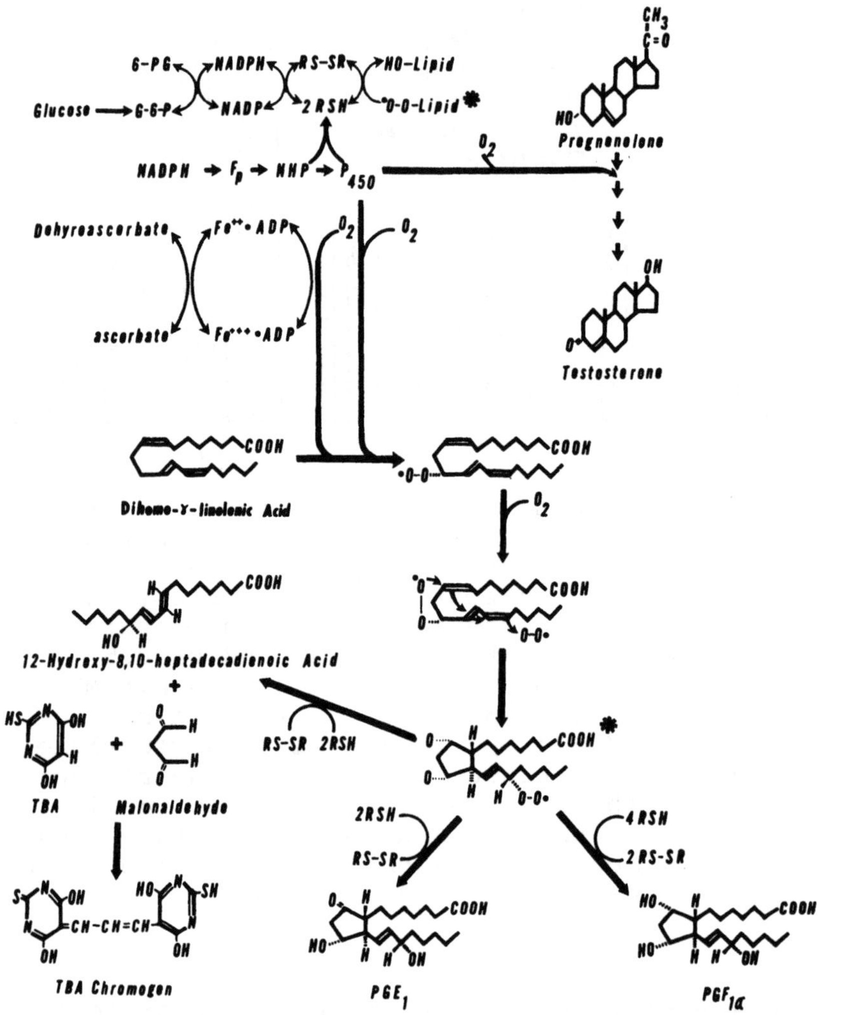

Figure 5. Proposed scheme showing the apparent routes for the synthesis of malonaldehyde, prostaglandin E(1) and prostaglandin F(a-alpha) and testosterone synthesis from pregnenolone. The asterisks are placed on the lipid peroxides to indicate their action on the conversion of sulfhydryl compounds to disulfide bridges. During lipid peroxidation, free radicals are formed. These free radicals are brought to the ground state by acdepting protons from other organic compounds. RSH is used in this scheme to represent either enzyme protein sylfhydryl groups, gluthathione sulfhydryl groups, or any organic RH2 moiety. The chain reaction of a free radical producing another free radical in this way can be stopped by the radical reacting with sulfhydryl groups or with antioxidants so that stable free radicals are formed. Once disulfide bridges are formed from sulfhydryl groups, the sulfhydryl groups may be regenerated through the

synthesis with serotonin acting to regulate the androstenedione:testosterone ratios during aging and the development of the testes. Thus, antoxidants appear important in regulating the flow of electrons and protons through the cells, thereby determining whether the energy is used in steroid hydroxylations, prostaglandin synthesis or malonoldehyde production. A summary of what is currently known concerning the utilization of NADPH by rat testicular tissue with respect to androgen synthesis, prostaglandin synthesis and molonaldehyde production is illustrated in Fig. 5. This scheme shows that the P450 step is required for lipid peroxidation and hence prostaglandin synthesis, but recent data indicate that this step is not mandatory (Takeguchi, 1971).

Another area where prostaglandins appear to function in the testes is through regulating contractile mechanisms. Unpublished data suggest that prostaglandins must be present for normal contractions of the testicular capsule and for oxytocin to act on the capsule. In aging studies of the rabbit testicular capsule, autorhythmic contractions were observed as early as one month postpartum when the testis weighed less than 0.2 g. Preparations from all age levels responded to prostaglandins, but the contractions in immature testis were smaller in amplitude and tension than those observed in mature testis.

---

glutathione-type reductase involving glucose-6-phosphate and NADPH as indicated at the upper-left-hand corner of the figure. NHP represents non-heme iron; P450 represents the iron containing pigment that develops an ultraviolet absorption spectrum with maximum absorption at 450 nanometers in the presence of carbon monoxide. The TBA chromogen is the pigment measured with thiobarbituric acid reagent as a measure of lipid peroxidation. Two methods are shown for the peroxidation of lipids. The first and predominant method is heat labile and enzymatic. This route involves NADPH, non-heme iron (NHP), and the P450 pigment. The second method involves ascorbic acid, ADP, and Fe^{2+}. It is heat stable but is much slower in its rate of synthesis of malonaldehyde. It is not known whetehr ascorbic acid acts as a proton donor in regenerating Fe^{2+} from Fe^{3+} or possibly as a complex with Fe^{2+}. (From Ellis and Baptista, 1969).

Clegg et al (1966) have described two effects of prostaglandins on the uterine wall: an immediate, direct effect on the membrane, and an "enhancement" noticed after several rinses, which is assumed to result from an effect of prostaglandin inside the cell. Our data appear to support the conclusions of these workers and suggests that the mechanism of action of the prostaglandins is the same for both preparations.

TABLE I

THE EFFECT OF HYPOPHYSECTOMY AND THE COMBINED TREATMENT OF GROWTH HORMONE (STH) AND ADRENOCORTICOTROPIC HORMONE (ACTH) ON LIPID PEROXIDATION BY RAT TESTICULAR PREPARATIONS *IN VITRO*, EXPRESSED ON A PER ANIMAL BASIS[1]

Treatment	Testes Wt. (g)	TBA O.D. Units
Control	2.40 ± 0.030	0.436 ± 0.016
Hypophysectomized	0.12 ± 0.008+	0.012 ± 0.003+
Hypophysectomized + STH + ACTH	0.16 ± 0.007*	0.029 ± 0.003**

* $P < 0.05$ when compared with the hypophysectomized group.
\+ $P < 0.001$ when compared with the control group.
**$P < 0.01$ when compared with the hypophysectomized group.
[1] Unpublished data.

TABLE II

THE QUANTITATIVE DISTRIBUTION OF LIPID PEROXIDATIVE ENZYME AND STEROID BIOTRANSFORMING ENZYME CAPACITIES FOR RAT TESTICULAR, TEASED-TUBULAR AND MINCED PREPARATIONS EXPRESSED ON A PER ANIMAL BASIS[1]

Steroid Isolated	Amount of Steroid Isolated nanoM	TBA O.D. Units per 30 min inc.
Mince		
Androstenedione	0.702 ± 0.059	
Testosterone	0.910 ± 0.054	0.239 ± 0.023
Total Androgens	1.612 ± 0.057	
Teased Tubules		
Androstenedione	0.112 ± 0.010	
Testosterone	0.006 ± 0.001	0.009 ± 0.002
Total Androgens	0.118 ± 0.007	

[1]Unpublished data

TABLE III

PROSTAGLANDIN SYNTHETASE ACTIVITY OF RAT TESTICULAR TISSUE AFTER HYPOPHYSECTOMY AND ADRENALECTOMY EXPRESSED ON A PER ANIMAL BASIS[1]

Treatment	(N)	Prostaglandin Synthetase Activity (O.D. Units)	P Value
Control	6	1.166 ± 0.133	
Hypophysectomy (3 weeks)	6	0.905 ± 0.149	>0.20
Hypophysectomy (5 weeks)	6	0.323 ± 0.029	<<0.001
Control	6	1.955 ± 0.158	
Adrenalectomy (10 days)	6	0.503 ± 0.130	<<0.001

[1]Unpublished data

REFERENCES

Anggard, E., Larsson, C., and Samuelsson, B., 1971, The distribution of 15-hydroxy prostaglandin dehydrogenase and prostaglandin-delta^{13}-reductase, Acta. Physiol. Scand. 81:396.

Burk, D., and Woods, M., 1963, Hydrogen peroxide, catalase, glutathione peroxidase, quinones, nordihydroquaiaretic acid, and phosphopyridine nucleotides in relation to x-ray action on cancer cells, Radiation Res. Suppl. 3:212.

Carpenter, M. P., and Wiseman, B., 1970, Prostaglandins of rat testes, Fed. Proc. 29:248.

Clegg, P. C., Hall, W. J., and Pickles, V. R., 1966, The action of ketonic prostaglandins on the guinea-pig myometrium, J. Physiol. 183:123.

Ellis, L. C., Inhibition of rat testicular androgen synthesis in vitro by melatonin and serotonin, Endocrinology, 89:193.

Ellis, L. D., 1970, Radiation effects, in "The Testis", (A. D. Johnson, W. R. Gomes, and N. L. Vandemark, eds.), Vol. III, Academic Press, New York, pp. 333-376.

Ellis, L. C., and Baptista, M. H., 1969, A proposed mechanism for the differential sensitivity of the immature rat testis, in "Radiation Biology of the Fetal and Juvenile Mammal" (M. R. Sikov and D. D. Mahlum, eds.), Proc. Ninth Annual Hanford Biology Symposium, Richland, Washington, May 5-8, 1969, pp. 963-974, U. S. Atomic Energy Commission, Division of Technical Information.

Ellis, L. C., and Berliner, D. L., 1963, The effects of ionizing radiation on endocrine cells. I. Steroid biogransformations and androgen production by tested from irradiated mice, Radiation Res. 20:549.

Ellis, L. C., and Berliner, D. L., 1965, Sequential biotransformation of 5-pregnololone-7-alpha-^{3}H and progesterone-4-^{14}C into androgens by mouse testes, Endocrinology 76:599.

Euler, U. S., and Eliasson, R., 1967, "Prostaglandins", pp. 1-159, Academic Press, New York.

Garattini, S., and Valzelli, L., 1965, in "Serotonin", pp. 148-168, Elsevier, Amsterdam.

Hargrove, J. L., Johnson, J. M., and Ellis, L. C., 1971, Prostaglandin E(1) induced inhibition of rabbit testicular contractions in vitro, Proc. Exptl. Biol. Med. 136:958.

Johnson, J. M., Hargrove, J. L., and Ellis, L. C., 1971, Prostaglandin F(1-alpha) induced stimulation of rabbit testicular contractions in vitro, Proc. Soc. Exptl. Biol. Med. 138:378.

Kitabachi, A. E., 1967, Inhibition of steroid C-21 hydroxylase by ascorbate: alterations of microsomal lipids in beef adrenal cortex, Steroids 10:567.

Kurzrok, R., and Lieb, C. C., 1930, Biochemical studies of human semen II. The action of semen on the human uterus. Proc. Soc. Exptl. Biol. Med. 28:268.

Matsuchita, S., Kabayashi, M., and Nitta, Y., 1970, Inactivation of enzymes by linoleic acid hydroperoxides and linoleic acid, Agri. Biol. Chem. 34:817.

Nakano, J., Montague, B., and Darrow, B., 1971, Metabolism of prostaglandin E in human plasma, uterus, and placenta, in swine ovary and in rat testicule, Biochem. Pharm. 9:2512.

Nugteren, D. H., 1970, Inhibition of prostaglandin biosynthesis by 8 cis, 12 trans, 14 cis-eicosatrienoic acid and 5 cis, 8 cis, 12 trans, 14 cis-eicosa-tetraenoic acid, Biochem. Biophys. Acta 210:171.

Samuelsson, B., Granström, E., and Hamberg, M., 1966, in "Prostaglandins" (S. Bergstrom and B. Samuelsson, eds.), pp. 31-44, Proc. Second Nobel Symposium, Stockholm, Interscience Publ., New York.

Savkovic, N., 1964, Effect of chemical protective agents (cysteamine and beta-amino-ethylisothiuranium (AET) cl HCl) and homologous testes-DNA on the death rate of the progeny of a first generation originating from male rats irradiated in the infantile period, Nature 203:1393.

Takeguchi, C., Kahno, E., and Sih, C. J., 1971, Mechanism of prostaglandin biosynthesis. I. Characterization and assay of bovine prostaglandin synthetase. Biochemistry 10:2372.

Van Dorp, D. A., 1967, Aspects of the biosynthesis of prostaglandins, Progr. Biochem. Pharmacol. 3:71.

THE ANTIHYPERTENSIVE AND NATRIURETIC ENDOCRINE FUNCTION OF THE KIDNEY: VASCULAR AND METABOLIC MECHANISMS OF THE RENAL PROSTAGLANDINS

J. B. Lee

State Univ. of New York, Buffalo School of Med.
Buffalo General Hospital
Buffalo, N. Y.

INTRODUCTION

The regulation of the extracellular and intracellular milieu in which all vital cell functions take place is a highly complex phenomenon involving intricate interreactions among various hemodynamic, humoral, metabolic and transport phenomena. The central role of the kidney in the maintenance of such homeostatic regulation has for years been widely appreciated. This is particularly reflected by the intense investigations which have been directed to the role of the kidney in the regulation of systemic arterial blood pressure and the mechanisms whereby the kidney, by virtue of its unique reabsorptive, excretory and metabolic activities, is capable of maintaining a remarkably constant internal environment. In the latter instance much attention has been given to the mechanism of sodium transport by the kidney since from an evolutionary standpoint the capacity for renal retention or elimination of sodium and water is considered to be a pivotal homeostatic function in the adaptation of terrestrial life to an environment in which exposure to sodium may be minimal or excessive.

The discovery that the kidney contains prostaglandins and that these compounds have profound effects on arterial blood pressure and sodium and water excretion as well as a host of other homeostatic actions has given rise to speculation that these compounds are importantly involved in the regulation of systemic blood pressure, plasma volume, and regional blood flow as well as the regulation of intrarenal hemodynamics and sodium and water excretion. It will be the purpose of this chapter to briefly review the evidence that an important relationship of the kidney to blood pressure regulation is represented by the so called antihypertensive renal endocrine function and that the renal prostaglandins may be important mediators of this activity. Secondly, an attempt will be made to outline the evidence that the renal prostaglandins may also function as a natriuretic "hormone" and may be important agents responsible, at least in part, for the activity of the kidney in the maintenance of sodium and water homeostasis. Lastly, new evidence, preliminary in nature, will be presented which demonstrates that the PGA compounds, but not the PGE compounds, have marked effects on the intermediary metabolism of the renal cortex and medulla *in vitro* which may be related in a significant way to the energetics involved in sodium transport by the renal tubule.

THE ANTIHYPERTENSIVE AND NATRIURETIC RENAL ENDOCRINE FUNCTION

The Antihypertensive Function

It has long been appreciated that the kidney is intricately involved in the development of various stages of hypertension, particularly experimental and renal human vascular hypertension and human essential hypertension. However, the great majority of investigations have been directed to uncovering the mechanisms of the well known pro-hypertensive functions of the kidney, most notably the renal renin-angiotensin system and the sympathetic nervous system. Although such pro-hypertensive activity is well documented there is also a growing body of experimental evidence which indicates that the kidney also has an antihypertensive function and that in the last analysis, certain states of hypertension may be the result of a deficiency of renal depressor systems rather than solely an increase in renal pressor mechanisms. Although many physiological abnormalities may be observed in hypertensive states, relative or absolute peripheral arteriolar vasoconstriction is believed to be in large part responsible for the observed elevation in blood pressure. Fig. 1b illustrates that "normal" arteriolar tone may result from vasodepressor mechanisms being equally offset by vasopressor mechanisms with the resultant state of normotension. States of hypertension (Fig. 1a) may be conceptually visualized as an absolute or relative deficiency of depressor systems giving rise to an absolute or relative increase in vasopressor systems with resultant peripheral arteriolar vasoconstriction and elevation in systemic arteriole blood pressure. Conversely, Fig. 1c shows that peripheral arteriolar dilation may be the result of a relative preponderance of depressor mechanisms associated with an absolute or relative deficiency of pressor agents which could lead to depression of systemic arterial blood pressure or shock.

The evidence for an antihypertensive renal function is derived from several sources of investigation. In the first place beginning with the early studies of Fasciolo (1938), it has long been known that a normally functioning kidney exerts a "protective" action during procedures designed to experimentally elevate blood pressure. Thus, partial unilateral renal arterial occlusion leads to a rise in systemic blood pressure, but usually only if the opposite kidney is removed, suggesting that the latter is exerting a protective antihypertensive function. Furthermore, once experimental renovascular hypertension is achieved by partial renal arterial occlusion with removal of the opposite kidney, transplantation of a normal kidney into the circulation promptly results in a fall in blood pressure to normotensive levels (Gomez et al, 1960). This protective antihypertensive activity of a normal kidney can be observed not only with unilateral renovascular hypertension but in the hypertension which occurs following the removal of both kidneys (renoprival hypertension) in which transplantation of a normal kidney in such animals (Kolff and Page, 1954; Muirhead et al, 1956) and man (Hume et al, 1955; Kolff et al, 1964) results in a lowering of blood pressure from hypertensive to normotensive levels.

The second line of evidence was obtained when Braun, Menendez and von Euler (1947) observed that removal of both kidneys resulted in a salt dependent renoprival hypertension which obviously could not be the result of any renal pro-hypertensive activity such as the renal renin-angiotensin system. The results of this experiment were believed by some to be non-specific due to the associated metabolic derangements occurring in the uremic stage which results from bilateral nephrectomy. However, Grollman et al (1949) showed that ureteral caval anastamosis, in which the same degree of uremia is obtained as with bilateral nephrectomy, is not associated with an elevation in blood pressure. This finding illustrated that the state of renoprival hypertension is indeed a specific entity and is not associated with the

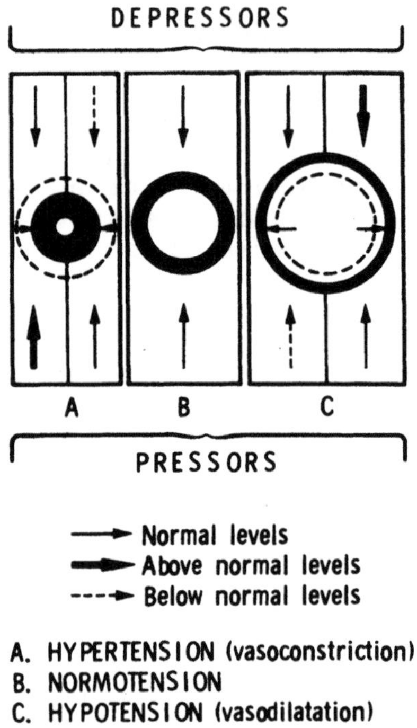

Figure 1. Schematical representation of the antihypertensive endocrine function. From Lee (1970). Reproduced with permission of the publisher.

excretory capacity of the kidney but rather with non-excretory, perfused renal tissue.

Thirdly, pressor agents have not been consistently detected in amounts to account for the observed degree of blood pressure elevation in animals with renovascular hypertension (Koletsky and Pritchard, 1963; Blaquier et al, 1960). Although partial constriction of the renal artery is associated with an enhanced elaboration of renin from the affected kidney, this phenomenon ceases after a few days despite continuation of blood pressure elevation. The failure to detect pressor agents in uncomplicated human essential hypertension has been interpreted as important negative evidence for the existence of an antihypertensive renal function.

Lastly, the kidney has long been suspected of containing potent vasodepressor and antihypertensive compounds. Grollman and his associates (1940) observed lowering of blood pressure following the administration of kidneys to hypertensive rats and humans. Similarly, Page

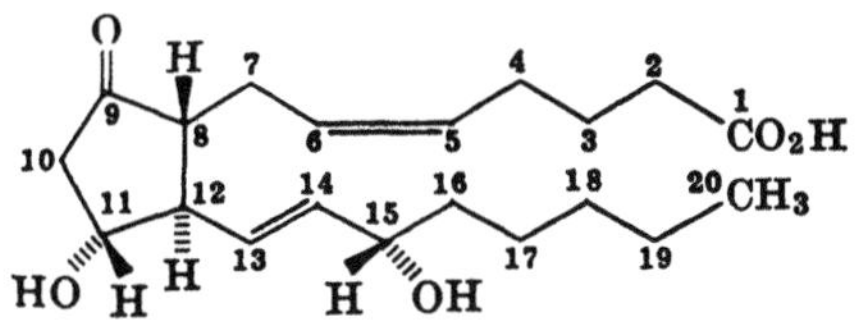

PGE_2 (compound 2)

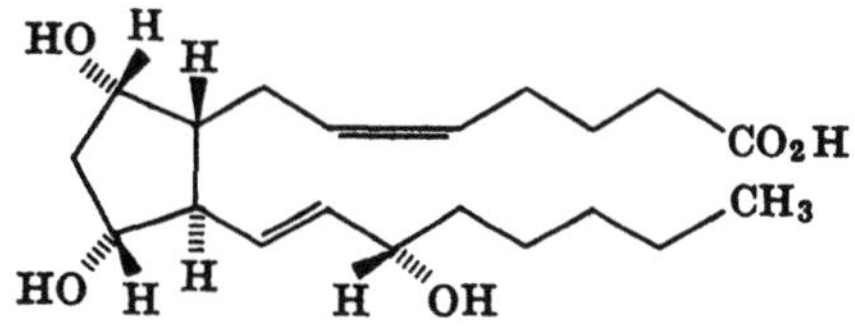

$PGF_{2\alpha}$ (compound 1)

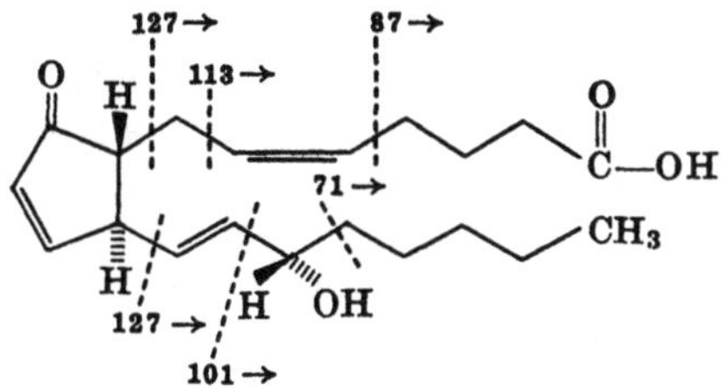

PGA_2 (medullin)

Suggested fragmentations for all three compounds are indicated in the structure of PGA_2

Figure 2. Structures of the renomedullary prostaglandins. From Lee et al (1967). Reproduced with permission of the publisher.

and his associates (1941) also found that extracts of whole kidney lowered blood pressure in hypertensive animals and humans. However, the results of intravenous experiments were believed to be non-specific because of the associated anaphylactoid reactions accompanying the administration of such crude kidney extracts. There have been two regions within the kidney which have been suggested as the source of the antihypertensive renal function. Hamilton and Grollman (1958) believe that the activity largely resides in renal cortex since hypotensive extracts which were dializable and water soluble were extracted from this region of the kidney but not from medulla. On the other hand, Muirhead and his associates (1960) observed that extracts of renal medulla (but not cortex) prevented the development of acute sodium dependent renoprival hypertension in the dog. Recently it has also been observed by Muirhead et al (1970) that transplantation of grafts of rabbit renal medulla into the subcutaneous tissue of rats with renovascular hypertension result in a fall in blood pressure to or toward normotensive levels, again suggesting that this region of the kidney contained antihypertensive agents. The active principles responsible for most of these vasodepressor and antihypertensive activities of the renal medulla may be the renomedullary prostaglandins, and their possible role as the antihypertensive "hormone" of the renal medulla will be discussed separately.

Natriuretic Function

The existence of a factor mediating the natriuresis accompanying salt administration has been suspected since DeWardener and his associates (1961) showed that such a natriuresis in the dog could occur independently of changes in glomerular filtration rate or aldosterone secretion. Despite continued interest in the possibility of the existence of a natriuretic hormone, efforts to isolate and identify the responsible material(s) have met with little success. Indeed, there is even conflicting evidence as to the existence of such a "hormone" and a variety of factors other than such an agent have been implicated as being chiefly responsible for the natriuresis accompanying saline administration. Some investigators (Early and Friedler, 1965, 1966; Brenner et al, 1969) have particularly favored the viewpoint that the natriuresis is primarily the result of intrarenal hemodynamic and physical factors which may preclude the necessity of postulating a natriuretic factor. Nevertheless, substances have been observed in the plasma and urine of salt loaded humans and sheep (Sealy and Laragh, 1971) and have been found to be of high molecular weight (10,000-50,000) and are natriuretic when injected into hydropenic or water loaded diabetes insipidus rats. On the other hand, material of low molecular weight (less than 1,000) has been observed in the plasma of saline loaded animals which inhibit sodium transport in the toad bladder (Buckalew et al, 1970) as well as para-amino-hippurate (PAH) uptake by the kidney cortical slice (Bricker et al, 1968).

In summary, therefore, there is evidence that a natriuretic factor exists although the source of the material and its chemical nature have proven elusive. The possibility that the renal prostaglandins may normally function as a natriuretic "hormone" has been entertained (Lee and Ferguson, 1969; Lee, 1970) but the evidence is entirely circumstantial. Nevertheless, the question as to whether these agents may normally have an intrarenal role in mediating the natriuretic capacity of the kidney will be discussed.

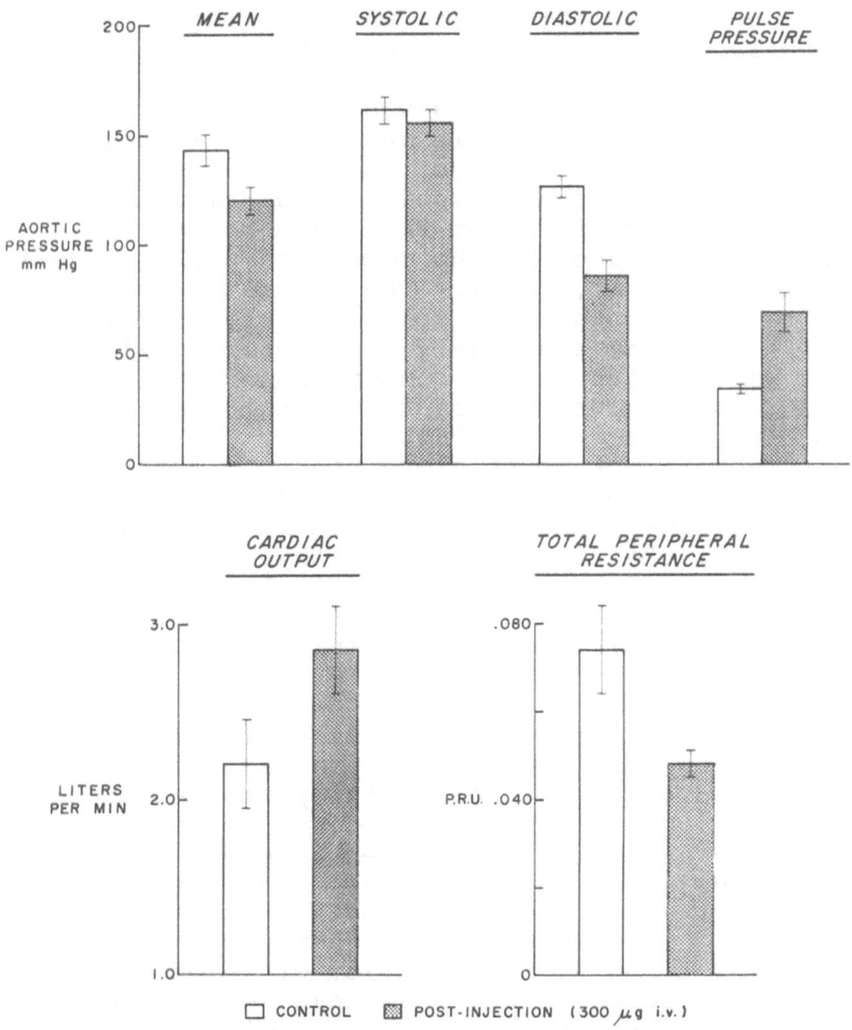

Figure 3. Hemodynamic effects of medullin - (PGA(2) - in the anesthetized dog. From Lee et al (1965). Reproduced with permission of the publisher.

THE RENAL PROSTAGLANDINS

Isolation and Identification

Since an antihypertensive renal function had long been suspected, an effort was made to determine whether the renal medulla or cortex possessed vasodepressor activity in the normotensive animal. Extracts of both rabbit kidney medulla and cortex were injected intravenously into the pentobarbitalized, pentolinium-treated, vagotamized rat and it was observed that the renal medulla, but not cortex, possessed potent sustained vasodepressor activities (Lee et al, 1962a). The compounds responsible for this vasodepressor activity were observed to be ethanol soluble and of a relatively low molecular weight. Furthermore, they could be differentiated by chromatographic and ultraviolet techniques as being different from such known vasodepressor compounds as the adenosine nucleotides and bradykinin (Lee et al, 1963). The vasoactive agents were observed to be polar acidic lipids (Hickler et al, 1964; Lee et al, 1965) and eventually three biologically active compounds were isolated by a combination of solvent extraction, column chromatography and thin layer chromatography (Lee et al, 1964b, 1965). Of the three isolates, one did not lower blood pressure but did result in contraction of non-vascular smooth muscle and was identified by non-classical means, including mass spectroscopy of the free acid and methyl ester, as PGF(2-alpha). The second compound which did lower blood pressure as well as contract non-vascular smooth muscle (Daniels et al, 1967; Lee et al, 1967) was identified as PGE(2) (Fig. 2).

However, the third compound called medullin was the first of what is now known as the prostaglandin A class to be isolated and characterized. It represented a new chemically and physiologically different class in that it was less polar and more unsaturated than PGE or PGF. Furthermore, medullin possessed only blood pressure lowering activity and did not stimulate non-vascular smooth muscle. The latter is a property of the prostaglandins E and F, the activity of which was originally discovered in sheep seminal vesicles and semen independently by von Euler (1934) and Goldblatt (1933) and further studied extensively by von Euler (1935, 1936) who characterized them as fatty acids. Their isolation (Bergstrom and Sjovall, 1957, 1960a, 1960b) and identification as PGE and PGF from sheep seminal vesicles and lung was achieved by Bergstrom and his associates (1962a, 1962b, 1962c, 1963).

Identification of the free acid and methyl ester of medullin by spectroscopic and mass spectral analysis (Lee et al, 1966, 1967) revealed its chemical characteristics to be due to elimination of a water molecule from the cyclopentanone ring of PGE(2) leading to the formation of an alpha/beta-unsaturated ketone having maximal ultraviolet absorption at 215 nanometers (Fig. 2). Confirmation of the structure of medullin as PGA(2) was

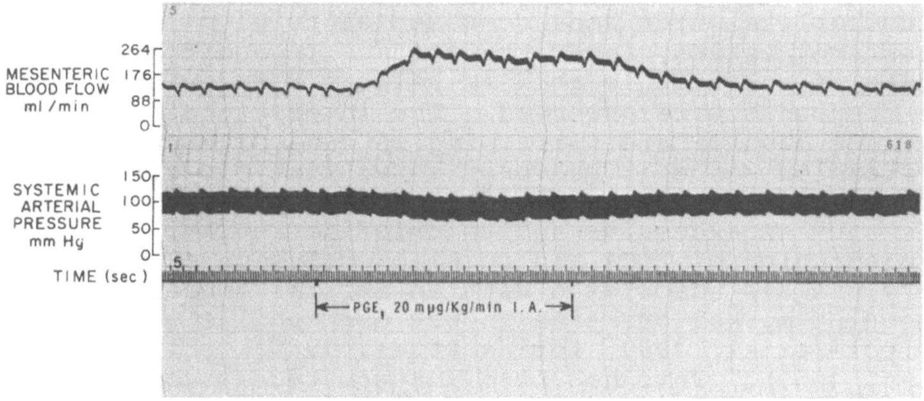

Figure 4. Effect of PGE(1) on mesenteric blood flow and systemic arterial blood pressure. Note the rise in mesenteric blood flow during PGE(1) infusion into the mesenteric artery with a coincident fall in systemic arterial blood pressure. Regional infusion of PGE(1) into other vascular beds has not been observed to be associated with a decrease in arterial blood pressure. From Covino, McMorrow and Lee, unpublished observations.

made by comparative thin layer chromatography and mass spectroscopy of PGA(2) obtained by dehydration of authentic PGE(2) which revealed them to be identical. PGA(1) as well as PGA(2) was also identified in human semen by Hamberg and Samuelsson (1966). The isolation of prostaglandin E(2) and A(2) with their powerful vasodepressor and antihypertensive properties in the kidney medulla constitutes added evidence for an antihypertensive renomedullary function.

Distribution and Metabolism

Despite intensive efforts to isolate prostaglandins from the renal cortex, extraction of large quantities of renal cortex failed to reveal biologically active prostaglandin. Thus, the renal prostaglandins are located primarily in the renal medulla with the highest concentration being in the inner medulla. The precise site of prostaglandin formation and/or storage in unknown, but there is accumulated evidence suggesting that the interstitial cells of the renal medulla which contain osmiophilic lipid granules may be the anatomic site of prostaglandin location. Nissen and Bojesen (1969) by ultracentrifugation of rat renal papilla, demonstrated lipid droplets which were composed of triglyceride cholesterol esters, and fatty acids. In addition, small amounts of arachidonic acid as well as PGE(2) and PGA(2)-like compounds were observed. The interstitial cells of the inner medulla are intriguing in regard to experimental hypertension since it has been shown that experiments designed to elevate the blood pressure by a variety of means such as excessive sodium chloride intake, administration of DOCA and partial occlusion of the renal artery are all associated with an initial increase and subsequent marked decrease in the osmiophilic granules (Muehrcke et al, 1969; Tobian et al, 1969; Ishii and Tobian, 1969). This has been interpreted as possibly being a reflection of increased synthesis and release of prostaglandins in response to the hypertensive stimulus. More recently, Muirhead et al (1970) have demonstrated that live grafts of renal medulla with viable interstitial cells when transplanted subcutaneously in animals with renovascular hypertension result in a reduction in blood pressure toward normotensive levels. Thus, if the interstitial cell granules contain prostaglandins they may indeed represent antihypertensive compounds of the kidney, and the hypothesis has been advanced that deficiency of interstitial cell prostaglandins may be associated with the development of sustained hypertension.

Until recently, relatively little attention has been given to the mechanism of the biosynthesis of the renal prostaglandins, although Hamburg (1969) has shown conclusively that biosynthesis takes place in rabbit renal medullary homogenates incubated with arachidonic acid. PGE(2) and PGF(2-alpha) were isolated with radioactivity derived from labelled precursor, but only traces of PGA(2) could be observed using this technique. Recently Anggard (1971) has found that the renomedullary PGE(2) is located exclusively in the supernatant, whereas its precursor, arachidonic acid, and biosynthesis is most active in the microsomal fraction. From this Anggard suggests that the renal medulla biosynthesis occurs by hydrolysis of microsomal precursor, probably esterified tc phospholipid. The arachidonate is then converted to PGA(2) by prostaglandin synthetase which is located in the microsomal fraction. Prostaglandin synthetase has been observed to be maximal in medulla and minimal in cortex, whereas the reverse is true for prostaglandin dehydrogenase activity which is maximal in cortex and almost completely absent in medulla (Anggard et al, 1971). Anggard concludes that PGE(2) formed in medulla may act locally or on juxtaglomerular cortical structures where

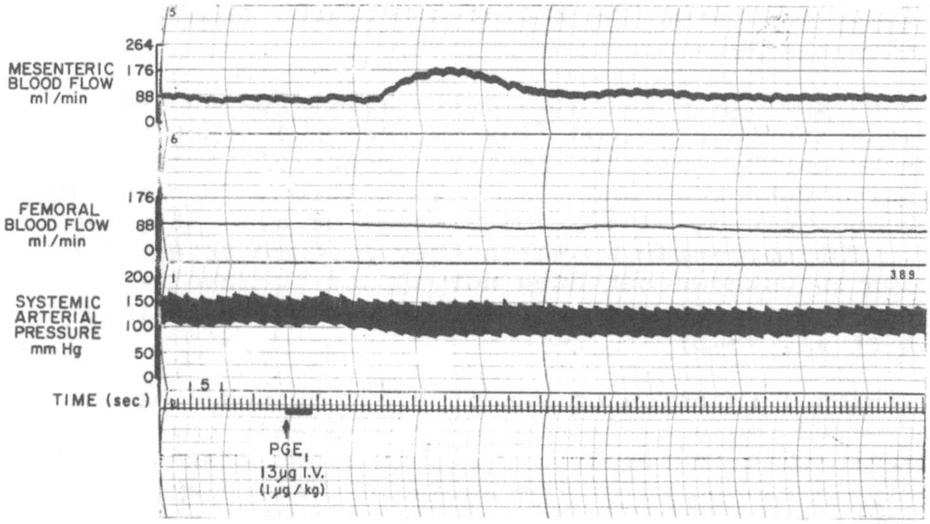

Figure 5. Effect of intravenous administration of PGE(1) on mesenteric blood flow and systemic arterial blood pressure. Again note the rise in mesenteric blood flow coincident with the fall in arterial blood pressure with a return of blood flow to control levels associated with a rise in arterial blood pressure to pre-injection values. Under these conditions there is no change in femoral blood flow. From Covino, McMorrow and Lee, unpublished observations.

they would be metabolized following their biological action on such cortical structures.

Cardiovascular Effects

<u>Systemic Hemodynamics</u>. Fig. 3 illustrates the overall mechanism of hypotensive action of PGA(2) in the normotensive dog. It is evident that following intravenous injection of PGA(2) there was a fall in mean blood pressure which was a result of greater fall in diastolic pressure than systolic pressure resulting in an increase in the pulse pressure. Cardiac output at the time of maximum vasodepression was elevated leading to a fall in the calculated total peripheral resistance. The increase in cardiac output was primarily secondary to a reflex increase in heart rate, probably occasioned by stimulation of the baroreceptor mechanisms in response to the fall in arterial blood pressure. This is supported by the fact that PGA(2) was devoid of any chronotropic or

inotropic effects on the isolated perfused rabbit heart (Lee et al, 1965).

When administered intra-arterially, PGA and PGE compounds increased blood flow to regional vascular beds of the coronary, carotic, femoral, brachial, mesenteric, pulmonary, cutaneous and renal circulations (Nakano, 1968; Nakano and McCurdy, 1967, 1968; Hauge et al, 1967; Lee, 1968). The mechanism by which they produce arteriolar dilation is unknown but does not appear to involve histaminergic, colonergic or adrenergic nerve endings (Smith et al, 1968).

Regional infusion of PGE or PGA compounds leading to an increased regional blood flow are usually not associated with any fall in systemic blood pressure with the notable exception of the splanchnic circulation (Covino et al, 1968). Fig. 4 shows that when PGE(2) is infused into the mesenteric artery there is a rise in splanchnic blood flow with a coincident fall in systemic blood pressure, suggesting that the splanchnic vascular bed is the major resistance system whose dilaticn by PGE(1) results in a decrease in arteriolar blood pressure. This is further confirmed by the fact that intravenous infusion of PGE or PGA compounds leads to a fall in blood pressure with unchanged or decreased regional blood flow to all vascular beds except the splanchnic vascular bed (Fig. 5). As with intra-arterial injection, the fall in systemic blood pressure resulting from intravenous infusion of prostaglandin is associated with a coincident rise in splanchnic blood flow.

Although PGA and PGE compounds are vasodepressor and act primarily on peripheral arteriolar smooth muscle, the important differences have recently been reported between the two classes (Kannegiesser and Lee, 1971). When PGE(1) or PGE(2) is administered high in the aorta of the anesthetized cat there is an immediate fall in blood pressure reaching a maximum within 20 sec with an immediate partial return toward control by 80 sec. There is then a relatively slow increase in blood pressure reaching control values over a period of 6-7 min (Fig. 6). In contrast, aortic injection of PGA(1) or PGA(2) leads to a gradual fall reaching a maximum at 80 sec with a slow return to normal over a period of 6-7 min coinciding with the slow phase of PGE(1) and PGE(2). The fall in blood pressure with PGA(1) or PGA(2) is of less magnitude than that observed with PGE(1) or PGE(2) and the time for maximum effect requires at least ten recirculations. This suggests that PGA compounds produce arteriolar dilations

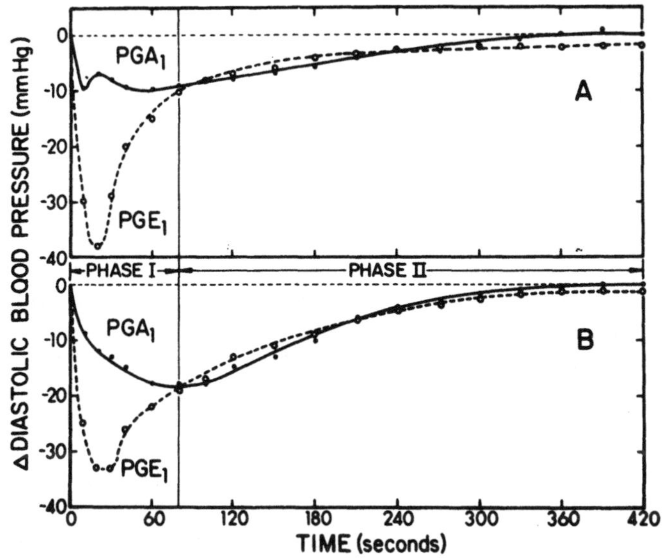

Figure 6. Hemodynamic differences between PGA(1) and PGE(1). A = non-pentolinium treated animals, B = pentolinium treated animals. Each point represents the mean of six animals. Prostaglandin dose: 30 nanograms intrathoracic aorta of the anesthetized cat. Note immediate fall in blood pressure with PGE(1) and the delayed response with PGA(1). Similar results were observed with PGA(2) and PGE(2). From Kannegiesser and Lee (1971). Reproduced with permission of the publisher.

through an indirect mechanism possibly by releasing vasoactive peptides. On the other hand, the PGE class immediately and probably directly dilate arteriolar smooth muscly distal to the site of injection which results in the prompt decrease in arterial blood pressure. The delayed effect of PGA compounds might be explained in part by the fact that these compounds are not actively metabolized by the lung and continue to circulate with biological activity for a period of time (McGiff et al, 1969). However, PGE compounds in the doses given are actively metabolized by the lung and after a single passage are not believed to circulate in a form that is biologically active. Thus, the entire early and delayed effect of PGE compounds can be attributed to events occurring between the thoracic aorta and the pulmonary circulation during the first 10 sec following injection.

<u>The Microcirculation</u>. The sight of action of the prostaglandins has been observed to be on all components of the microcirculation with dilation of the arterioles,

metarterioles, pre-capillary sphincter and venules (Kadar and Sunahara, 1969; Weiner and Kaley, 1969). This is associated with a rapid flow throughout the capillary circulation together with an increase in the capillary permeability rather than in an increased number of patent capillaries since Horton (1963) and Solomon et al (1968) have demonstrated elevation in the extrusion of capillary fluid using dye techniques.

Although it is known that the prostaglandins directly dilate arteriolar smooth muscle, the precise biochemical method whereby such actions take place is unknown. Strong and Bohr (1967) demonstrated that the spontaneous contractions of isolated mesenteric artery showed a biphasic response to PGA and PGE compounds with relaxation in small concentrations and contraction at high concentrations. Prostaglandin induced relaxation or contraction was not blocked by alpha- or beta-adrenergic blockade, atropine, lysergic acid or histamine. In *in vivo* studies it has been shown that the increased regional blood flow to the extremity of the anesthetized dog following PGE(1) administration is not blocked by atropine, propranolol or tripelennamine (Smith et al, 1968) suggesting that the vasodilator action of PGE(1) is not mediated by histaminergic, cholinergic, or adrenergic nerve endings. Although PGE(1) inhibition of spontaneous contractions of isolated dog mesenteric artery were uninfluenced by pretreatment with atropine or propranolol, Kadar and Sunahara (1969) observed that norepinephrine-induced contractions were enhanced at a low concentration of potassium and diminished at a high concentration. Furthermore, the effect of PGE(1) was abolished following treatment of the tissue with ouabain. From these results the authors concluded that PGE(1) elicits its action on vascular smooth muscle through Na^{+}-K^{+} ATPase. Thus, one mechanism whereby PGE and PGA compounds might lower systemic arterial blood pressure would be by direct post-synaptic myogenic relaxation of vascular smooth muscle leading to generalized peripheral arteriolar dilation and a fall in systemic blood pressure.

A second mechanism could be that these compounds interact with the well established pressor systems since a variety of investigations have shown that PGA and PGE compounds appear to act as potent adrenergic antagonists. However, they are not specific as such since it has been shown by Holmes et al (1963) and Weiner and Kaley (1969) that pressor responses to angiotensin and vasopressin are also blunted. The interaction of renal prostaglandins with the pressor systems of the kidney is of particular

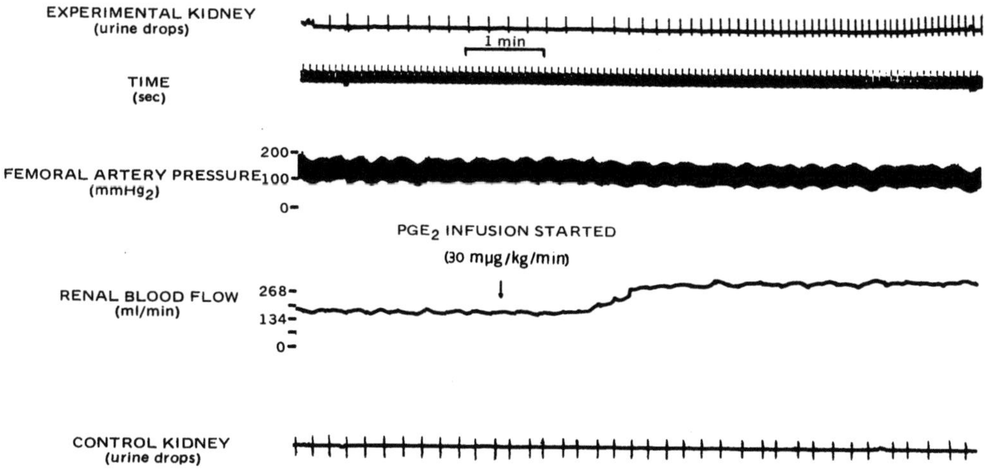

Figure 7. Effect of PGE(2) infusion in the renal artery in renal hemodynamics and urine flow. Renal blood flow was measured by electromagnetic flowmeter and urine formation by photoelectric transducer adapted for drop counting. Note increase in renal blood flow associated with an elevation in urine formation in the infused kidney but not the contralateral control kidney. The elevation in renal blood flow was not associated with any change in systemic arterial blood pressure. From Lee (1968). Reproduced with permission of the publisher.

interest with relation to a possible antihypertensive endocrine function. It has been observed that PGE-like material was released from kidneys made ischemic from unilateral arterial constriction, renal nerve stimulation or following infusion of angiotensin or norepinephrine (McGiff et al, 1970a, 1970b; Dunham and Zimmerman, 1970). Continued infusion of the pressor substances was associated with a spontaneous loss of renovasoconstriction with a return of renal blood flow toward pre-infusion levels. This increase in renal blood flow was associated with a release of prostaglandin-like material suggesting that there might be antagonism of renal pressor systems by prostaglandins released within the kidney in response to varied renovasoconstrictors.

Renal Effects

<u>Hemodynamics</u>. Although it has been indicated that PGA and PGE compounds result in generalized peripheral arteriolar dilation and an increased blood flow to most regional vascular beds, their effect on renal blood flow

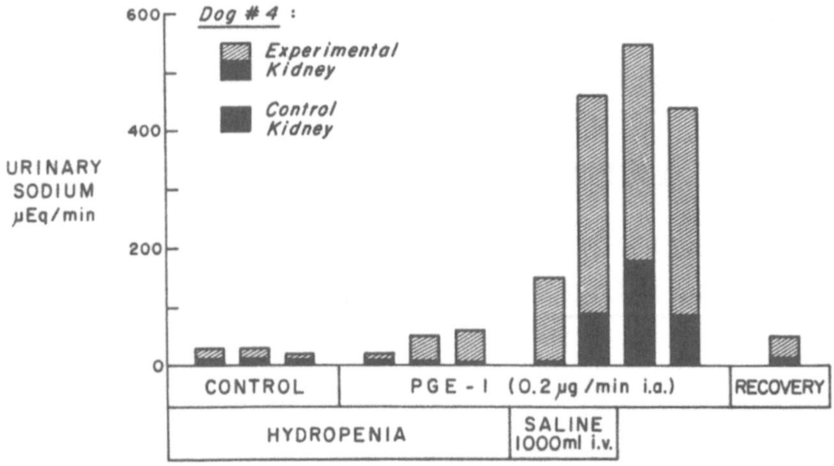

Figure 8. Natriuretic effect of PGE(1) infused into the renal artery of the anesthetized dog. There is a facilitated increase in sodium excretion from the experimental kidney when compared to the control kidney when isotonic saline is administered intravenously during continued PGE(1) infusion. From Lee (1967). Reproduced with permission of the publisher.

is complex and unique. When PGE(2) or PGA(2) is infused into the renal artery of the anesthetized dog there is approximately a 33% increase in renal blood flow to the infused kidney but not to the contralateral kidney (Fig. 7). This increase in renal blood flow is unaccompanied by any change in systemic blood pressure suggesting that the renal resistance per se does not contribute significantly to maintenance of systemic arterial blood pressure. The increase in renal blood flow is accompanied by an enhanced elaboration of urine from the infused kidney but again not from the control, non-infused kidney. It has been shown by a number of investigators that the increased urine flow occasioned by prostaglandin administration is accompanied by a marked increase in the excretion of sodium, potassium and chloride (Johnston et al, 1967; Vander, 1968). Barger and Herd (1966) utilizing krypton-85 washout techniques confirmed by radioautography and Silastic injection before and following infusion of PGA(2) found that the increase in total renal blood flow is almost entirely the result of an increase in cortical flow accompanied by a decrease in outer medullary blood flow. The effects on renal blood flow can be observed with concentrations of prostaglandins as low as 0.003 microgram/kg/min, an infusion rate at which there is no effect on systemic blood pressure when administered intravenously. The redistribution of blood flow from outer medulla to cortex is not a phenomenon specific to prostaglandins but has been observed under many conditions which induce a renal salt-losing state. These include administration of ethacrynic acid, furosemide, acetylcholine and, of great interest, following the infusion of saline intravenously (Jones and Herd, 1970). Since the elevation in renal blood flow during PGA or PGE administration is not associated with a corresponding elevation in glomerular filtration rate, a decrease in filtration fraction takes place and it thus appears that these prostaglandins act on the renal circulation primarily by dilation of efferent cortical arterioles.

In summary, therefore, PGA(2) and PGE(2) which are normally present in the kidney medulla result in an increased cortical renal blood flow when administered into the renal artery. This is primarily the result of preferential efferent arteriolar vasodilation, the mechanism of which remains obscure.

Water and Electrolyte Excretion. PGA and PGE compounds are among the most potent naturally occurring natriuretic factors known. Fig. 8 shows that when PGE(2) is infused into the renal artery of the hydropenic anesthetized dog at rates in which there is a redistribution of blood flow from medulla to cortex, natriuresis was observed from the infused kidney while antinatriuresis occurred in the opposite kidney. Superimposition of a saline load given intravenously resulted in a marked increase in sodium excretion from 50 microEq/min to 540 microEq/min from the infused kidney with a smaller rise from 10 microEq/min to 150 microEq/min in the opposite kidney. This facilitation of the ability of the kidney to increase excretion of sodium in the face of an elevation in filtered load was accompanied by a marked kaliuresis and an increase in osmolal and (to a much lesser extent) free water clearance. Although it is possible that the increase in free water clearance may be the result of inhibition of vasopressin by PGE(1), probably through inhibition of adenylate cyclase (Grantham and Orloff, 1968), a major factor in the elevation of free water clearance may be the marked reduction in outer and inner medullary blood flow under the influence of prostaglandin.

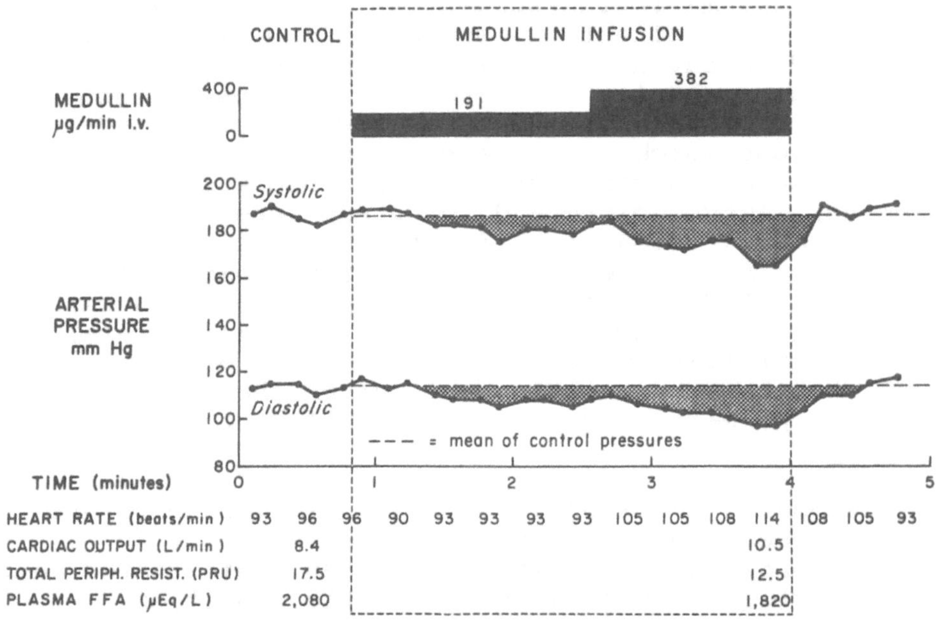

Figure 9. Hemodynamic effects of PGA(2) (medullin) in a patient with essential hypertension. From Lee (1967). Reproduced with permission of the publisher.

The changes in renal hemodynamics and renal handling of water and electrolytes suggest that such compounds may be acting by decreasing intrarenal renin release. However, Vander (1968) has observed no change in renal venous renin concentration during infusion of PGE(1). Similarly, in patients with essential hypertension, administration of PGA(1) results in an increase of renal blood flow with an elevated or unchanged renal venous renin content (Lee et al, 1971b). These studies suggest that at least under acute conditions, the elevation in renal blood pressure is not immediately mediated by the renal renin-angiotensin system. The effect of PGA and PGE compounds on renal hemodynamics and urinary electrolyte and water excretion are summarized in Table I.

Effects on Renal Intermediary Metabolism. The intermediary metabolism of the rabbit renal cortex *in vitro* is characterized by a high rate of oxygen consumption (Warburg, 1927) which is primarily the result of oxidation of endogenous and exogenous fatty acids (Lee et al, 1962b). In addition, the kidney cortex is capable of a high degree of oxidative deamination (Krebs, 1935) and gluconeogenesis (Benoy and Elliott, 1937) but a relatively small degree of glycolysis. In contrast, an extremely high rate of glycolytic activity which is the sole source of energy occurs in the inner medulla (Kean et al, 1962; Lee and Peter, 1969), a site incapable of significant oxidative metabolism, being an area of the kidney of very low oxygen tension. On the other hand, the outer medulla exhibits both properties, that is, it is capable of a high rate of glucose and fatty acid oxidation but is also an area of the kidney in which glycolysis is a major pathway (Davis, 1964). This is particularly true when outer medullary slices are incubated under low oxygen tension where glycolysis is capable of providing a large part of the energy which otherwise results from oxidative pathways (Lee and Peter, 1969). In addition, it has been shown by many investigators that sodium and potassium markedly alter oxidative metablism in both kidney cortex and outer medulla. In this regard it is of interest that a large degree of the energy formed from the oxidation of various substrates is believed to be applied to iso-osmotic sodium transport in the cortical proximal tubule (Thurau, 1961; Kiil et al, 1961; Deetjen and Kramer, 1961) and to hyperosmotic sodium transport by the ascending limb of the loop of Henle in the outer medulla (Abcdeely and Lee, 1971). Furthermore, Na^{+}-K^{+} ATPase is present both in the kidney cortex and in outer medulla, being two to three times as high in the latter tissue (Jorgenson and Skou, 1969; Hendler et al, 1969; Martinez-Maldonaldo et al, 1969). It has been shown that in addition to the increased oxygen consumption which occurs when osmolality is increased by the addition of sodium chloride, there is a parallel rise in Na^{+}-K^{+} ATPase activity both reaching a maximal value at a medium osmolality of 800 mOsm/kg H2O (Abodeely and Lee, 1971; Alexander and Lee, 1970). Both oxygen consumption and Na^{+}-K^{+} ATPase were inhibited by ouabain at an identical Ki (10^{-5}). From these results it has been postulated that changes in the osmolality of the medium result in an enhanced substrate oxidation which may be in large part mediated through increased activity of Na^{+}-K^{+} ATPase. Thus peritubular osmolality may be an important regulator of such enzyme activity in the tubule cells of the ascending limb which in turn may be an important determinant for the rate of substrate oxidation necessary for production of high energy requirements for hyperosmotic active sodium transport in this part of the nephron.

Although the PGE compounds have been found to have varied and rather profound metabolic effects in a variety of tissues, biological activity of the PGA class to date has been primarily circulatory in nature with the majority

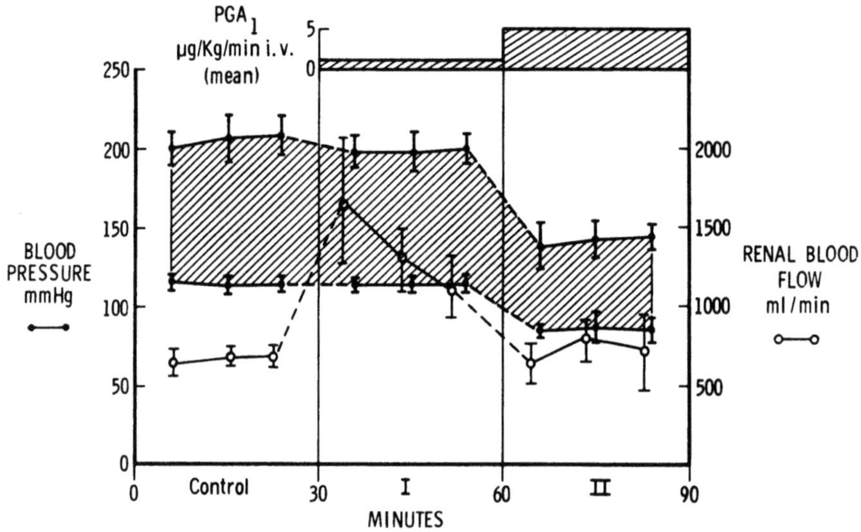

Figure 10. Effect of PGA(1) on blood pressure and renal blood flow in six patients with essential hypertension. Each point represents observations over a 10 min clearance collection. Note the immediate rise in renal blood flow in Period I before blood pressure decreases followed by a lowering in blood pressure to normotensive levels in Period II associated with a return in renal blood flow to control values. From Lee et al (1971). Reproduced with permission of the publisher.

of observed effects being the result of effects on vascular smooth muscle. To determine if a metabolic mechanism might be operative, slices of rabbit renal cortex and outer medulla were incubated in Krebs-tris buffer, glucose-U-^{14}C, 10 mM, 38°C, 90 min with and without PGA(2) and PGE(2) (10^{-3} M). These studies were done in collaboration with Mr. John Lafferty and Dr. Hartmut Kannegiesser at Saint Louis University School of Medicine and Dr. Charles Parker at Washington University School of Medicine. Measurements were made of QO2, glucose to CO2, Na^+-K^+ ATPase and cyclic AMP. Table II indicates that PGA(2) resulted in a 50% fall in QO2 which was the result, in part, of a decrease in glucose incorporation into CO2. There was a highly significant fall in Na^+-K^+ ATPase from 45.1 to 7.1 microM Pi/mg protein/60 min. Control cyclic AMP was 32±4 microM and was uninfluenced by PGA(2). Significantly, PGE(2) was without effect on cortical oxidative metabolism and ATPase whereas PGE(2) was devoid of such activity. The only observed difference was that in medullary slices there was a slight rise in medullary cyclic AMP during PGA(2), but not PGE(2), incubation. The metabolic differences between PGA and PGE compounds together with other known differences (Table III) may ultimately provide clues as to the role and mode of action of each class.

The concentrations at which PGA(2) produced metabolic effects were high (10^{-3} M) but it is of interest that PGE(2) was without effect at this concentration, whereas ethacrynic acid and furosemide, potent natriuretic agents, inhibit renal cortical oxidative metabolism to the same degree and at the same concentration (10^{-3} M). Lastly, the concentration needed to observe metabolic effects *in vitro* may be artificially high because of the poor perfusion characteristics of the slice preparation.

Although the results are quite preliminary, it is of interest that in these studies PGA(2) and PGE(2) had either no effect or caused a very slight rise in cyclic AMP. Although the prostaglandins have been shown to interact with the cyclic AMP system, the absence of a significant effect on renal cyclic AMP is noteworthy. However, present studies were carried out during a 90 min incubation and it is possible that a time-dose relationship study would uncover effects on cyclic AMP which may be obscured by the present experimental design. Thus, further studies will have to be undertaken to conclusively show that PGA(2) and PGE(2) are without definite effects on renal cortical or renal medullary cyclic AMP.

These results suggest that in addition to cardiovascular mechanisms PGA(2) has marked metabolic effects manifested primarily by inhibition of oxidative metabolism which may lead to a decrease in energy production available for active iso-osmotic sodium transport. In outer medulla there may be a similar mechanism of action whereby hyperosmotic sodium transport by thick ascending limb may be inhibited leading to the well established decrease in free water reabsorption resulting from PGA(2) administration *in vivo*. In both instances, the decrease in oxidative metabolism may be the result of inhibition of Na^{+}-K^{+} ATPase which would lead to a diminution of high energy intermediates available for sodium transport and may in part explain the natriuresis induced by the PGA class.

STUDIES IN HUMAN ESSENTIAL HYPERTENSION

Systemic Hemodynamic Effects

Further information on the effects of the PGA compounds on renal function and their relationship to the hypertensive state have been obtained from human studies. The first prostaglandin to be infused into humans with essential hypertension was PGA(2) (isolated from the kidney as medullin). Fig. 9 shows that intravenous administration (382 microgram/min) to a patient with fixed diastolic hypertension resulted in a fall in blood pressure from 185/115 to 165/95 mm Hg. As in the animal studies the decrease in blood pressure was associated with an increase in cardiac output which was primarily due to an increase in heart rate from 96 to 114 beats/min. The increase in heart rate followed the lowering of blood pressure, suggesting a tachycardia of reflex origin probably secondary to enhanced baroreceptor activity. The net result was a fall in total peripheral resistance from 17 to PRU. During this first study there were no observable side effects such as facial flushing, diarrhea or headache although interestingly, there was a marked diuresis which occurred during the infusion of PGA(2).

More recently investigations have been carried out in twenty hypertensive patients with a closely related analog, PGA(1) derived from dehydration of PGE(1), obtained biosynthetically from sheep seminal vesicles (Daniels et al, 1968). Fig. 10 shows that at an infusion rate of 1 microgram/kg/min, PGA(1) administered intravenously usually had no effect on systemic pressure during the first 15-30 min shown as Period I. Subsequently, there was a decline in blood pressure from an elevated control of 200/110 to 140/85 mm Hg during Period II. This fall in blood pressure occurred whether the prostaglandin was administered at a constant rate of 0.25 microgram/kg/min (Lee et al, 1971a) or 1-2 microgram/kg/min (Westure et al, 1970) or whether the infusion rate was increased to 5 microgram/kg/min as shown in Fig. 10. Again during the period of blood pressure fall, there was a rise in cardiac output which was primarily the result of an associated increase in heart rate from 80 to 100 beats/min (Table V). This resulted in a significant fall in the peripheral resistance. Similar observations have been made by Carr (1970) who observed an

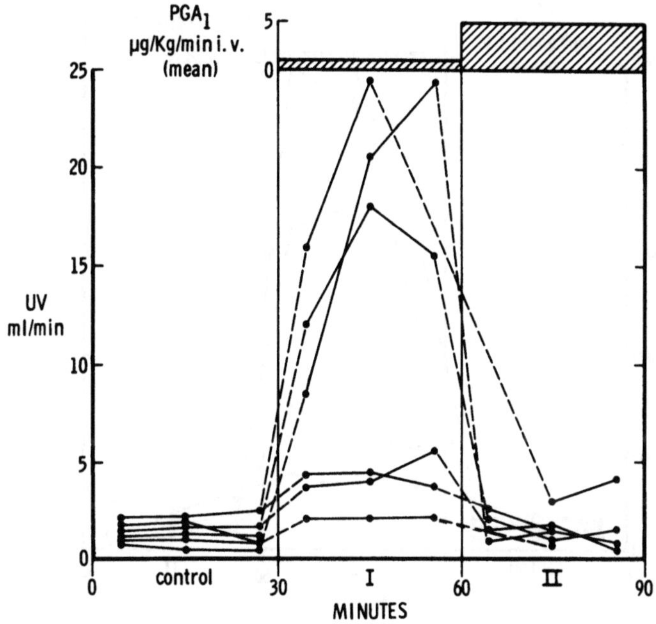

Figure 11. Effect of PGA(1) on urinary flow in patients with essential hypertension. Same conditions as Fig. 9, except each point represents an individual patient. The rise in urine flow during Period I occurs before any change in arterial blood pressure. Again, when blood pressure falls to normotensive levels urinary flow returns to control values. From Lee et al (1971). Reproduced with permission of the publisher.

increased cardiac output from 4.2 to 5.4 l/min when PGA(1) was infused into patients with mild essential hypertension. Thus it would appear that the mechanism of effect of PGA(1) and PGA(2) is primarily the result of peripheral arteriolar vasodilation leading to a fall in total peripheral resistance and a secondary, probably reflex, increase in heart rate which accounts in large part for the elevation in cardiac output.

Renal Effects

Renal Hemodynamics. The earliest and most impressive effect of PGA(1) in patients with essential hypertension was an almost immediate rise in blood flow, before any significant fall in arterial pressure. Fig. 10 illustrates that renal blood flow rose from 600 to 1690

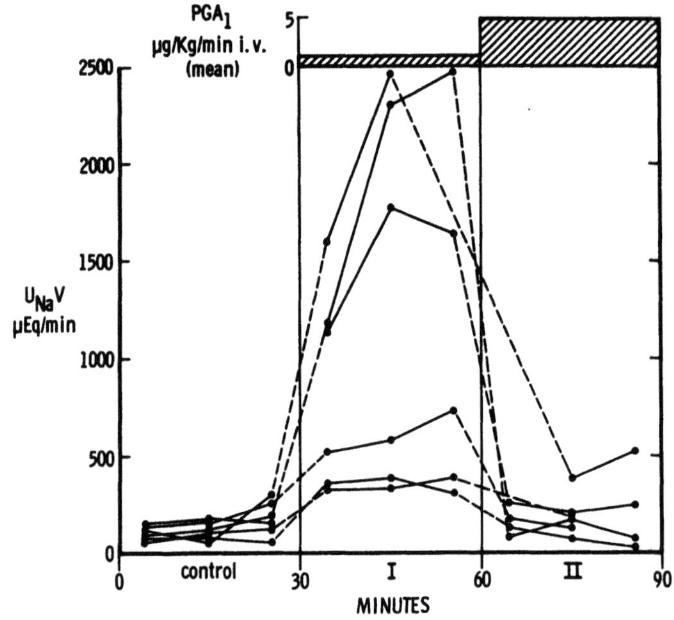

Figure 12. Effect of PGA(1) on sodium excretion in patients with essential hypertension. Same conditions as Fig. 10. The rise in sodium excretion during Period I with a subsequent fall to control values in Period II parallel with the observations made with urine flow. From Lee et al (1971). Reproduced with permission of the publisher.

ml/min at the beginning of Period I when blood pressure remained elevated at 200/110 mm Hg. Subsequently, there was a progressive fall in renal blood flow which reached control levels when blood pressure fell to normotensive levels during Period II.

The rise in renal blood flow was associated with an increase in glomerular filtration rate during Period I which was again transitory, falling toward control when blood pressure became normalized. The increase in renal blood flow was greater than the rise in glomerular filtration rate so there was a significant fall in filtration fraction suggesting preferential efferent cortical arteriolar dilation.

Urinary Flow and Electrolyte Excretion. Fig. 11 illustrates that accompanying the elevation in renal blood flow and glomerular filtration rate there was a rise in urinary flow, in this instance, from a control of approximately 1 ml/min to as high as 24 ml/min. The highest rates of urinary flow occurred in the patients who had the greatest increase in renal blood flow and glomerular filtration rate. Again the pattern of return of urinary flow to control values was observed during Period II when blood pressure fell.

The changes in renal blood flow, glomerular filtration rate and urinary flow were associated by similar alterations in sodium excretion. Natriuresis began almost immediately following PGA(1) infusion with a mild rate of sodium excretion seen in some patients and a marked elevation in others (Fig. 12). As with other renal functions the elevation in sodium excretion during Period I fell to control levels during Period II when normotension occurred. Identical patterns were also observed for potassium excretion and osmolal and free water clearance. Similar changes were also reported by Carr (1970) who observed an elevation in the excretion of calcium and uric acid as well as the previously mentioned electrolytes. An increase in phosphorus excretion was also observed by Carr during studies in hypertensive patients during a water diuresis. The natriuresis and diuresis which occurred in the initial phases of PGA(1) administration resulted in a decrease in plasma volume of approximately 10% (Table V) which accounted, at least in part, for a fraction of the blood pressure decrease. There was also a decrease in serum potassium from 3.9 ± 0.05 to 3.0 ± 0.08 mEq/l at the end of Period II. In five of six patients studied there was an elevation in the renal venous renin content during Period I and in two patients continued elevation during Period II (Lee et al, 1971b).

Side Effects. As with the studies of PGA(2) none of the patients undergoing infusion with PGA(1) exhibited side effects previously observed with PGE(2), that is facial flushing, abdominal cramps, headaches and diarrhea. Two patients showed a severe bradycardia when the infusion rate of PGA(1) exceeded 5 microgram/kg/min. This was alleviated by elevation of the extremities and the administration of atropine together with cessation of PGA(1) administration. No changes have been observed in hematocrit, white count, platelet count, transaminases or BSP retention (Carr, 1970). However, an increased serum amylase was observed on several occasions on the day following PGA(1) administration (Carr, 1970).

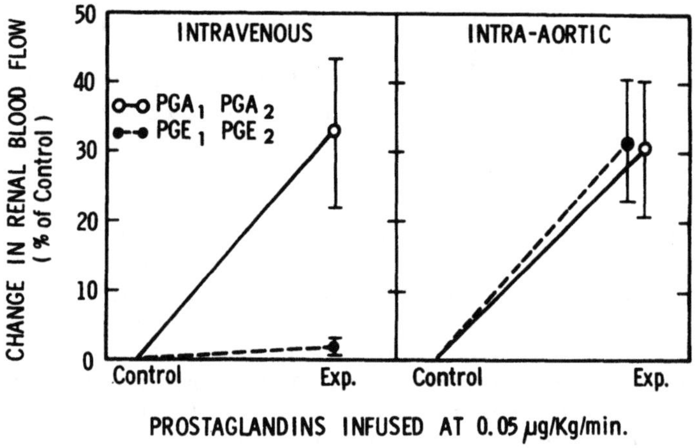

Figure 13. Selective passage of PGA(1) and PGA(2) across dog lung. Intra-aortic infusion of prostaglandins 0.05 microgram/kg/min above the renal arteries results in a prompt rise in renal blood flow both with the PGA and the PGE compounds. In contrast, intravenous administration of these prostaglandins results in an increase in renal blood flow with the PGA compounds but not the PGE compounds indicating metabolism of the latter by the lung. From McGiff et al (1969). Reproduced with permission of the publisher.

PROSTAGLANDINS AS MEDIATORS OF THE ANTIHYPERTENSIVE AND NATRIURETIC RENAL ENDOCRINE FUNCTION

Systemic Mechanisms of Action

There are at least two systemic mechanisms of action whereby prostaglandins might mediate either an antihypertensive or a natriuretic renal function. In the first place, in response to an appropriate stimulus such as an increased volume expansion secondary to increased sodium chloride load, or to behavioral stimuli which tend to raise blood pressure (emotion, intellectual activity, etc.), renal medullary prostaglandins might be synthesized and released into the renal venous blood and reach the systemic arterial circulation where they could (1) exert natriuretic actions by dilation of the renal cortical vasculature or (2) antihypertensive effects by extrarenal arteriolar dilation. If such a systemic release of prostaglandins occurred in the kidney, the compounds responsible for any circulatory or natriuretic effects would have to be the PGA compounds since, as Fig. 13

shows, the PGA compounds are not metabolized by the lung in concentrations which are known to produce vasoactive effects. In contrast, at such concentrations PGE compounds are actively degraded to inactive metabolites.

An alternative systemic hypothesis, not involving the kidney, would be that PGA or PGE compounds are released locally in various regional vascular beds to dilate peripheral arterioles. Following local synthesis and action they might be metabolized near the sight of their synthesis or be degraded systemically by liver, lung and renocortical tissue. Although the systemic hypothesis is not implausible, it is noteworthy that prostaglandins have not been detected in blood in sufficient concentrations to explain their physiological action except in certain specific instances such as medullary carcinoma of the thyroid and in the blood of pregnant women at term. Nonetheless, highly specific and sensitive assays are only now becoming available and with such methods a systemic role for the prostaglandins should be able to be defined in the relatively near future.

Intrarenal Mechanism of Action

Antihypertensive Function. The evidence that PGA(2) or PGE(2) are antihypertensive hormones is at present largely circumstantial and stems from the following considerations. First, PGA(2) and PGE(2) are present normally in the renal medulla and they are among the most potent renal cortical vasodilators causing increases in renal blood flow in concentrations as low as 0.0001 microgram/ml renal blood (McGiff et al, 1969). Secondly, they are physiological antagonists of such pressor mechanisms as angiotensin and norepinephrine which are recognized as being in some fashion importantly involved in blood pressure regulation. Thirdly, from human studies PGA compounds appeared to function as "ideal" antihypertensive agents in that their mechanism of action is by vasodilation of the peripheral arterioles resulting in a decreased peripheral resistance accompanied by a fall in blood pressure. Importantly, this is associated with a reflex increase in cardiac output so that the observed hypotension induced by prostaglandins does not occur by depression in cardiac contractility. It is also noteworthy that the renal circulation actively participates in the lowering of blood pressure induced by PGA compounds so that when normotension is ultimately established, renal function is well maintained. Presumably this latter mechanism results when PGA

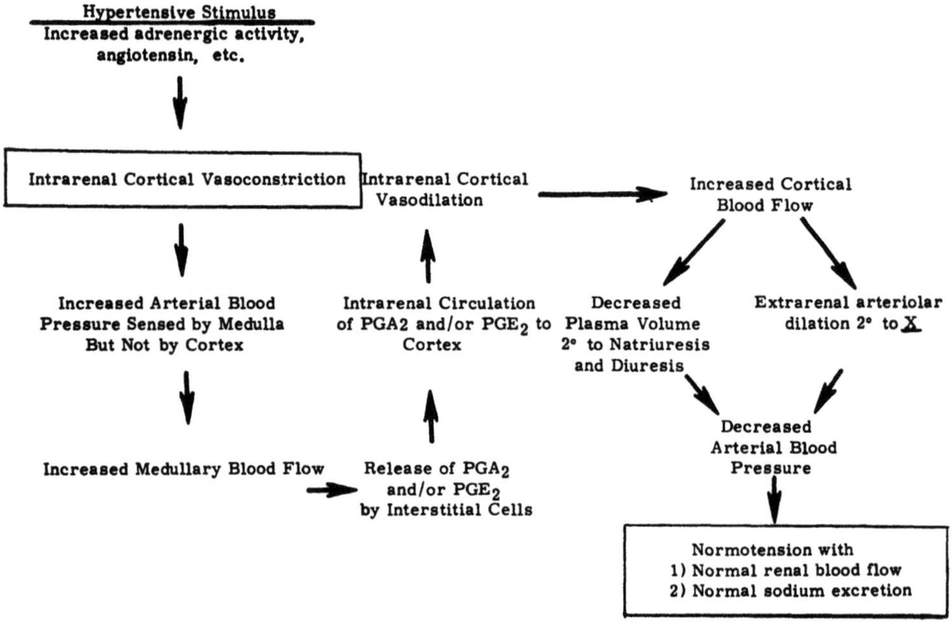

Figure 14. Schematic representation for a possible intrarenal role of the renomedullary prostaglandins in the regulation of systemic blood pressure and maintenance of normotension.

initially causes an increased renal blood flow without a change in systemic blood pressure. Following this there is a fall in systemic blood pressure due to extrarenal arteriolar dilation so that renal arterial perfusion falls. This offsets the continued renal vasodilating action of PGA so that renal blood flow returns from elevated to normal levels. In essence, therefore, PGA(1) mediated antihypertensive action results in the establishment of normotension with normal renal blood flow and normal sodium and water excretion. This is rather unique to the prostaglandins since most antihypertensive agents currently in use may result in a compromise in renal blood flow secondary to a fall in renal arterial perfusion pressure.

Recently, Guyton and Coleman (1969) utilizing a computer involving all the factors known to be important in the establishment of hypertension have concluded that the renal resistance is pivotal in the production of hypertension. This may be especially significant when one considers that in patients with essential hypertension an increase in renal resistance is one of the earliest detectable abnormalities (Bradley et al, 1950; Gomez, 1951). Since PGA(1) and PGA(2) reduce blood pressure in association with relief of intrarenal vasoconstriction, the hypothesis is appealing that they may normally maintain normotension through their actions on the renal circulation.

A proposed mechanism of prostaglandin mediated antihypertension which does not involve elaboration of renomedullary prostaglandins into the systemic circulation is indicated in Fig. 14. Behavioral stimuli - alarm, emotion, intellectual activity, etc. - are known to cause elevations in blood pressure which are transient but are associated with intrarenal vasoconstriction. Within each increase of systemic blood pressure there is a corresponding constriction of renal cortical arterioles so that renal blood flow remains constant despite the elevation in blood pressure. This so called "autoregulation" is not believed to take place in renal medulla so that this area of the kidney "senses" the elevation in blood pressure (Thurau, 1964) and an increase in renomedullary blood flow results. This might result in release of PGA(2) or PGE(2) by the interstitial cells since these cells are in close apposition to the renomedullary vascular tree. Experimental evidence for such PGA(2) or PGE(2) release is derived from data showing that various states of experimental hypertension are associated with a decrease in the osmiophilic granules in these cells which are believed by some to contain renomedullary prostaglandins.

If PGA(2) or PGE(2) are released by such mechanisms it is hypothesized they may circulate possibly via the vasa rectae or the lymphatic vessels to the cortex where they are capable of producing intrarenal cortical vasodilation leading to an increased cortical blood flow. This could lead to two effects which have been observed experimentally: First, a decrease in plasma volume occurs secondary to the associated natriuresis and diuresis, the latter resulting from such factors as increased peritubular capillary pressure and, as suggested in the present study, by decreased ATPase activity and energy production from inhibition of oxidative metabolism. Secondly, extrarenal arteriolar dilation may take place, the mechanism of which is unknown but probably indirect, and is designated in the schema in Fig. 13 as X. Both of these phenomena can contribute to the observed decrease in blood pressure resulting from the fall in total peripheral resistance.

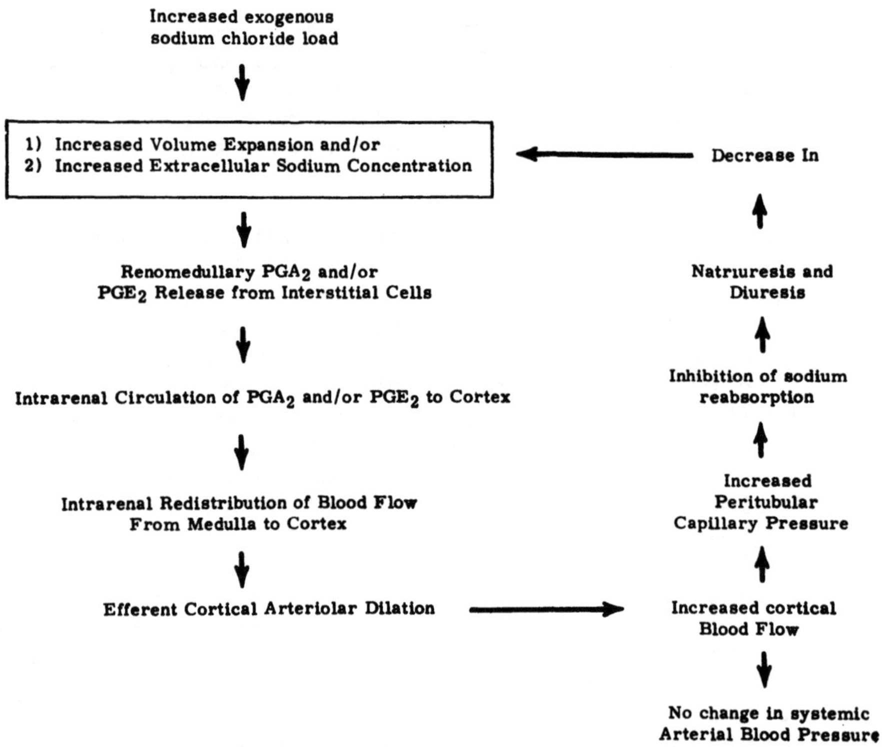

Figure 15. Schematic representation of a possible intrarenal role of the renomedullary prostaglandins in the regulation of extracellular fluid volume and fluid and electrolyte homeostasis.

Since the renal prostaglandins promote cortical vasodilation, an hereditary or acquired deficiency of PGA(2) or PGE(2) in patients with essential hypertension would be expected to result in intrarenal vasoconstriction, a documented finding in patients with essential hypertension. In this fashion blood pressure control would reside within the kidney itself and elaboration of renal prostaglandins into the general circulation would not be necessary for them to exert a hemodynamic regulatory function.

Inconsistent with this hypothesis, is the fact that patients with essential hypertension demonstrate an exaggerated natriuresis when given an endogenous sodium chloride load. The reverse would be expected to be true if there were a deficiency of renal prostaglandins. However, one explanation might be that, from the known differences in biological and metabolic activity between PGE(2) and PGA(2), one compound might be principally involved in antihypertensive activity. Thus a deficiency of PGA(2) formation might lead to systemic hypertension with preservation of natriuretic activity from PGE(2) or vice versa. An alternative explanation for the exaggerated natriuresis is that the hypertensive state itself may have important effects on the renal handling of sodium and water independent of prostaglandin involvement.

Natriuretic Function. As has been mentioned, a number of investigators have suggested that there is a factor or factors which act as natriuretic "hormones". Although none of these particularly indicate that the responsible agents are prostaglandin, there is a striking similarity between the renal effects of renomedullary prostaglandins and saline administration. Both result in a marked increase in renal sodium, water and potassium excretion and increased renal blood flow with redistribution of flow from outer medulla to cortex. The natriuresis resulting from either stimuli is associated with an increase in osmolal clearance and to a lesser degree free water clearance, together with an increase in glomerular filtration rate as well as a decrease in the extraction of PAH and filtration fraction. Less circumstantial evidence in favor of the possibility of prostaglandin mediated natriuresis is the observation that plasma from saline loaded animals inhibits renal cortical slice uptake of PAH *in vitro* (Bricker et al, 1968) while it has been demonstrated that incubation of cortical slices with PGA or PGE compounds significantly inhibits the slice to medium ratio of PAH (Lee and Ferguson, 1969).

Fig. 14 presents an hypothesis based on current experimental evidence, that naturally occurring prostaglandins may mediate saline induced natriuresis. It is known that an increased exogenous sodium chloride load leads to an increased extracellular fluid volume expansion at times associated with an increase in extra-cellular sodium concentration. Intrarenal release of PGE(2) or PGA(2) occasioned by such stimuli, may lead to redistribution of blood flow from outer medulla to cortex similar to that postulated for the antihypertensive effects of these compounds. However in normotensive humans, PGA(1) administration is not associated with any lowering of blood pressure (Patel, N. C., personal communication). Nevertheless, the increased peritubular capillary pressure may lead to natriuresis, diuresis and a decrease in the expanded extracellular fluid volume.

The lack of effect of blood pressure changes in normotensive humans is an important observation in that it may help explain the divergent effects of intrarenal cortical vasodilation in normotensive ahd hypertensive humans. In normotensive but volume-expanded individuals, prostaglandin-induced increase in renal blood flow may lead to inhibition of sodium reabsorption leading to natriuresis without affecting blood pressure. However, the major effect of prostaglandin in hypertensive states, as has been shown, is to lower blood pressure, a phenomenon associated with only a transient increase in renal blood flow and natriuresis.

CONCLUSIONS

Evidence has been presented that the rencmedullary prostaglandins, PGA(2) and PGE(2) may normally participate in an antihypertensive and natriuretic function and that the effect of PGA(2) on the intermediary metabolism of the kidney may be related to the energetics applied to sodium reabsorption in the renal tubule. Obviously, proof of any such hypotheses for the renomedullary prostaglandins will have to await the measurement of these compounds by highly specific and sensitive assays. Although direct experimental proof is still to be obtained, the antihypertensive hypothesis is particularly appealing since PGA(2) and PGE(2) normally present in the kidney exert important effects on renal resistance, total peripheral resistance, sodium balance, extracellular fluid volume, and indirectly on cardiac output leading to a lowering of arterial blood pressure. All of these factors are known to be intimately involved in the development of the state of essential hypertension in humans, as well as experimental, renoprival and renovascular hypertension. The role of the renal prostaglandins as regulators of regional blood flow and arterial blood pressure will almost certainly involve interactions with the various humoral, metabolic, and hemodynamic factors involved in local tissue perfusion and extracellular and intracellular homeostasis.

TABLE I

EFFECT OF PGA AND PGE COMPOUNDS ON RENAL FUNCTION IN NORMOTENSIVE DOG

Increase	Decrease	Unchanged
Total renal blood flow	PAH extraction	Glomerular filtration rate
Renal cortical blood flow	Outer medullary blood flow	Blood pressure
Sodium excretion	Filtration fraction	
Potassium excretion		
Free water clearance		
Osmolal clearance		
Urine flow		

TABLE III

DIFFERENCES BETWEEN PGA(2) and PGE(2) COMPOUNDS*

PGA(2)	PGE(2)
Delayed arteriolar dilation	Immediate arteriolar dilation
Delayed blood pressure fall	Immediate blood pressure fall
No effect on non-vascular smooth muscle	Stimulation of non-vascular smooth muscle
Not metabolized by the lung	Metabolized by the lung
Decreased renal QO2	No effect on renal QO2
Glucose to CO2	Glucose to CO2
Na+-K+ ATPase	Na+-K ATPase

*Similar differences have been observed with PGA(1) and PGE(1) with the exception of Na+-K ATPase and cyclic AMP which have not been measured.

TABLE II

EFFECT OF PGA(2) AND PGE(2) ON THE INTERMEDIARY METABOLISM OF RABBIT RENAL CORTICAL SLICES

	Control	PGA(2)	P	n	Control	PGE(2)	P	n
CO_2 microliters	2109 ±95	1117 ±101	<0.001	6	1823 ±82	1612 ±70	N.S.	6
Glucose to CO_2 microM	4.99 ±0.32	3.82 ±0.34	<0.05	6	4.00 ±0.63	3.81 ±0.50	N.S.	6
Na^+-K ATPase microM	45.1* ±4.4	7.1 ±3.9	<0.001	4	45.1* ±4.4	43.4 ±6.0	N.S.	4
3',5'-cyclic AMP picoM	32+ ±4	33 ±11	N.S.	4	32 ±4	46 ±8	N.S.≠	4

Results expressed/gm initial wet wt/90 min except ATPase which is expressed/mg protein/60 min. Each result represents the mean ±SEM. N.S. = non significant. Results analyzed statistically by Student's *t* test. *Same controls. +Same controls ≠P = <0.025 when analyzed by paired data analysis.

TABLE IV

EFFECTS OF PGA(1) ON CARDIOVASCULAR HEMODYNAMICS IN ESSENTIAL HYPERTENSION

	Control	PGA(1)*
Blood pressure, mm Hg	175/100	160/90
Heart rate (beats/min)	83 ±4	103 ±2
Cardiac output ($l/min/m^2$)	2.98 ±0.57	3.69 ±0.61
Peripheral resistance (dynes sec/cm^5)	2018 ±433	1420 ±256

*PGA(1) infused at 0.25 microgram/kg/min and results obtained 30 min following infusion. From Lee et al (1971a). Reproduced with permission of the publisher.

TABLE V

EFFECT OF PGA(1) ON PLASMA VOLUME IN ESSENTIAL HYPERTENSION

Patient	Control	PGA(1)*	Change	Percent
J. H.	5565	5124	-441	8
M. N.	3458	3169	-289	8
R. F.	3630	3245	-385	11
P. S.	5238	4806	-432	8
Mean	4472 ±541	4086 ±511	-386 ±34	9

*Infused at 0.25 microgram/kg/min. Plasma volume determined 60 min after infusion. Results in ml/patient. From Lee et al (1971a). Reproduced with permission of the publisher.

ACKNOWLEDGEMENTS

The author thanks Dr. Natoo Patel and the Upjohn Company for gifts of PGA(1) and Ms. Carole Atallah for secretarial assistance. These studies were supported by Grants AM 13036-03 and AM 13036-04 from the National Institute of Arthritis and Metabolic Diseases.

REFERENCES

Abodeely, D. A., and Lee, J. B., 1971, The fuel of respiration of outer renal medulla: regulation of intermediate metabolism by extracellular sodium, Am. J. Physiol. 220:1693.

Alexander, J. C., and Lee, J. B., 1970, The effect of osmolality on Na^+-K^+ ATPase in outer renal medulla, Am. J. Physiol. 219:1742.

Anggard, E., 1971, Studies on the analysis and metabolism of the prostaglandins, Ann. N. Y. Acad. Sci. 180:200.

Anggard, E., Boman, L. O., Griffith, J. E., III, Larsson, C., and Maunsbach, A., 1971, Subcellular localization of the prostaglandin system in the rabbit renal papilla, Acta Physiol. Scand. (in press, personal communication).

Barger, A. C., and Herd, J. A., 1966, Study of renal circulation in the unanesthetized dog with inert gases: External counting, Proc. Third International Congress of Nephrology, Karger, Basel, p. 174.

Benoy, M. P., and Elliot, K. A. C., 1937, The metabolism of lactic and pyruvic acids in normal and tumor tissues: V. Synthesis of carbohydrate, Biochem. J. 31:1268.

Bergstrom, S., and Sjovall, J., 1957, The isolation of prostaglandin, Acta Chem. Scand. 11:1086.

Bergstrom, S., and Sjovall, J., 1960a, The isolation of prostaglandin F from sheep prostate glands, Acta Chem. Scand. 14:1693.

Bergstrom, S., and Sjovall, J., 1960b, The isolation of prostaglandin E from sheep prostate glands, Acta Chem. Scand. 14:1701.

Bergstrom, S., Dressler, F., Krabisch, L., Ryhage, R., and Sjovall, J., 1962a, The isolation and structure of a smooth muscle stimulating factor in normal sheep and pig lungs, Ark. Kemi 20:63.

Bergstrom, S., Dressler, F., Ryhage, R., Samuelsson, B., and Sjovall, J., 1962b, The isolation of twc further prostaglandins from sheep prostate glands, Ark. Kemi 19:563.

Bergstrom, S., Ryhage, R., Samuelsson, B., and Sjovall, J., 1962c, The structure of prostaglandin E, F(1) and F(2), Acta Chem. Scand. 16:501.

Bergstrom, S., Ryhage, R., Samuelsson, B., and Sjovall, J., 1963, The structures of prostaglandin E(1), F(1-alpha) and F(1-beta), J. Biol. Chem. 238:3555.

Blaquier, P., Bohr, D. F., and Hoobler, S. W., 1960, Evidence against an increase in circulating pressor material in renal hypertensive rats, Am. J. Physiol. 198:1148.

Bradley, S. E., Bradley, G. P., Tyson, C. J., Curry, J. J., and Blake, W. D., 1950, Renal function in renal diseases, Am. J. Med. 9:766.

Braun-Menendez, E., and von Euler, U. S., 1947, Hypertension after bilateral nephrectomy in the rat, Nature (London) 160:905.

Brenner, B. M., Falchuk, K. H., Keimowitz, R. I., Berliner, R. W., 1969, Relationship between peritubular capillary protein and fluid reabsorption by the renal proximal tubule, J. Clin. Invest. 48:1519.

Bricker, N. S., Klahr, S., Purkerson, M., 1968, In vitro assay for a humoral substance present during volume expansion and uremia, Nature 219:1058.

Buckalew, V. M., Jr., Martinez, F. J., and Green, W. E., 1970, Effect of dialystates and ultrafiltrates of plasma of saline-loaded dogs on toad bladder sodium transport, J. Clin. Invest. 49:929.

Carr, A. A., 1970, Hemodynamic and renal effects of a prostaglandin, PGA(1), in subjects with essential hypertension, Am. J. Med. Sci. 259:21.

Covino, B. G., Lee, J. B., and McMorrow, J. V., 1968, Circulatory effects of prostaglandins, Circulation 38:(Suppl. VI):60A.

Daniels, E. G., and Pike, J. E., 1968, Isolation of prostaglandins, In: Prost. Symp. Worc. Found. Exper. Biol., (P. W. Ramwell and J. E. Shaw, Eds.), Interscience Publishers, New York, p. 379.

Davis, R. P., 1964, Glycolytic Pathways and Cation Transport, In: Renal Metabolism and Epidemiology of Some Renal Diseases, Proc. 15th Ann. Conf. on the Kidney, Jack Metcoff, M. D., Ed. p. 114.

Deetjen, P., and Kramer, K., 1961, Abhangigkeit des O2-Verbrauchs der Niere von der Na-Ruckresorption, Pflugers Arch. Ges. Physiol. 273:636.

DeWardener, H. E., Mills, I. M., Clapham, W. F., and Hayter, C. J., 1961, Studies on the efferent mechanism of sodium diuresis which follows the administration of intravenous saline in the dog, Clin. Sci. 21:249.

Dunham, E. W., and Zimmerman, B. G., 1970, Release of prostaglandin-like material from dog kidney during nerve stimulation, Am. J. Physiol. 219:1279.

Earley, L. E., and Friedler, R. M., 1965, Studies on the mechanism of natriuresis accompanying increased renal blood flow and its role in the renal response to extracellular volume expansion, J. Clin. Invest. 44:1857.

Earley, L. E., and Friedler, R. M., 1966, The effects of combined renal vasodilation and pressor agents on renal hemodynamics and the tubular reabsorption of Na, J. Clin. Invest. 45:542.

Fasciolo, J. C., 1938, Accion del rinon sano sobre la hipertension arteriol por isqemia renal, Rev. Soc. Argent. Biol. 14:15.

Goldblatt, M. W., 1953, Depressor substance in seminal fluid, Chem. Ind. 52:1056.

Gomez, D. M., 1951, Evaluation of renal resistance with special reference to changes in essential hypertension, J. Clin. Invest. 30:1143.

Gomez, A. H., Hoobler, S. W., and Blaquier, P., 1960, Effect of addition and removal of kidney transplant in renal and adrenocortical hypertensive rats, Circ. Res. 8:464.

Grantham, J. J., and Orloff, J., 1968, Effect of prostaglandin E(1) on the permeability response of the isolated collecting tubule to vasopressin, adenosine 3',5'-monophosphate and theophylline, J. Clin. Invest. 47:1154.

Grollman, A., Williams, J. R., Jr., and Harrison, T. R., 1940, Reduction of elevated blood pressure by administration of renal extract, J. A. M. A. 115:1169.

Grollman, A., Muirhead, E. E., and Vanatta, J., 1949, Role of the kidney in pathogenesis of hypertension as determined by a study of the effects of bilateral nephrectomy and other experimental procedures on the blood pressure of the dog, Am. J. Physiol. 157:21.

Guyton, A. C., and Coleman, T. G., 1969, Quantitative analysis of the pathophysiology of hypertension, Circ. Res. 24:(Suppl. I):I-1.

Hamburg, M., 1969, Biosynthesis of prostaglandins in the renal medulla of rabbit, FEBS Letters 5:127.

Hamberg, M., and Samuelsson, B., 1966, Prostaglandins in human seminal plasma, J. Biol. Chem. 241:257.

Hamilton, J. G., and Grollman, A., 1958, The preparation of renal extracts effective in reducing blood pressure in experimental hypertension, J. Biol. Chem. 233:528.

Hauge, A., Lunde, P. K. M., and Waaler, B. S., 1967, Effects of prostaglandin E(1) and adrenaline on the pulmonary vascular resistance (PVR) in isolated rabbit lungs, Life Sciences 6:673.

Hendler, E. K., Toretti, J., Weinstein, E., and Epstein, F. H., 1969, Functional significance of the distribution of Na-K ATPase within the kidney, J. Clin. Invest. 48:37a.

Hickler, R. B., Lauler, D. P., Saravis, C. A., Vagnucci, A. I., Steiner, G., and Thorn, G. W., 1964, Vasodepressor lipid from the renal medulla, Can. Med. Assoc. J. 90:280.

Holmes, S. W., Horton, E. W., and Main, I. H. M., 1963, The effect of prostaglandin E(1) on responses of smooth muscle to catecholamines, angiotensin and vasopressin, Brit. J. Pharm. 21:528.

Horton, E. W., 1963, Action of prostaglandin E(1) on tissues which respond to bradykinin, Nature 200:892.

Hume, D. M., Merrill, J. P., Miller, B. F., and Thorn, G., 1955, Experiences with renal homotransplantation in the human: Report of nine cases, J. Clin. Invest. 34:327.

Ishii, M., and Tobian, L., 1969, Interstitial cell granules in renal papilla and the solute composition of renal tissue in rats with Goldblatt hypertension, J. Lab. Clin. Med. 74:47.

Johnston, H. H., Herzog, J. P., and Lauler, D. P., 1967, Effect of prostaglandin E(1) on renal hemodynamics, sodium and water excretion, Am. J. Physiol. 213:939.

Jones, L. G., and Herd, J. A., 1970, Intrarenal distribution of blood flow during saline diuresis, Fed. Proc. 29:398a.

Jorgenson, P., and Skou, J. C., 1969, Preparation of highly active Na-K ATPase from the outer medulla of rabbit kidney, Biochem. Biophys. Res. Commun. 37:39.

Kadar, D., and Sunahara, F. A., 1969, Inhibition of prostaglandin effects by ouabain in the canine vascular tissue, Can. J. Physiol. Pharm. 47:871.

Kannegiesser, H., and Lee, J. B., 1971, Difference in haemodynamic response to prostaglandin A and E, Nature 229:314.

Kean, E. L., Adams, P. H., Winters, R. W., and Daview, R. C., 1961, Energy metabolism of the renal medulla, Biochem. Biophys. Acta 54:474.

Kiil, F., Aukland, K., and Refsum, H. E., 1961, Renal sodium transport and oxygen consumption, Am. J. Physiol. 201:511.

Koletsky, S., and Pritchard, W. H., 1963, Vasopressor material in experimental renal hypertension, Circ. Res. 13:552.

Kolff, W. J., and Page, I. H., 1954, Blood pressure reducing function of the kidney; reduction of renoprival hypertension by kidney perfusion, Am. J. Physiol. 178:75.

Kolff, W. J., Nakamoto, S., Poutasse, E. F., Straffon, R. A., and Figueroa, J. E., 1964, Effect of bilateral nephrectomy and kidney transplantation on hypertension in man, Circulation 30:(Suppl. II):23.

Krebs, H. A., 1935, Metabolism of amino acids: III. Deamination of amino acids, Biochem. J. 29:1620.

Lee, J. B., 1967, Chemical and physiological properties of renal prostaglandins with emphasis on the cardiovascular effects of medullin in essential human hypertension, In: Prostaglandin II Nobel Symposium, (S. Bergstrom and B. Samuelsson, Eds.), Stockholm, Almqvist and Wicksell; New York, Interscience, p. 197.

Lee, J. B., 1968, Cardivascular implications of the renal prostaglandins, In: Prostaglandin, Symp. Worc. Found. Expt'l. Biol., Wiley, p. 131.

Lee, J. B., 1970, Prostaglandins, The Physiologist, 13:379.

Lee, J. B., Hickler, R. B., Saravis, C. A., and Thorn, G. W., 1962a, Sustained depressor effect of renal medullary extract in the normotensive rat, Circulation 26:747, Part II.

Lee, J. B., Vance, V. K., and Cahill, G. F., Jr., 1962b, Metabolism of C-14 labelled substrates by rabbit kidney cortex and medulla, Am. J. Physiol. 203:27.

Lee, J. B., Hickler, R. B., Saravis, C. A., and Thorn, G. W., 1963, Sustained depressor effects of renal medullary extract in the normotensive rat, Circ. Res. 13:359.

Lee, J. B., Mazzeo, M. A., and Takman, B. H., 1964a, The acidic lipid characteristics of sustained rencmedullary depressor activity, Clin. Res. 12:254.

Lee, J. B., Takman, B. H., and Covino, B. G., 1964b, The isolation and chemical characteristics of renomedullary depressor substances, The Physiologist 7:188.

Lee, J. B., Covino, B. G., Takman, B. H., and Smith, E. R., 1965, Remomedullary vasodepressor substance, Medullin: Isolation, chemical characterization and physiological properties, Circ. Res. 7:57.

Lee, J. B., Gougoutas, J. Z., Takman, B. H., Caniels, E. G., Grostic, M. F., Pike, J. E., Hinman, J. W., and Muirhead, E. E., 1966, Vasodepressor and antihypertensive prostaglandins of PGE type with emphasis on the identification of medullin as PGE(2)-217, J. Clin. Invest. 45:1036.

Lee, J. B., Crowshaw, K., Takman, B. H., and Gougoutas, J. Z., 1967, The identification of PGE(2), PGF(2-alpha), and PGA(2) from rabbit kidney medulla, Biochem. J. 105:1251.

Lee, J. B., and Ferguson, J. F., 1969, The effect of renal prostaglandin on PAH uptake by kidney cortex, Nature 222:1185.

Lee, J. B., and Peter, H. M., 1969, The effect of oxygen tension on glucose metabolism in rabbit kidney cortex and medulla, Am. J. Physiol. 217:464.

Lee, J. B., Kannegiesser, H., O'Toole, D., and Westura, E. E., 1971a, Hypertension and the renomedullary prostaglandins: A human study of the antihypertensive effects of PGA(1), Ann. N. Y. Acad. Sciences, 180:218.

Lee, J. B., McGiff, J. C., Kannegiesser, H., Aykent, Y. Y., Mudd, J. G., and Frawley, T. F., 1971b, Antihypertensive renal effects of prostaglandin A(1) in patients with essential hypertension, Ann. Int. Med. 74:703.

Martinez-Maldonado, M., Allen, J. C., Knoyan, G. W., Suki, W., and Schwartz, A., 1969, Renal concentrating mechanism: Possible role for sodium potassium activated adenosine triphosphatase, Science 165:807.

McGiff, J. C., Terragno, N. A., Strand, J. C., Lee, J. B., Lonigro, A. J., and Ng, K. K. F., 1969, Selective passage of PG's across the lung, Nature 223:742.

McGiff, J. C., Crowshaw, K., Terragno, N. A., Lonigro, A. J., 1970a, Renal prostaglandins: Possible regulators of the renal actions of pressor hormones, Nature 227:1255.

McGiff, J. C., Crowshaw, K., Terragno, N. A., Lonigro, A. J., Strand, J. C., Sr., Williamson, M. A., Lee, J. B., and Ng, K. K. F., 1970b, Prostaglandin-like substances appearing in canine renal venous blood during renal ischemia: Their partial characterization by pharmacologic and chromatographic procedures, Circ. Res. 27:765, 1970.

Muehrcke, R. C., Mandal, A. K., Epstein, M., and Volini, F. E., 1969, Cytoplasmic granularity of the renal medullary interstitial cells in experimental hypertension, J. Lab. Clin. Med. 73:299.

Muirhead, E. E., Stirman, J. A., Lesch, W., and Jones, F., 1956, The reduction of postnephrectomy hypertension by renal homotransplant, Surg., Gynec. Obst. 103:673.

Muirhead, E. E., Jones, F., and Stirman, J. A., 1960, Antihypertensive property in renoprival hypertension of extract from renal medulla, J. Lab. Clin. Med. 56:167.

Muirhead, E. E., Brown, G. B., Germain, G. S., and Leach, B. E., 1970, The renal medulla as an antihypertensive organ, J. Lab. Clin. 76:641.

Nakano, J., 1968, Effects of prostaglandins E(1), A(1) and F(2-alpha) on the coronary and peripheral circulations, Proc. Soc. Exper. Biol. Med. 127:1160.

Nakano, J., and McCurdy, J. R., 1967, Cardiovascular effects of prostaglandin E(1), J. Pharm. Exp. Therap. 156:538.

Nakano, J., and McCurdy, J. R., 1968, Hemodynamic effects of prostaglandins E(1), A(1), and F(2-alpha) in dogs, Proc. Soc. Exper. Biol. Med., 128:39.

Nissen, H. M., and Bojesen, I., 1969, On lipid droplets in renal interstitial cells, IV. Isolation and identification, Z. Zellforsch. 97:274.

Page, I. H., Helmer, O. M., Kohlstaedt, K. G., Kempf, G. F., Gambill, W. D., and Taylor, R. D., 1941, The blood pressure reducing property of extracts of kidneys in hypertensive patients and animals, Ann. Int. Med. 15:347.

Sealey, J. E., and Laragh, J. H., 1971, Further studies of a natriuretic substance occurring in human urine and plasma, Circ. Res. 28:(Suppl. II):32.

Smith, E. R., McMorrow, J. V., Covino, B. G., and Lee, J. B., 1968, Studies on the vasodilator action of prostaglandin E(1), In: Prost. Symp. Worc. Found. Exper. Biol., (P. W. Ramwell and J. E. Shaw, Eds.), Interscience Publishers, New York, p. 259.

Solomon, L. M., Juhlin, L., and Kirschenbaum, M. B., 1968, Prostaglandin on cutaneous vasculature, J. Invest. Derm. 51:280.

Strong, C. G., and Bohr, D. F., 1967, Effects of prostaglandins E(1), E(2), A(1), and F(1-alpha) on isolated vascular smooth muscle, Am. J. Physiol. 213:725.

Thurau, K., 1961, Renal Na-reabsorption and oxygen-uptake in dogs during hypoxia and hydrochlorothiazide infusion, Proc. Soc. Exper. Biol. Med. 106:714.

Thurau, K., 1964, Renal hemodynamics, Am. J. Med. 36:698.

Tobian, L., Ishii, M., and Duke, M., 1969, Relationship of cytoplasmic granules in renal papillary interstitial cells to "postsalt" hypertension, J. Lab. Clin. Med. 73:309.

von Euler, U. S., 1934, Zur Kenntnis der pharmakologischen Wirkungen von nativsekreten und extrakten mannlicher accessorischer Geschlectsdrusen, Arch. Exptl. Pathol. Pharmakol. 175:78.

von Euler, U. S., 1935, Uber die spezifische blutdrucksenke Substanz des menschlichen Prostata und Samenglasenkretes, Klin. Wochschr. 14:1182.

von Euler, U. W., 1936, On the specific vasodilating and plain muscle stimulating substances from accessory genital glands in man and certain animals (prostaglandin and vesiglandin), J. Physiol. 81:65.

Warburg, O., 1927, Uber die Klassifizierung eierischer gewebe nach ihrem stoffwechsel, Biochem. Ztschr. 184:484.

Weiner, R., and Kaley, G., 1969, Influence of prostaglandin E(1) on the terminal vascular bed, Am. J. Physiol. 217:563.

Westura, E. E., Kannegiesser, H., O'Toole, J. D., and Lee, J. B., 1970, Antihypertensive effects of prostaglandin A(1) in essential hypertension, Circ. Res. 27:(Suppl. I):131.

DISCUSSION

KESSLER: I would like to show a few slides which are complementary to Dr. Lee's.

Dr. Lee has been investigating the effects of prostaglandins on the excretion of sodium and water in hypertensive states. We have looked at the other side of the coin, so to speak, studying the effects of saline infusion, in hypotensive states, on urinary excretion. Endotoxin shock in the dog is our model. In Table 1 is a summary of the effects of saline in normal animals. You observe the expected increase in urine flow and sodium excretion with a fall in urine osmolality down to iso-osmotic levels. The difference in water reabsorption between the control and experimental is not significant. There was no change in blood pressure. In the endotoxin experiments the drop in blood pressure is clear. During the administration of saline the increase in urine flow is less than in normal dogs, but exceeds pre-endotoxin levels. Another thing that is clear is the failure of the kidney to increase the sodium excretion beyond pre-endotoxin control levels. Even more striking is the fall in water reabsorption so that free water appears at low rates of solute clearance. These results were quite puzzling, and we considered it possible that there was something in plasma which might cause these aberrant responses. We therefore induced saline diuresis in other animals and abruptly changed the infusing solution to normal plasma, or endotoxin plasma, or hemorrhagic shock plasma, without changing the infusion rate. With both shock plasmas, but not normal plasma, we saw an increase in urine flow, a decrease in water reabsorption, a fall in urine osmolality, and a fall in the filtration fraction. These observations suggested that there was some material in the shock plasmas which was vasodilator and inhibited water reabsorption.

Fig. 1: We began to extract plasma with ethyl acetate at pH 3 from normal dogs and following the induction of shock. This slide shows the mean values after conversion to PGB compounds with potassium hydroxide. There is a marked difference in the optical density of the shock plasmas from the normals. Quite regularly we obtained peaks around 278 nanometers following lipid extraction of the shock plasma, but not normal plasma.

Fig. 2: We have compared individual control plasmas to the shock plasmas in the same animal. The peaks at 278 nanometers are shown in the middle column after extraction and elution from silicic acid columns. In every instance shock plasma extracts showed higher peaks than control. We also noted peaks in shock plasma extracts at 217 nanometers compared to normal plasma extracts. In four of the five instances we obtained peaks at 238 nanometers as well, following thin layer chromatography.

Fig. 3: We can show material that looks like PGA(2) in the shock plasma. Occasionally we find a spot on TLC in normal plasma extracts which is comparable to the shock plasma and PGA(2), but these spots are never as intense as they are in the shock plasma extracts. The final point is that the extracts eluted from the columns are vasodepressor in the rat. Therefore, one wonders whether in the shock process, prostaglandins are being elaborated, possibly, in the kidney, which predispose the endotoxin shock animal to react in the peculiar way it does to saline infusion. Using the shock model to compare with the hypertensive model, there may be something basically common in these models. If, as Dr. Lee suspects, hypertension may be associated with a deficiency of prostaglandin in the kidney, it seems possible that shock may be associated with excess prostaglandin production.

BLOCH: Do you have any comments, Dr. Lee?

LEE: It will certainly be consistent with the possibility that locally or intrarenally there might be a release of vasoactive prostaglandins which would participate in generalized arteriolar dilation or shock.

BLOCH: Your paper should allow us to ascertain the precise sites in the kidney where some of these controls exist. This is a problem which is not very well settled.

LEE: You are right, Dr. Bloch, because the kidney offers a unique model to study the synthesis and degradation of the prostaglandins. We believe we have a model where we can study the prostaglandins being synthesized at a site different from where they act, whereas in other tissues, it is possible that the prostaglandins could be synthesized, have their effect and be metabolized all at the same site.

The latter situation, of course, makes investigations of their mode of operation quite difficult. In the case of the kidney, anatomic separation of these functions may afford clues as to unifying mechanisms of action. In this instance the kidney may well be an extremely useful model applicable to many other systems in the body. We have hypothesized that one unifying mechasism of PGE and PGA action might be local regulation of blood flow to various organ systems. The appropriate response would be inherent and specific to the organ or tissue involved. Thus, increases in renal cortical flow mediated by prostaglandins may lead to changes in renal function specifically related to the kidney, i.e., natriuresis and blood pressure regulation. In other organs it might lead to responses inherent to the particular tissue (such as the effect of PGF(2-alpha) on luteal blood flow leading to luteolysis and accompanying reproductive effects). Of course, this oversimplification does not explain many of the non-vascular effects of the PGE and PGF class, nor the renal metabolic effects of the PGA compounds. Nevertheless, local prostaglandin release triggered by the appropriate stimulus appears, at this time, to be the best starting point for future investigations.

Table 1.
Effects of saline infusion in endotoxin shock.

	Mean AP	$\frac{V}{Cin}$x100	$\frac{CNa}{Cin}$x100	$\frac{T^cH_2O}{Cin}$x100	Cin	Cpah	Filt. Fraction
Control	150	2.4	2.9	1.9	49.4	138	.37
Saline (10)	152	20.0	18.3	-0.9	57.1	161	.37
Cont.	136	3.8	4.3	1.6	69	142	.53
Endo & Saline (7)	95	8.4	3.1	-4.4	42	130	.37

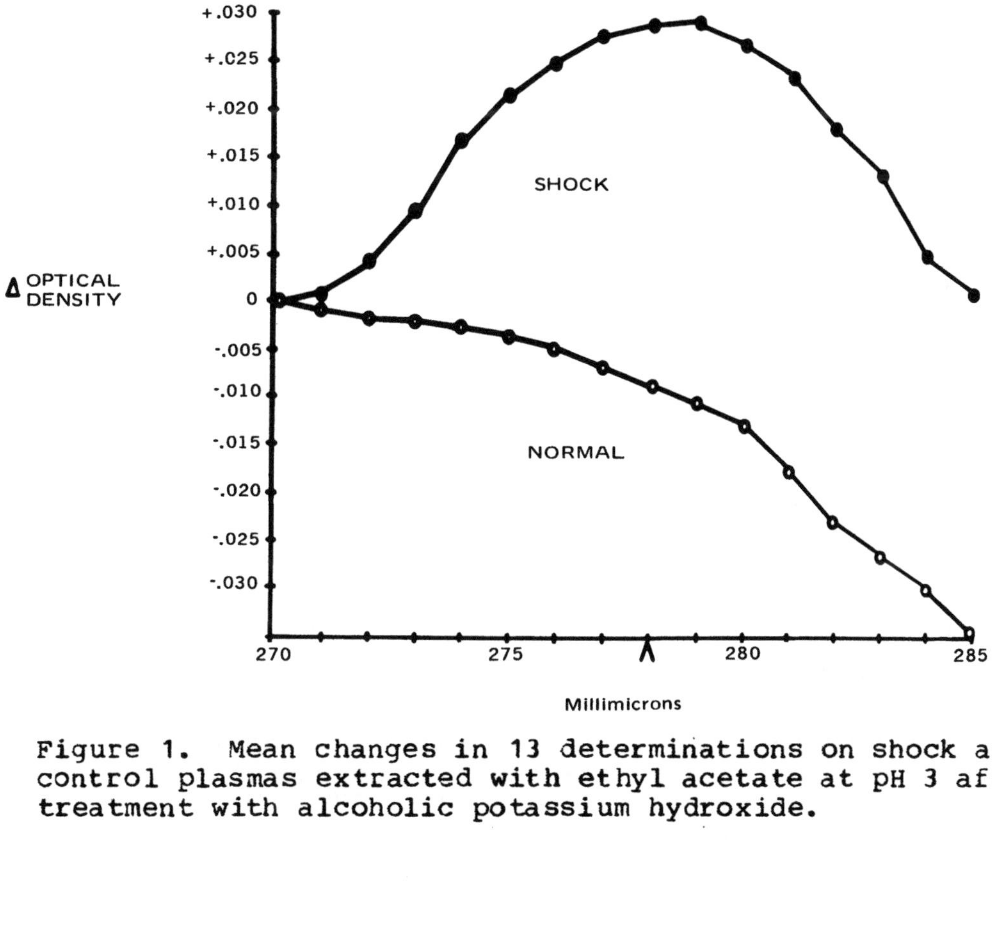

Figure 1. Mean changes in 13 determinations on shock and control plasmas extracted with ethyl acetate at pH 3 after treatment with alcoholic potassium hydroxide.

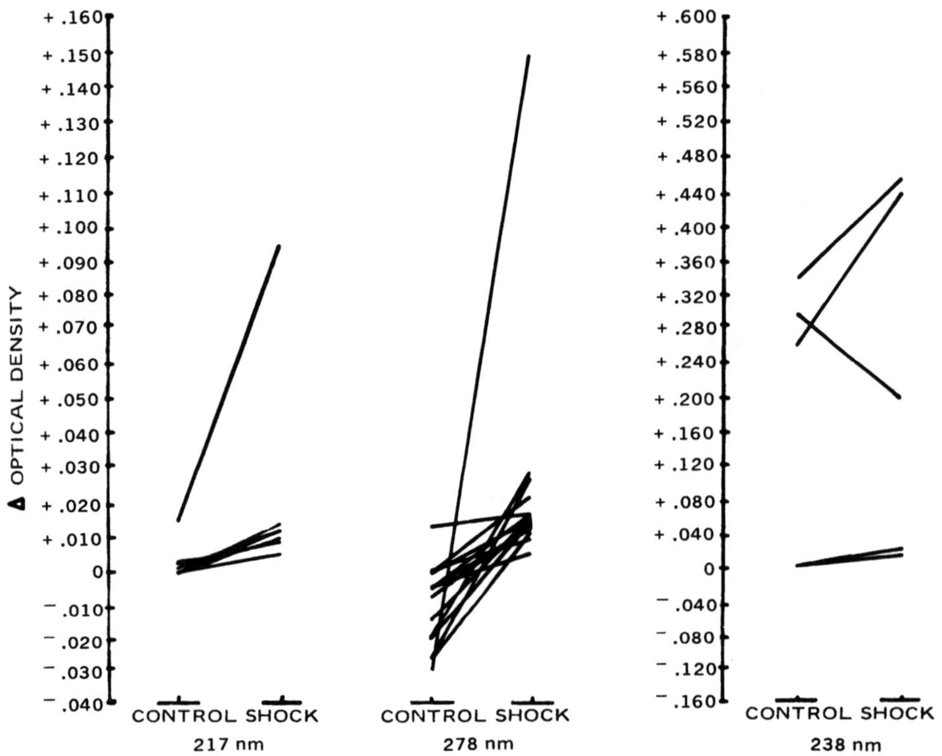

Figure 2. Comparison of changes in optical density of control and shock plasma extracts in individual dogs at 217, 278 and 238 nanometers using 210, 270 and 230 nanometers as baselines, respectively.

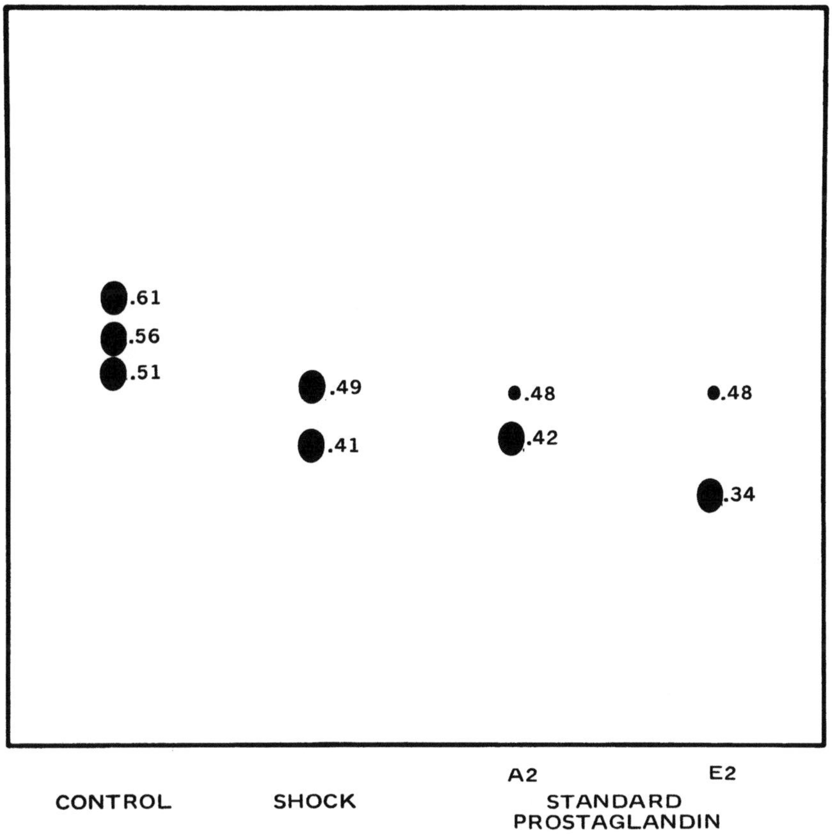

Figure 3. Thin layer chromatographic separation of plasma extracts in a solvent system consisting of 2,2,4-trimethylpentane:isopropyl alcohol:glacial acetic acid, 135:45:2.

PROSTAGLANDINS AND THE MICROVASCULAR SYSTEM: PHYSIOLOGICAL AND HISTOCHEMICAL CORRELATIONS

G. R. Siggins

Laboratory of Neuropharmacology
Division of Special Mental Health Research, NIMH
Saint Elizabeth's Hospital, Wash., D. C.

INTRODUCTION

It is well known that the prostaglandins markedly alter cardiovascular activity.. However, the nature of the vascular response to exogenously administered prostaglandins depends not only upon the type and dose of prostaglandin studied but also upon the test system used. In general, a myriad of reports have shown that prostaglandins E(1), E(2), A(1), A(2), F(1-alpha) and F(2-alpha) are potent vasodilator substances. All of these prostaglandins, and especially PGF(2-alpha), have also been reported to produce vasoconstriction in some vascular preparations (DuCharme and Weeks, 1967; Jackson and Stovall, 1968; Nakano, 1968; Strong and Bohr, 1967; Sunahara and Kadar, 1968; Davies and Withrington, 1969; Mark et al, 1971). At least a few of the discrepancies in effect reported for the prostaglandins may result from the general employment of indirect methods, such as estimations of systemic blood pressure, flow and vascular resistance, to estimate changes in vessel diameter. The few exceptions (Kaley and Weiner, 1967; Strong and Bohr, 1967; Sunahara and Kadar, 1967; Berman and Siggins, 1968; Viguera and Sunahara, 1969; see also Kaley; Strong; and Winkelmann, this volume) have used either isolated blood vessels or thin membrane preparations to directly observe living microvascular elements and their response to prostaglandins.

An early observation was that prostaglandins can antagonize the pressor action of catecholamines (von

Euler, 1939; Steinberg et al, 1963). Although it has been directly shown that certain prostaglandins have the ability to block the vasoconstrictor action of several vasoactive agents such as the catecholamines and, sometimes, angiotensin and vasopressin (Kaley and Weiner, 1968; Strong and Bohr, 1967; Viguera and Sunahara, 1969), the exact cellular site of these interactions was not determined. Consideration of the site of action is important with respect to potential functional interactions of prostaglandins with the sympathetic nervous system, because von Euler, Hedqvist and colleagues (see Hedqvist, 1970) have shown that PGE(1) and PGE(2) in non-vascular tissues are capable of blocking junctional release of catecholamines as well as their postsynaptic actions. Moreover, Nakano and MuCurdy (1967), Kaplan et al (1969) and Lavery et al (1970) reported that prostaglandins can also exert strong cardiovascular actions mediated by the central nervous system. Similarly, since angiotensin is known to exert at least part of its pressor action through the sympathetic nervous system (cf. Severs et al, 1966; Deuben and Buckley, 1970; Peach et al, 1970), studies of possible prostaglandin-angiotensin interactions at pre- and post-junctional levels would also be helpful.

Some of these problems were explored in the present study by utilizing _in vivo_ microscopy of the microvasculature of two thin membrane preparations. The frog retrolingual membrane is ideal for neurotropic drug studies because its thinness permits easy visualization of terminal vasoconstrictor and vasodilator nerves, which may be selectively stimulated by microelectrode (Siggins, 1967; Berman and Siggins, 1968). Five different prostaglandins, PGE(1), PGE(2), PGF(1-alpha), PGF(2-alpha) and PGA(2) were tested on this preparation. For comparison of prostaglandin effects in a mammal, the somewhat thicker hamster cheek pouch (Fulton et al, 1947) was also studied. Finally, for further assessment of the possible functional significance of the extremely potent effects seen with topically applied prostaglandins, a newly described histochemical technique for staining prostaglandin 15-hydroxy dehydrogenase (PGDH) (Nissen and Anderssen, 1968) was applied to the two membrane preparations. This dehydrogenase is known to provide a major pathway for biological inactivation of the prostaglandins (Granstrom and Samuelsson, 1971; Samuelsson et al, 1971), and determination of the cellular localization of the enzyme could aid in understanding the sites of prostaglandin catabolism, as well as providing insights into potential regions of prostaglandin elaboration.

METHODS

Preparation of the Frog Retrolingual Membrane

Male and female frogs (*Rana pipiens*, 28-55 g body weight) were immobilized by immersion in 50-150 ml of a 1:1000 solution of MS-222 (tricaine methanesulfonate, Sandoz) in water or, less frequently, by single pithing of the brain. Each retrolingual membrane was prepared as described elsewhere (Pratt and Read, 1930; Fulton and Lutz, 1942; Siggins, 1967). This structure is the thin (20-70 microns) epithelial surface of a large lymph sac at the base of the tongue. After the underlying tissue was dissected away, leaving only the retrolingual membrane, the structure was trans-illuminated and observed with a modified Bausch and Lomb microscope at 100 or 200 X.

Terminal arterioles (15-60 micron outer diameter) were located within the membrane along with the occasional fine (1-8 micron diameter) nerves which innervate these arterioles (Berman and Siggins, 1968; Siggins and Bloom, 1970). These vasomotor nerves, or the wall of the arteriole itself, were stimulated with a monopolar platinum microelectrode (maneuvered into the microscope field by micromanipulator) which was insulated with glass to the 10 micron diameter tip. Stimulating current was generated by a Grass S-4 stimulator and monitored on a Tektronix oscilloscope. Controls and methods of selectively stimulating either vasodilator or vasoconstrictor nerves have been described by Siggins (1967) and Berman and Siggins (1968).

Drug solutions were administered topically to the arterioles from tuberculin syringes; to avoid differential penetrations of some drugs through either surface of the retrolingual membrane and to better insure that the drugs always reached effector cells, the solutions were applied to both ventral (epithelial) and dorsal (internal) surfaces of the membrane arteriole. Responses to drugs, Ringer's solution alone or electrical stimulations were documented and measured as described below. All experiments were carried out at room temp (20-23°C).

Preparation of the Hamster Cheek Pouch

Golden hamsters (*Mesocricetus auratus*, 105-145 g body weight) were anesthetized by chloral hydrate (350 mg/kg). The cheek pouch was everted, pinned out over a lucite block, and prepared according to the "single membrane preparation" of Fulton et al (1947), in which a portion of one side of the pouch is dissected away. After careful removal of several layers of connective tissue from the remaining underlying side of the pouch, the structure was trans-illuminated and observed microscopically as with the frog retrolingual membrane. Since in this thicker tissue terminal vasomotor nerves are rarely visible *in vivo*, and connective tissue over arteriolar walls is abundant, no attempt was made to activate perivascular nerves by microelectrode. Drug solutions were warmed to 37°C and applied topically as with the retrolingual membrane, although only to the exposed upper surface of the membrane.

Denervation of Retrolingual Membranes

In order to determine if the action of drug solutions was exerted pre- or post-junctionally, it was deemed necessary in some experiments to prevent the release of neurotransmitters by terminal vasomotor nerves. Such a functional denervation of arterioles was attempted only in the retrolingual membrane. A combination of two methods were used. First, chronic chemical sympathectomy was accomplished by topical application to exposed retrolingual membranes of 0.5 mg/ml of 6-hydroxydopamine HCl for 2 hr on 2 successive days. This regimen has been shown to destroy all adrenergic nerve terminals in the retrolingual membrane for at least 21 days after the last treatment (Siggins and Bloom, 1970). Second, to disrupt the cholinergic vasodilator nerves (Siggins and Weitsen, 1971) in the membrane, the large mixed nerve bundles seen to enter the tongue bilaterally were sectioned at the base of the tongue.

Arterioles were studied 5 to 10 days after the combined surgery and chemical sympathectomy; denervation was evidenced by the lack of vasodilator or vasoconstrictor responses of arterioles to nerve stimulation.

Quantitating Vasomotor Responses by a Digitizing Optical Analyzer

Images of the blood vessels of either the frog retrolingual membrane or the hamster cheek pouch were passed through a beam splitter in the microscope, which was coupled to a Grass Kymograph camera. In this manner, responses of the arterioles to drugs or electrical stimulation were photomicrographed on Kodak 35 mm Tri-X film at rates of 0.1 to 0.27 frames/sec. The total magnification of the optical system was calibrated by a stage micrometer.

The film was then bulk-developed on a Pakoro CTX processor (Pako Corp) and mounted in a Benson-Lehner Oscar Model F optical analyzer, which increased the magnification of the 35 mm image by 9.1 X. The Oscar was programmed to punch on paper tape (in ASCII code) relative digital values in each film frame of arteriolar diameter, taken from the sides of a chosen fixed point on an arteriolar wall. The punch tape of the complete vasomotor response was then fed into the teletype tape reader of a Digital PDP-12 computer, which was programmed to type out the absolute values of arteriolar outer diameter (O.D.), as well as the circumference which was determined mathematically. Simultaneously, these values were stored on a magnetic data tape and displayed as a digital curve, with respect to time, on the oscilloscope face of the computer (see Fig. 1). These response curves were then photographed.

Histochemical Procedures

Histochemical localization of prostaglandin 15-hydroxy dehydrogenase was carried out according to the technique of Nissen and Anderssen (1968), as slightly modified by Siggins et al (1971). Frog retrolingual membranes and hamster cheek pouches were prepared as for physiological experiments. Arterioles were identified and mapped by hand-drawings and the relevant portion of the membrane excised and stretch-mounted on a microscope slide. After air drying for 5-15 min the unfixed tissue was exposed for 20 to 60 min to an incubation mixture containing diphosphopyridine nucleotide 0.5 mg/ml (as hydrogen carrier), nitroblue tetrazolium 0.5 mg/ml (as hydrogen indicator), and prostaglandin E(1), 1 mg/ml (as substrate) in Tris buffer at pH 8.3 and 33-37°C. Agar (0.25-0.5%) was often added to the incubation mixture to impede diffusion of the enzyme and/or reaction product.

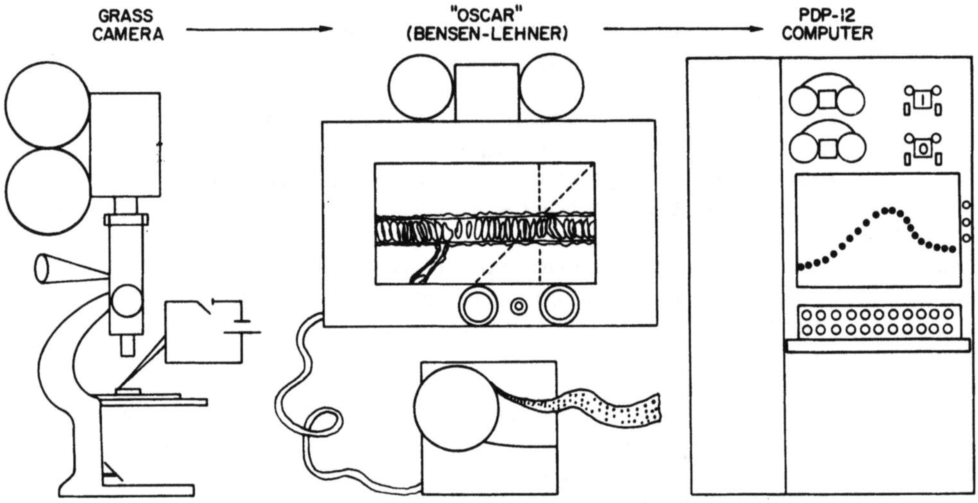

Figure 1. Schematic diagram of the method used for quantifying vasomotor responses. Arterioles are time-lapse photographed on 35 mm film with a Grass Kymograph camera. After magnifying the film in an Oscar Analyzer, the coordinates of the arteriolar wall are located by the "cross hairs" and punched out on paper data tape. The data tape is then analyzed by a Digital PDP-12 computer (see text).

For controls, paired membranes or portions of the same membrane were identically exposed to an incubation mixture containing the same substances but without the substrate PGE(1). As an additional control, stretch-mounted membranes were stained for succinic dehydrogenase by incubation for 20 to 60 min in a solution containing Na succinate (1 mg/ml), and nitroblue tetrozolium (0.5 mg/ml) in Tris buffer at pH 8.5 and 33-37°C. After incubation in experimental or control media, the tissues were fixed for 10 min in 5% phosphate-buffered formaldehyde, infiltrated with glycerin and covered with a cover-slip.

Solutions and Drugs Used

All drugs were administered in Ringer's physiological solution, which for frogs consisted of the following (mM): NaCl, 112; NaHCO3, 3; KCl, 2; and CaCl2, 1; for the hamster the solutions were made up in (mM): NaCl, 147; NaHCO3, 4; KCl, 4; and CaCl2, 3. The drugs used were 1-norepinephrine bitartrate (Levophed, Winthrop), 1-epinephrine bitartrate (Nutritional Biochemical), angiotensin II (Calbiochem), 6-hydroxydopamine HCl (Aldrich), PGE(1) (Upjohn), PGE(2) (Alza), PGF(1-alpha) (Upjohn), PGF(2-alpha) (Alza) and PGA(2) (Upjohn). Drug concentrations are expressed generally as g of the base per V of vehicle.

RESULTS

Direct Effects of the Prostaglandins

As outlined in Table I, all the prostaglandins tested had extremely potent vasodilator effects on both arterioles and precapillary sphincters (but not venules) in both the frog retrolingual membrane and the hamster cheek pouch (Fig. 2 and 3). Although low doses of PGF(2-alpha) always evoked vasodilation, vasoconstriction of arterioles of the hamster (but not frog) was often produced by PGF(2-alpha) (10^{-6} to 10^{-7} g/ml) (see Table I). Sometimes low doses of PGF(2-alpha) (10^{-8} to 10^{-9} g/ml) produced a noticeable dilation while 10^{-7} to 10^{-6} g/ml evoked no response in the hamster. Hamster venules never responded to PGF(2-alpha) in either direction. Arterioles of the cheek pouch seemed less responsive than frog arterioles, perhaps as a result of either: (1) the poorer penetration of prostaglandin to hamster arterioles, owing to the greater thickness of the cheek pouch, or (2) the greater dilatation of hamster arterioles seen in the untreated state, which (by the law of initial values) might diminish observation of small, threshold changes in diameter. It was often found necessary tc first preconstrict hamster arterioles with norepinephrine to then see a masked prostaglandin dilatory action. Furthermore, the exact threshold response to the individual prostaglandins was sometimes obscured by the occasional and unexplained slight dilatory action of Ringer's solution alone. However, statistical evaluation (student's T-test) of digitized response curves (Fig. 4) indicate that the most effective prostaglandins in the retrolingual membrane are PGE(1) and PGE(2), both with thresholds in the frog in the range of 10^{-12} to 10^{-11} g/ml (Table I). Since about 0.1 ml was delivered, this amounts to a total of 10^{-12} to 10^{-13} g. The order of potency of the other prostaglandins in the retrolingual membrane is PGA(2) > PGF(2-alpha) > PGF(1-alpha). In the cheek pouch, PGA(2), PGE(1), and PGE(2) all seem about equipotent as vasodilators, with thresholds in the region of 10^{-10} g/ml. PGF(1-alpha) and PGF(2-alpha) were 1/10 as effective in producing vasodilation.

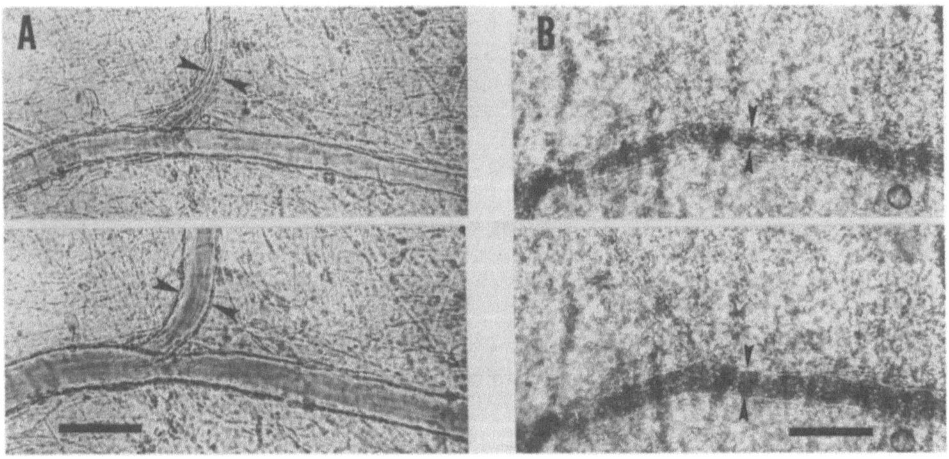

Figure 2. Time-lapse photomicrographs of the maximal vasodilations of arterioles of the frog retrolingual membrane (A) and hamster cheek pouch (B). Top panels taken just before prostaglandin E(2), 10^{-8} g/ml (in A) and F(2-alpha), 10^{-6} g/ml (in B) applied. Bottom panels taken from the maximum response of each arteriole tc the respective prostaglandin, about 45 sec after application. Arrows show points measured for the data of Fig. 3 and 4. Calibration mark = 100 microns.

The half-maximal dilatory response to the E group of prostaglandins occurred with doses of 10^{-9} to 10^{-10} g/ml in the retrolingual membrane and about 10^{-8} g/ml in the hamster. Maximal responses (Fig. 2, 3, and 4), amounting sometimes to more than a twofold increase in diameter in the frog (much less in the hamster), required 10^{-8} to 10^{-9} g/ml of PGE(1) or E(2). The latency from the time of topical application to the onset of response (ranging from 5 to 45 sec) was often highly variable from test to test with the same dose, as well as with different doses, prostaglandins and preparations. The topical method of application may be responsible in part for this observation, although underlying spontaneous rhythmic vasomotion, especially in the retrolingual membrane, also often obscured exact measurement of the response latency. In this regard, it is interesting that in the retrolingual membrane prostaglandin often appeared to increase the extent, if not the frequency, of rhythmic vasomotion during the time of vasodilation (see oscillations in response curve of Fig. 3A-E).

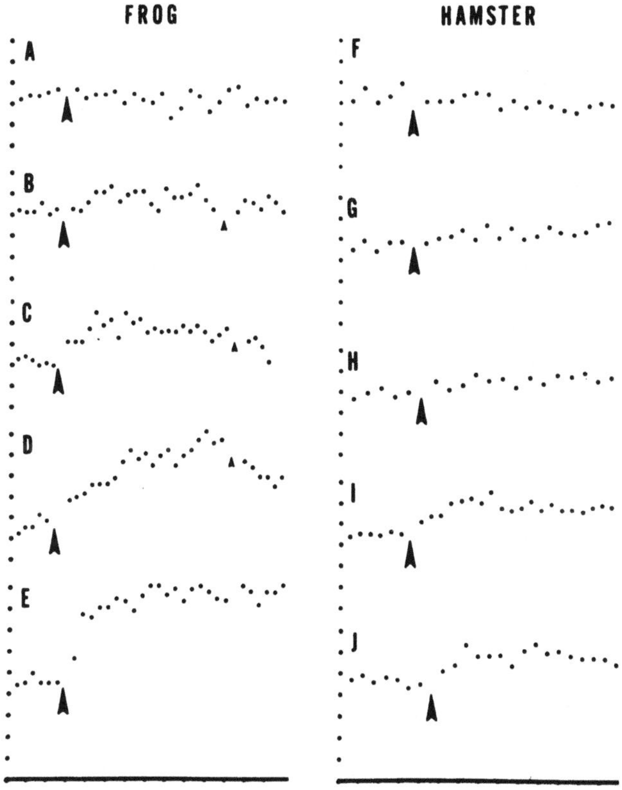

Figure 3. Digitalized response curves of a precapillary sphincter of the retrolingual membrane (A-E) and an arteriole of the cheek pouch (F-J) in response to different concentrations of PGE(2) and PGF(1-alpha), respectively. Same blood vessels as illustrated _in vivo_ in Fig. 2. First large arrow in each curve shows point of topical drug application as follows: A = PGE(2), 10^{-13}; B = 10^{-12}; C = 10^{-10}; D = 10^{-9}; E = 10^{-8} g/ml; for the hamster arteriole, F = PGF(1-alpha), 10^{-10}; G = 10^{-9}; H = 10^{-8}; I = 10^{-7}; J = 10^{-6} g/ml. The second (small) arrow in some panels indicates the time of rinse with Ringer's solution. There are 3.8 sec between each _data_ point in every panel. In panels A-E, each _ordinate_ point is 5 microns; in panels F-J each ordinate point is 7 microns. The control diameters in the frog ranged between 20-25 microns; those in the hamster 29-35 microns. All dilatations in curves B-E and G-J are statistically significant at the 0.0001 level (student's T-test).

Tachyphylaxis (desensitization of response) during a single application of prostaglandin was not seen in either preparation with any of the five prostaglandins, since a single 0.5 to 0.15 ml suprathreshold application of prostaglandin to a membrane (without a succeeding washout) usually produced vasodilation persisting in full extent for 10-25 min. However, in the frog there did appear occasional desensitizations in the same arteriole from one application to another. Thus, the response to a normally maximal dilatory dose of prostaglandin was sometimes reduced to a near-threshold response by prior administration of a half-maximal dose.

The vasodilatory response to the various prostaglandins in the frog was unabated by prior chemical sympathectomy with 6-OHDA, combined with surgical denervation. In fact, the threshold doses of the prostaglandin (as well as angiotensin and norepinephrine) were usually even lower after sympathectomy and denervation; for example with PGE(1) thresholds were a remarkable 10^{-12} to 10^{-13} g/ml.

During the maximal vasodilations produced by the prostaglandins, blood flow through arterioles always appeared to increase greatly. Stasis of flow, sometimes seen with histamine (unpublished observations), was never seen after prostaglandin. Likewise, the prostaglandins did not produce petechial hemorrhages, thrombi, "sludged blood" or other apparent forms of endothelial, white or red cell "sticking".

Interaction of Prostaglandins with Other Vasoactive Amines and Autonomic Nerves

In the frog retrolingual membrane several prostaglandins, PGE(1), PGE(2), PGF(2-alpha), and to a small extent PGA(2-alpha), were capable of dramatic and prolonged antagonism of the vasoconstrictor action of norepinephrine (NE) and epinephrine (E). PGE(1), PGE(2), and PGF(2-alpha) were almost equipotent in this regard; PGE(1) (10^{-8} to 10^{-9} g/ml) was usually capable of totally blocking a maximally constricting dose of NE (about 10^{-6} g/ml). Constriction with submaximal concentrations of NE were blocked at lower doses (10^{-9} to 10^{-11} g/ml). In fact, NE often evoked a noticeable vasodilation after PGE(1) treatment. PGA(2) (10^{-8} to 10^{-7} g/ml) only partially blocked the effect of catecholamines.

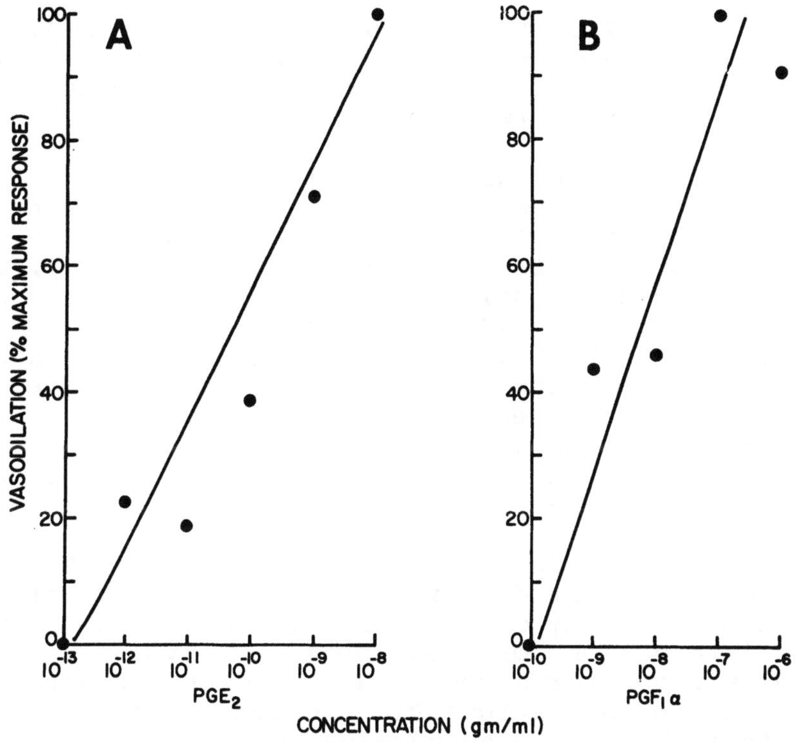

Figure 4. Dose-response curves of the frog precapillary sphincter (A) and hamster arteriole (B) illustrated in Fig. 2, in response to PGE(2) and PGF(1-alpha), respectively. Each point was derived by averaging five or more successive stable points both before and after prostaglandin application and calculating the percent mean change in diameter from mean control diameter; these values are then expressed as percent of the maximum response obtained with each prostaglandin, corrected for the slight vasoconstriction measured with application of vehicle alone, or subthreshold concentrations of prostaglandin (see Fig. 3A and F).

The most remarkable feature of the interaction between the prostaglandins and catecholamines is the extremely prolonged time-course of the blockade by a single application of PGE(1); after PGE(1) 10^{-9} g/ml, at least 1 hr was required for complete recovery of response to a previously maximal concentration of NE. This blockade persisted even after repeated bathing of the retrolingual membrane with Ringer's solution to rinse away residual prostaglandin, and long beyond the subsequent return of the arterioles to their initial diameters. Thus, the catecholamine-blocking effect of prostaglandin cannot be due merely to a mechanical antagonism as a result of summation of two opposite effects.

The specificity of prostaglandin blockade of catecholamines is further attested to by studies with PGE(1) and angiotensin II. Although the constrictor action of angiotensin is also antagonized by PGE(1), the following observations point to a lesser specificity in this interaction: (1) 100 to 1000 fold greater concentrations of prostaglandin are required than those needed to block the catecholamines; (2) the blockade only endures for 3-5 min after the administration of prostaglandin; (3) the blockade of the angiotensin response is not seen after the arteriole returns to its initial control caliber.

Prostaglandin blockade of responses to catecholamines or angiotensin was not diminished by prior denervation and chemical sympathectomy of retrolingual membranes. Thus this drug interaction does not appear to involve perivascular vasomotor nerves, such as might be expected if catecholamines or angiotensin exerted their constrictor effects either by releasing terminal vasoconstrictor agents or by blocking release of dilator agents.

In the hamster cheek pouch, only PGE(1) reproducibly blocked the action of NE or E, although PGE(2) (10^{-5} to 10^{-6} g/ml) often slightly reduced arteriolar responses to NE. The concentrations of PGE(1) required for total blockade of a just-maximal dose of NE ranged from 10^{-6} to 10^{-7} g/ml; complete recovery of the adrenergic response usually required 15-30 min after PGE(1) application, in spite of repeated rinsing of the cheek pouch with Ringer's solution. The effects of angiotensin have not yet been tested in the cheek pouch.

PGE(1), PGE(2), and PGF(2-alpha) actions on sympathetic nerve-induced vasoconstriction in the retrolingual membrane exactly paralleled their effects on the catecholamines. Thus, when terminal vasomotor nerves were stimulated by a micro-electrode with parameters which favored vasoconstriction, maximal responses were completely blocked and sometimes reversed to vasodilation by about 10^{-8} g/ml of PGE(1) or PGE(2) (Fig. 5), or 10^{-7} g/ml of PGF(2-alpha). Similarly, the longitudinally conducted vasoconstriction seen with direct stimulation of arteriolar walls, thought to be due solely to activation of perivascular nerves (Siggins and Bloom, 1970), was also blocked by these three prostaglandins, leaving only a localized constriction due to direct activation of the smooth muscle. Low concentrations of PGE(1) or PGE(2) (10^{-10} to 10^{-9} g/ml) acted to reduce the submaximal vasoconstrictor responses to nerve and mural stimulation; also, the stimulating currents required to evoke threshold vasoconstriction were markedly elevated by 10^{-10} to 10^{-9} g/ml of PGE(1) or PGE(2).

As with blockade of the catecholamines, the inhibition of nerve-induced vasoconstriction persisted for 45-60 min after a single prostaglandin administration (despite repeated rinsing of the preparation) and at least 35 min beyond the time at which the arteriole returned to control diameter. Thus, a mere mechanical effect of the vasodilation is not responsible for the antagonism. Although it was difficult to test the effect of the prostaglandins on the cholinergic (cf. Siggins and Weitsen, 1971) vasodilator nerves in the retrolingual membrane during the early PG-evoked vasodilation, stimulation of these nerves after the arteriole had regained some tone (4-10 min after prostaglandin administration) revealed no significant diminution of the neurogenic vasodilator response nor elevation of threshold currents.

Histochemical Localization of Prostaglandin Dehydrogenase

In whole stretch-mounts of the frog retrolingual membrane, marked staining for prostaglandin dehydrogenase was observed only in the skeletal muscle fibers (sparsely scattered throughout the membrane), arterioles (Fig. 6A), and precapillary sphincters. The staining was observed in all arterioles. It appears to be selective primarily for prostaglandin dehydrogenase of the smooth muscle cells and not for endothelial, nerve or connective tissue cells, for the following reasons: (1) varicose perivascular networks such as seen with the formaldehyde-fluorescence, acetylcholinesterase or methylene blue staining techniques were not observed; (2) staining was not seen in venules or capillaries, which in the retrolingual membrane have no smooth muscle cells; (3) the reaction product was most intense in bands around the arterioles (Fig. 6) corresponding to the circular or spiral wrapping of the smooth muscle cells.

The reaction product in arterioles and precapillary-sphincters appears to be specific for prostaglandin

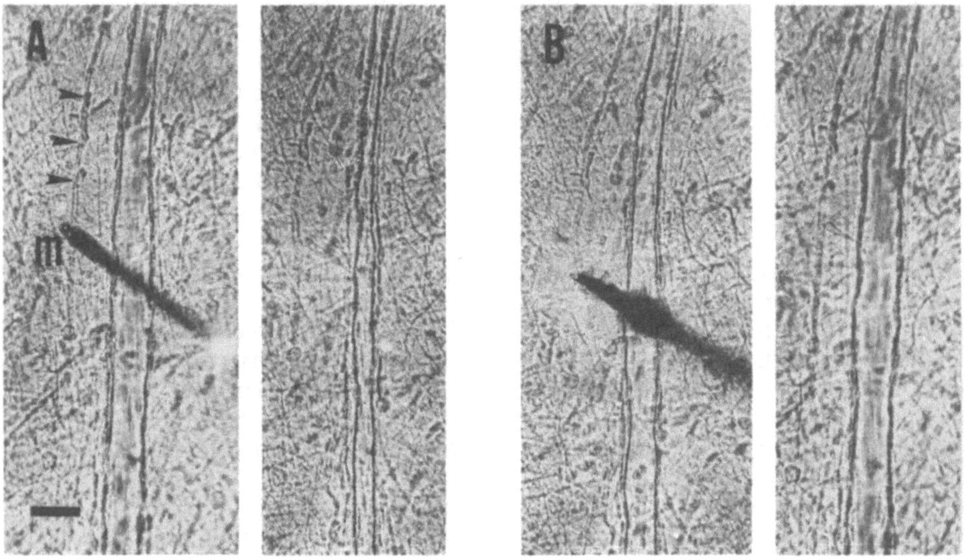

Figure 5. Effect of prostaglandin E(2) on arteriolar response to nerve stimulation; retrolingual membrane arteriole. First panel in (A) (control) taken just before stimulation of vasomotor nerve (arrows) with microelectrode (m). Second panel shown just-maximal control constriction evoked at 18 sec after delivery of a single shock of 0.8 mA to the nerve. Both panels in (B) taken 17 min after PGE(2), 10^{-8} g/ml topically applied to retrolingual membrane. In spite of the return of arteriolar tone (left panel) (due to repeated rinsing of the membrane with Ringer's solution), slight vasodilation only is now evoked by a previously supramaximal shock of 1.7 mA (right panel, 18 sec after stimulus). Calibration mark = 50 microns for all panels.

dehydrogenase, since no staining of these structures was observed when PGE(1) in the incubation mixture was removed ("nothing" dehydrogenase), or replaced with Na succinate. In the latter control, staining (for succinic dehydrogenase) appeared only in the striated muscle fibers.

In the hamster cheek pouch staining for prostaglandin dehydrogenase was substantially the same as in the frog. However, the greater thickness of whole mounts of the cheek pouch often obscured staining in the finer arterioles. Greater definition of arteriolar smooth muscle was obtained by incubation of unfixed cryostat

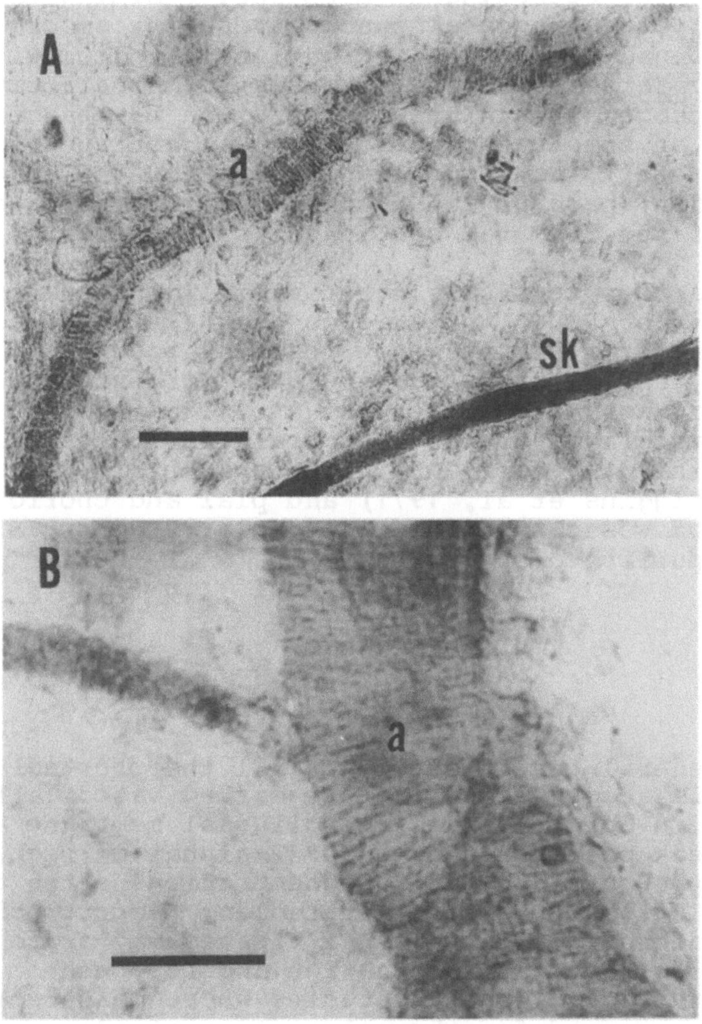

Figure 6. Histochemical localization of prostaglandin dehydrogenase (PGDH) in frog retrolingual membrane (A) and hamster cheek pouch (B). In both preparations, most PGDH staining is seen in arterioles (a) and skeletal muscle (sk). Note the intense staining of the circular smooth muscle of both arterioles. (A) is a stretch mount preparation; (B) is a cryostat section of 50 microns thickness. Calibration bar in (A) = 100 microns, in (B) = 50 microns.

sections (50 microns thick) of the cheek pouch (Fig. 6B); here again circular or spiral bands of smooth muscle, but not perivascular nerves, are readily apparent. In contrast to the retrolingual membrane, staining was occasionally seen in cheek pouch arteriolar smooth muscle after prolonged (> 45 min) incubation in medium without substrate. This might result either from the activity of other dehydrogenases in the tissue wiich have access to low levels of some unknown endogenous substrate (lactate, succinate, etc.), or it could be indicative of endogenous prostaglandin, perhaps released by the freeze-thawing sequence of the cryostat sectioning procedure.

For further controls, other mammalian structures were also tested for PGDH. Arterioles of the stretch-mounted hamster mesocecum did not show reaction product, although the adipocytes in this structure did. Cryostat sections of rat brain were also observed (cf. Siggins et al, 1971); here the vasculaure of the deep brain substance did not stain consistently for PGDH, although certain neuronal elements (Siggins et al, 1971) and pial and chorioid plexus blood vessels stained intensely (Siggins, Bloom and Hoffer, unpublished observations).

DISCUSSION

Direct Prostaglandin Effects

The general direct effect of all the prostaglandins used in this study was to produce marked vasodilation of arterioles in both the frog retrolingual membrane and the hamster cheek pouch, although PGF(2-alpha) in high, perhaps supra-physiological, concentrations often evoked vasoconstriction of hamster arterioles. Reports of PG-evoked vasoconstriction in other tissues may arise from the methods of drug administration and response measurement used. Thus, in studies where measurements of blood flow are used for the calculation of flow resistance in large vascular beds, intravascular injection of prostaglandins could produce rheological changes in the blood (thus altering flow) which could mask changes in vascular diameter. Furthermore, excessive manipulation or trauma in isolation of the tissue or in drug administration could evoke the release of sufficient endogenous prostaglandin (cf. Piper and Vane, 1971) or other vasodilator substances to produce a vasodilation that would obscure the dilator effect of subsequent prostaglandin administration. The use of direct observation of arteriolar responses to topical prostaglandin administration to natural thin membranes, which require little or no artificial manipulation for their preparation, avoids some of these problems.

Perhaps the most striking result of this study is the great sensitivity of the retrolingual membrane arterioles to all the prostaglandins tested. The concentrations of topical prostaglandin required for threshold vasodilation are generally lower than reported for other preparations. This could result from the extreme thinness of the membrane, from the rather sensitive method of quantifying arteriolar responses, or it may truly reflect a greater responsiveness of these arterioles to topical prostaglandins. Arterioles of the much thicker hamster cheek pouch required 10 to 100 fold greater concentrations of prostaglandin for threshold vasodilation than the frog. However, membrane thickness may not be the major limiting factor in the cheek pouch; these arterioles often appear nearly maximally dilated in the control state and further small threshold increases might escape detection. Alternatively, these arterioles may not be highly sensitive to the prostaglandins. Regardless of the underlying factors, the difference in the responsiveness of the arterioles of the two preparations suggests that caution should be used in drawing parallels from one tissue or species to another.

The prostaglandin-evoked vasodilation might have been mediated by perivascular vasomotor nerves, either by suppression of tonic sympathetic activity, or by release of acetylcholine or other vasodilating substances from nerve terminals. However, the prostaglandins were, if anything, even more potent vasodilators after denervation and chemical sympathectomy than before. This increased responsiveness to the prostaglandins may be akin to the "denervation supersensitivity" seen with certain neurotransmitters; still, the physiological significance of this phenomenon is unclear.

Prostaglandin-Adrenergic Interactions

Although destruction of perivascular nerves does not reduce prostaglandin-evoked vasodilation, suppression of catecholamine evoked vasoconstriction occurs with low concentrations of the prostaglandins (10^{-6} to 10^{-10} g/ml) in both normal and denervated preparations. In the frog retrolingual membrane, three prostaglandins, PGE(1), PGE(2), and PGF(2-alpha), inhibit the effects of NE and E, whereas only PGE(1) is effective in the hamster cheek pouch.

The great similarities between the blocking action of prostaglandins on the effects of catecholamines and on vasoconstrictor nerve stimulation in the retrolingual membrane strongly support the thesis (Berman and Siggins, 1968; Siggins and Bloom, 1970) that these nerves are adrenergic in nature. It would also seem that their blockade by prostaglandin is predominantly exerted at post-synaptic sites. Thus, the effects of both the catecholamines and stimulation of constrictor nerves are suppressed by similar concentrations of prostaglandin for equally prolonged periods (45 min to 1 hr). Furthermore, the prostaglandins are capable of actually reversing to vasodilation the normal constrictor effects of both the catecholamines and nerve stimulation. This reversal is identical to that seen with treatment of retrolingual membranes with alpha-adrenergic blocking agents (Siggins, 1967; Berman and Siggins, 1968). It is thus possible that the prostaglandins may either inhibit alpha-adrenergic receptors in vascular smooth muscle, and/or allow beta-receptor activation by exogenous or endogenous catecholamines to predominate over alpha-receptors. A possible intermediary role of cyclic AMP or Ca^{++} (Strong and Bohr, 1967) could be involved in this phenomenon.

The suppression of the effects of angiotensin II by prostaglandin requires some comment. Since this blockade persists only for the brief duration of the maximal dilatory period of prostaglandin (3-5 min) and requires 100 to 1000 fold the concentration of prostaglandin needed to block the effects of NE or E, it might be argued that the suppression of the action of angiotensin is non-specific. A similar type of short-acting, non-specific blockade of several diverse types of pressor agents has been reported for histamine and beta-histidine (Baez et al, 1971), which are also potent vasodilators. Nonetheless, it is possible that the differential sensitivity of the catecholamines and angiotensin II subserves some physiological function. There could be, for example, a graded release of the prostaglandins, with the amount released depending upon different types of physiological or pathological stimuli (sympathetic activation, trauma, anaphylactic or immune responses, etc.). Thus, with mild stimuli the release of small amounts of prostaglandin would exert adrenergic blockade; in more drastic circumstances, enough prostaglandin could be released to bring about angiotensin blockade.

In this respect it is of more than phenomenological interest that very low concentrations of prostaglandin (10^{-8} to 10^{-9} g/ml) totally block the maximal effects of NE, E or stimulation of sympathetic nerves in the retrolingual membrane for very prolonged periods. This finding might lead to the assumption that, under normal resting circumstances, very low concentrations of prostaglandin are released spontaneously near adrenergic effector sites; otherwise, adrenergic stimuli would never be effective. This assumption is in accord with several reports showing spontaneous release of nanogram amounts of prostaglandins or prostaglandin-like substances from various tissues (cf. Laity, 1969; Bartels et al, 1970; Dunham and Zimmerman, 1970; McGiff et al, 1970; Horton et al, 1971; Piper and Vane, 1971). It is possible, however, that in times of physiological emergency such as cardiovascular or anaphylactic shock, reactive hyperemia or inflammation, the amounts of prostaglandins released (cf. Piper and Vane, 1971) are sufficient to inactivate adrenergic (or even polypeptidal) responses for prolonged periods. The inflammatory action of some prostaglandins has been described by Kaley and Weiner (1968), and is the subject of a portion of this Symposium.

Cytochemical Localization of Prostaglandin Dehydrogenase

Of course, tissue levels of prostaglandin might well be subject to other mechanisms of regulation than by synthesis and/or release; namely by enzymatic degradation. Since prostaglandin 15-hydroxy dehydrogenase is currently thought to be one of the major degradative enzymes for inactivation of the prostaglandins (Samuelsson et al, 1971; Granstrom and Samuelsson, 1971) its histochemical localization could indicate sites specialized for prostaglandin inactivation.

With this consideration, it is interesting that in all the arteriolar preparations studied which stained for PGDH, only the muscular arteriolar coat, (and skeletal muscle) and not perivascular nerves, motor nerves or endothelial or connective tissue cells, showed significant reaction products. This may be relevant to recent suggestions that prostaglandins are released (in response to stretch or irritation) from smooth muscle cells or other post-junctional sites, rather than nerve terminals (Gilmore et al, 1968; Piper and Vane, 1971). However, the significance of high PGDH activity in striated muscle fibers is unclear, as no functional role has yet been delegated for the prostaglandins in skeletal muscle (cf. e.g., Ginsborg and Hirst, 1971).

The post-junctional localization of PGDH in arterioles could of course subserve the important function of terminating the vasodilator effect of released prostaglandins, or perhaps of preventing excessive spontaneous release of intracellular prostaglandin. It might be argued that in the retrolingual membrane, the first suggestion is weakened by the relative lack of desensitization to the prostaglandins in arterioles which show marked PGDH activity. However, it is likely that with topical prostaglandin administration, in which the volume of drug solution is many times greater than the total volume of muscular tissue in the retrolingual membrane, the substrate has saturated the PGDH enzyme.

Finally, the regional differences in brain PGDH localization call for some speculation. It was noted in this preliminary study that while many "peripheral" and meningeal blood vessels showed dramatic staining for PGDH, there was little reaction product in the vasculature of deep brain structures. Significantly, these regional differences parallel those of the innervation of intracerebral blood vessels: adrenergic (Falck et al, 1968) and cholinergic (Lavrentieva et al, 1968) perivascular nerves are confined largely to meningeal vessels. The relative lack of PGDH in parenchymal vessels may be of some functional significance with respect to the influence of circulating prostaglandins on the central nervous system, since it could afford easier access of active prostaglandins to central neurons than would occur with high PGDH levels. This suggestion is compatible with previous findings of exogenous brain prostaglandin and of behavioral effects (Horton and Main, 1967; Holmes and Horton, 1968; Potts and East, 1971) and centrally-mediated cardiovascular responses (Nakano and McCurdy, 1967; Kaplan et al, 1969; Lavery et al, 1970) produced by parenteral administration of prostaglandins. Since central parenchymal blood vessels might also respond (presumably by dilation) to circulating or locally-released prostaglandin in a more intense and prolonged fashion with low PGDH activity in the vessels than with high PGDH levels, the observed low levels could contribute to the etiology of several brain vascular phenomena such as migraine headaches. Conversely, the low cerebrovascular PGDH activity might allow release of significant amounts of prostaglandins from brain structures (cf. Holmes and Horton, 1968). However, verification of these speculations will require direct analysis of prostaglandin release and of prostaglandin actions on the central vasculature.

CONCLUSIONS

The results of this study suggest that both the dilatory and the adrenergic blocking effects of various prostaglandins in the microvascular system are exerted at the post-junctional level. However, these two phenomena do not appear to be irrevocably linked, as the latter effect can outlast the former by 30 min or more. Thus, prostaglandin antagonism of adrenergic stimuli does not result from the passive summation of two opposite effects. Indeed, two different post-junctional "receptors" for the dilatory and adrenergic blocking effects of certain prostaglandins might be postulated. The lack of adrenergic blockade seen with some prostaglandins which are nonetheless potent vasodilators fortifies this proposal. Finally, the histochemical localization of significant prostaglandin dehydrogenase activity postjuctionally in the same arterioles correlates well with these in vivo studies, indicating that the dramatic arteriolar responses to the prostaglandins are not merely pharmacological observations, but have functional significance.

TABLE I

COMPARISON OF THE APPROXIMATE CONCENTRATIONS OF SEVERAL PROSTAGLANDINS REQUIRED FOR THRESHOLD VASODILATION OF ARTERIOLES IN TWO THIN MEMBRANES

	Average Threshold Concentrations (g/ml) for Dilation of Arterioles of:	
Prostaglandin	Retrolingual Membrane	Cheek Pouch
PGE(1)	8×10^{-12}	3×10^{-10}
PGE(2)	1×10^{-11}	2×10^{-10}
PGA(2)	5×10^{-11}	1×10^{-10}
PGF(2-alpha)	2×10^{-9}	1×10^{-9}*
PGF(1-alpha)	1×10^{-8}	1×10^{-9}

*High concentrations of PGF(2-alpha) (10^{-6} to 10^{-7} g/ml) often evoked vasoconstriction in cheek pouch arterioles.

ACKNOWLEDGEMENTS

The author is deeply indebted to Professor Herbert J. Berman for his collaboration in preliminary physiological studies with PGE(1) on retrolingual membrane arterioles, during the author's predoctoral traineeship at Boston University. Thanks are also due Drs. Floyd Bloom, Barry Hoffer and James Wedner for their assistance in the histochemical studies and for helpful criticism of the manuscript. Valuable technical assistance was furnished by Mrs. Jeannie Alwine; Mrs. Odessa Colvin provided speedy typing of the manuscript, and H. Poole and R. Coates prepared the photographs.

The prostaglandins used in the studies were kindly supplied by Dr. J. E. Pike of Upjohn and Dr. P. W. Ramwell of Alza Corporation.

REFERENCES

Baez, S., Orkin, L. R., and Lagisquet, J. A. L., 1971, Antagonism of some vascular smooth muscle agonists by histamine and beta-histidine, Microvascular Res. 3:170.

Bartels, J., Kunze, H., Vogt, W., and Wille, G., 1970, Prostaglandin: liberation from and formation in perfused frog intestine, Naunyn-Schmiedebergs Arch. Exp. Pathol. Pharmakol. 266:199.

Berman, H. H., and Siggins, G. R., 1968, Neurogenic factors in the microvascular system, Fed. Proc. 27:1384.

Davies, B. N., and Withrington, P. G., 1969, Actions of prostaglandins A(1), A(2), E(1), E(2), F(1-alpha) and F(2-alpha) on splenic vascular and capsular smooth muscle and their interactions with sympathetic nerve stimulation, catecholamines and angiotensin, in "Prostaglandins, Peptides and Amines," (P. Mantegazza and E. W. Horton, eds.) pp. 53-56, Academic Press, New York.

Deuben, R. R., and Buckley, J. P., 1970, Identification of a central site of action of angiotensin II., J. Pharmacol. Exp. Therap. 175:139.

Ducharme, D. W., and Weeks, J. R., 1967, Cardiovascular pharmacology of prostaglandin F(2-alpha), a unique pressor agent, in "Nobel Symposium 2: Prostaglandins," (G. Bergstrom and B. Samuelsson, eds.), pp. 173-182, Interscience, New York.

Dunham, E. W., and Zimmerman, B. G., 1970, Release of prostaglandin-like material from dog kidney during nerve stimulation, Am. J. Physiol. 219:1279.

Falck, B., Nielsen, K. C., and Owman, C., 1968, Adrenergic innervation of the pial circulation, Scand. J. Clin. Lab. Invest., Suppl. 102:96.

Fulton, G. R., and Lutz, B. R., 1942, Smooth muscle motor units in small blood vessels, Am. J. Physiol. 135:531.

Fulton, G. R., Jackson, R. G., and Lutz, B. R., 1947, Cinephotomicroscopy of normal blood circulation in the cheek pouch of the hamster, Science 105:361.

Gilmore, N., Vane, J. R., and Wyllie, J. H., 1968, Prostaglandins released by the spleen, Nature 218:1135.

Ginsborg, B. L., and Hirst, G. D. S., 1971, Prostaglandin E(1) and noradrenaline at the neuromuscular junction, Brit. J. Pharmacol. 42:153.

Granstrom, E., and Samuelsson, B., 1971, On the metabolism of prostaglandin F(2-alpha) in female subjects, J. Biol. Chem. 246:5254.

Hedqvist, P., 1970, Studies on the effect of prostaglandins E(1) and E(2) on the sympathetic neuromuscular transmission in some animal tissues, Acta Physiol. Scand., Suppl. 345:1.

Holmes, S. W., and Horton, E. W., 1968, The distribution of tritium-labelled prostaglandin E(1) injected in amounts sufficient to produce central nervous effects in cats and chicks, Brit. J. Pharmacol. 34:32.

Horton, E. W., and Main, I. H. M., 1967, Further observations on the central nervous actions of prostaglandins F(2-alpha) and E(1), Brit. J. Pharmacol. Chemother. 30:568.

Horton, E., Jones, R., Thompson, C., and Poyser, N., 1971, Release of prostaglandins, Ann. N. Y. Acad. Sci. 180:351.

Hyman, A. L., 1969, The active responses of pulmonary veins in intact dogs to prostaglandin F(2-alpha) and E(1), J. Pharmacol. Exp. Therap. 165:267.

Jackson, R. T., and Stovall, R., 1968, Vasoconstriction of nasal blood vessels induced by prostaglandin, in

"Prostaglandin Symposium of the Worcester Foundation for Experimental Biology," (P. W. Ramwell and J. W. Shaw, eds.), pp. 329-334, Interscience, New York.

Kaley, G., and Weiner, R., 1968, Microcirculatory studies with prostaglandin E(1), in "Prostaglandin Symposium of the Worcester Foundation for Experimental Biclogy," (P. W. Ramwell and J. E. Shaw, eds.), pp. 321-328, Interscience, New York.

Kaplan, H. R., Gregg, G. J., Sherman, G. P., and Buckley, J. P., 1969, Central and reflexogenic cardiovascular actions of prostaglandin E(1), Intern. J. Neuropharmacol. 8:15.

Laity, J. L. H., 1969, The release of prostaglandin E(1) from the rat phrenic nerve-diaphragm preparation, Brit. J. Pharmacol. 37:698.

Lavery, H. A., Lowe, R. D., and Scroop, G. C., 1970, Cardiovascular effects of prostaglandins mediated by the central nervous system of the dog, Brit. J. Pharmacol. 39:511.

Lavrentieva, N. B., Mchedlishvili, G. I., and Plechkova, E. K., 1968, Distribution and activity of chclinesterase in the nervous structures of the pial arteries (a histochemical study), Biull. Eksper. Biol. Med. (USSR) 64:110.

Mark, A. L., Schmid, P. G., Eckstein, J. W., and Wendling, M. G., 1971, Venous responses to prostaglandin F(2-alpha), Am. J. Physiol. 220:222.

McGiff, J. C., Crowshaw, K., Terragno, N. A., and Lonigro, A. J., 1970, Release of a prostaglandin-like substance into renal venous blood in response to angiotensin II, Circulation Res. 26 and 27(1):1.

Nakano, J., 1968, Effect of prostaglandins E(1), A(1) and F(2-alpha) on cardiovascular dynamics in dogs, in "Prostaglandin Symposium of the Worcester Foundation for Experimental Biology" (P. W. Ramwell and J. E. Shaw, eds.), pp. 201-213, Interscience, New York.

Nakano, J., and McCurdy, J. R., 1967, Cardiovascular effects of prostaglandin E(1), J. Pharmacol. Exp. Therap. 156:538.

Nissen, H. M., and Anderssen, H., 1968, On the localization of a prostaglandin dehydrogenase activity in the kidney, Histochemie 14:189.

Peach, M. J., Cline, W. H. Jr., Davila, D., and Khairallah, P. A., 1970, Angiotensin-catecholamine interactions in the rabbit, European J. Pharmacol. 11:286.

Piper, P., and Vane, J., 1971, The release of prostaglandins from lung and other tissues, Ann. N. Y. Acad. Sci. 180:363.

Potts, W. J., and East, P. F., 1971, The effect of prostaglandin E(2) on conditioned avoidance response performance in rats, Arch. Intern. Pharmacodyn. Therap. 191:74.

Pratt, F. H., and Reid, M. A., 1930, A method for working on the terminal nerve-muscle unit, Science 72:431.

Samuelsson, B., Granstrom, E., Green, K., and Hamberg, M., 1971, Metabolism of prostaglandins, Ann. N. Y. Acad. Sci. 180:138.

Severs, W. B., Daniels, A. E., Smookler, H. H., Kinnard, W. J., and Buckley, J. P., 1966, Interrelationship between angiotensin II and the sympathetic nervous system, J. Pharmacol. Exp. Therap. 153:530.

Siggins, G. R., 1967, Nervous control of the arterioles in the retrolingual membrane of the frog (Rana pipiens), Doctoral Dissertation, Boston University Graduate School.

Siggins, G. R., and Bloom, F. E., 1970, Cytochemical and physiological effects of 6-hydroxydopamine on periarteriolar nerves of frogs, Circulation Res. 27:23.

Siggins, G. R., Hoffer, B. J., and Bloom, F. E., 1971, Prostaglandin-norepinephrine interactions in brain: Microelectrophoretic and histochemical correlates, Ann. N. Y. Acad. Sci. 180:302.

Siggins, G. R., and Weitsen, H. A., 1971, Cytochemical and physiological evidence for cholinergic, neurogenic vasodilation of amphibian arterioles and pre-capillary sphincters. I. Light microscopy, Microvasc. Res. 3:308.

Smith, E. R., McMorrow, J. V. Jr., Covino, B. G., and Lee, J. B., 1968, Studies on the vasodilator action of prostaglandin E(1), in "Prostaglandin Symposium of the

Worcester Foundation for Experimental Biology," (P. W. Ramwell and J. E. Shaw, eds.), pp. 259-266, Interscience, New York.

Steinberg, D., Vaughan, M., Nestel, P. J., and Bergstrom, S., 1963, Effect of prostaglandin E opposing thcse of catecholamines on blood pressure and on tryglyceride breakdown in adipose tissue, Biochem. Pharmacol. 12:764.

Strong, C. G., and Bohr, D. F., 1967, Effects of prostaglandins E(1), E(2) and F(1-alpha) on isolated vascular smooth muscle, Am. J. Physiol. 213:725.

Sunahara, F. A., and Kadar, D., 1968, Effects of ouabain on the interaction of autonomic drugs and prostaglandins on isolated vascular tissue, in "Prostaglandin Symposium of the Worcester Foundation for Experimental Biclogy," (P. W. Ramwell and J. E. Shaw, eds.), pp. 247-258, Interscience, New York.

Viguera, M. G., and Sunahara, F. A., 1969, Microcirculatory effects of prostaglandins, Can. J. Physiol. Pharmacol. 47:627.

von Euler, U. S., 1939, Weitere untersuchungen uber prostaglandin, die physiologisch aktive substanz gewisser genitaldrusen, Skand. Arch. Physiol. 81:65.

von Euler, U. S., and Hedqvist, P., 1969, Inhibitory action of prostaglandin E(1) and E(2) on the neuromuscular transmission in the guinea pig vas deferens, Acta Physiol. Scand. 77:510.

Weiner, R., and Kaley, G., 1969, Influence of prostaglandin E(1) on the terminal vascular bed, Am. J. Physiol. 217:563.

DISCUSSION

SIGGINS: We find an interesting parallel between Dr. Lee's latency with PGA(1) and PGA(2). In our system the delay for vasodilation can be up to 1 min or longer.

NAKANO: We studied PGDH activity in rat cerebral cortex and cerebellum homogenates (Nakano and Prancan, 1971, J. Pharm Pharmacol. 23:231). We could not find the high activity that occurs in kidney or lung. However, this may be due to variations in the locations of the dehydrogenase even in the same tissue, as you have stated previously. Did you see any LDH activity in similar areas where you found the high PGDH activities?

SIGGINS: We only studied this with succinic dehydrogenase. In the arterials, succinic dehydrogenase appears in the thin membranes primarily in the skeletal muscle preparations. Lactic dehydrogenase appears primarily in the granular cell layer of the cerebellum. Prostaglandin dehydrogenase appears in the cerebellum primarily in the Purkinje cells, which is a very small population of the cells and which might account for the fact that you don't see very high levels in host cerebellum biochemically. In the thin membranes we find an interesting parallel to the long latent periods seen by Dr. Lee with PGA(1) and PGA(2). In our system the delay for vasodilation can be up to 1 min after PGA(2) application.

NAKANO: We studied PGDH activity in the brain of rats and we could not find high activity as occurs in kidney or lungs. However, this may be due to variations in the cellular locations of the dehydrogenase in the tissue, as you have stated previously. Did you see any alcohol dehydrogenase activity in similar areas of PGDH activities?

SIGGINS: In the microvascular preparations I have only studied this with succinic dehydrogenase, which is probably a better control. Succinic dehydrogenase appears in the thin membrane preparations primarily in the skeletal muscle fibers, with little staining in arterioles. In the cerebellum, lactic and succinic dehydrogenase appear in the granular and molecular cell layers of the cerebellum (Siggins et al, 1971), whereas prostaglandin dehydrogenase appears primarily in the Purkinje cells, which is a very small percent of the total population of the cerebellar cells. In fact, the Purkinje cells probably represent an even smaller percent of total wet weight of the cerebellum, which might account for the fact that you don't see very high PGDH levels in the cerebellum by biochemical analysis.

EFFECT OF PROSTAGLANDINS ON ADRENERGIC NEUROTRANSMISSION TO VASCULAR SMOOTH MUSCLE

P. J. Kadowitz, C. S. Sweet, and M. J. Brody

Departments of Pharmacology
College of Medicine, Univ. of Iowa, Iowa City,
and Tulane Medical School, New Orleans, La.

INTRODUCTION

Prostaglandins are released into the circulation from several organs when the sympathetic nerves are stimulated, or vasoconstrictors are infused (Bergstrom et al, 1968; Davies et al, 1968; Gilmore et al, 1968; McGiff et al, 1970; Sweet et al, 1971; Dunham and Zimmerman, 1970). Prostaglandins of the E and A series are potent vasodilators in most vascular beds and the vasodilator action is not blocked by classical pharmacological blocking agents such as atropine, propranolol, or antihistamines (Nakano and McCurdy, 1967; Dougherty, 1971; Strong and Bohr, 1967; Nakano, 1968). PGE(1) has been reported to decrease vasoconstrictor responses to catecholamines in several vascular beds in the dog (Kadowitz et al, 1971a; Kadowitz et al, in press; Hedwall et al, 1971), the cat (Holmes et al, 1963), and the rat (Weiner and Kaley, 1969; Viguera and Sunahara, 1969). PGE(2) has been reported to inhibit constrictor responses to nerve stimulation and norepinephrine in the isolated cat spleen (Hedqvist, 1970) but in high concentrations enhances the response to nerve stimulation in the cutaneous vascular bed of the dog (Kadowitz et al, 1971a). PGE(2) is released into the circulation along with PGF(2-alpha) by the kidney and spleen when the sympathetic nerves are stimulated or vasoconstrictors are infused (Davies et al, 1968; Gilmore et al, 1968; McGiff et al, 1970; Dunham and Zimmerman, 1970).

PGF(2-alpha) on the other hand increases the blood pressure when injected into the dog (DuCharme et al, 1968). The increase in pressure is thought to be due to an increase in cardiac output brought about by venoconstriction and an increased venous return (DuCharme et al, 1968). PGF(2-alpha) enhances responses to norepinephrine and nerve stimulation in the saphenous vein (Kadowitz et al, 1971b) and to sympathetic nerve stimulation in the dog hindpaw (Kadowitz et al, 1971a). PGA(1) is a potent vasodilator which escapes degradation by the lung and could theoretically act as a circulating hormone (McGiff et al, 1969; Ferreira and Vane, 1967). This prostaglandin is of special interest because of its potential use in the treatment of hypertension (Westura et al, 1970).

In view of these findings, the present study was designed to compare the effect of a wide range of concentrations of prostaglandins E(1), E(2), A(1), and F(2-alpha) on vasoconstrictor responses to norepinephrine and sympathetic nerve stimulation in the cutanecus vascular bed of the dog under conditions of constant flow. Furthermore, the effect of PGE(1) on adrenergic responses was compared in skin and skeletal muscle vessels and the effect of PGF(2-alpha) on adrenergic transmission to resistance and capacitance vessels was compared in the hindpaw. Finally, the effects of PGE(1) and PGA(2) were compared in the hindlimb and the effect of neurohumors released during acute renal ischemia was evaluated in the hindpaw perfused with renal venous blood.

METHODS AND MATERIALS

Dogs (11-20 kg) of either sex were anesthetized with pentobarbital sodium (30 mg/kg IV) and after cannulation of the trachea, were respired artificially with room air using a Harvard respirator. Arterial blood pressure was measured through a catheter inserted into the carotid artery. Systemic injection of drugs was made through a catheter in the jugular vein. The left hindpaw was perfused according to the method of Zimmerman and Gomez (1965). After administration of heparin sodium (5 mg/kg IV), the cranial tibial artery was cannulated and the hindpaw perfused at constant flow by a Sigmamotor pump model T8 with blood supplied from the left iliac artery. Flow was initially set to provide a perfusion pressure that approximated systemic pressure and was not altered during the experiment. Flow averaged 37 ± 2 (S.E.) ml/min in 73 experiments. Paw perfusion pressure was monitored from a T-fitting in the tubing between the pump and artery. Vascular isolation was confirmed by stopping the Sigmamotor pump and observing the residual pressure in the paw. This pressure ranged between 10 and 20 mm Hg and since it approached small vein pressure was considered good evidence for complete vascular isolation. The lumbar sympathetic chain was approached through a left flank incision and a Harvard electrode was placed on the left chain between L5 and L6. The lumbar chain was ligated and crushed with a hemostat several millimeters above the electrode. The nerve was stimulated with square wave pulses, 2 msec duration, 8-16 volts, variable frequency for 10 sec periods with a Tektronix assembly (160 series waveform and pulse generators). The left gracilis muscle was perfused by the method described by Zimmerman and Whitmore (1967). After administration of heparin sodium, the gracilis muscle was perfused by a Sigmamotor pump with blood from the left iliac artery through a catheter in the left femoral artery with all arteries ligated except the main vessel supplying the muscle. Flow averaged 16 ± 2 (S.E.) ml/min in 17 experiments. Perfusion pressure was measured in the same manner as in hindpaw experiments. All pressures were measured with Statham transducers and recorded on a Beckham Dynograph.

Mean pressures were obtained from the pulsatile signal by electrical integration. The lumbar nerves were stimulated as in hindpaw experiments with the exception that stimuli were applied for 15 sec periods. In experiments in which the effect of PGF(2-alpha) on adrenergic venomotor responses was studied, the dorsal branch of the lateral saphenous vein was perfused according to the method of Webb-Peploe and Shepherd (1968). The vein was cannulated below the ankle and after administration of heparin sodium, was perfused at constant flow by a Sigmamotor pump with blood from the left iliac artery. Flow was initially set to provide a perfusion pressure of approximately 20 mm Hg and was not altered during the experiment. Flow averaged 22 ± 5 (S.E.) ml/min in 13 experiments. Perfusion pressure was measured from a sidearm in the tubing between the pump and vein and the nerves were stimulated in the same manner as in gracilis muscle experiments.

For hindlimb perfusion the left hindlimb was perfused by a Sigmamotor pump with blood obtained from the abdominal aorta through a catheter in the left iliac artery in dogs previously given heparin. Flow averaged 160 ml/min in these experiments and systemic arterial and

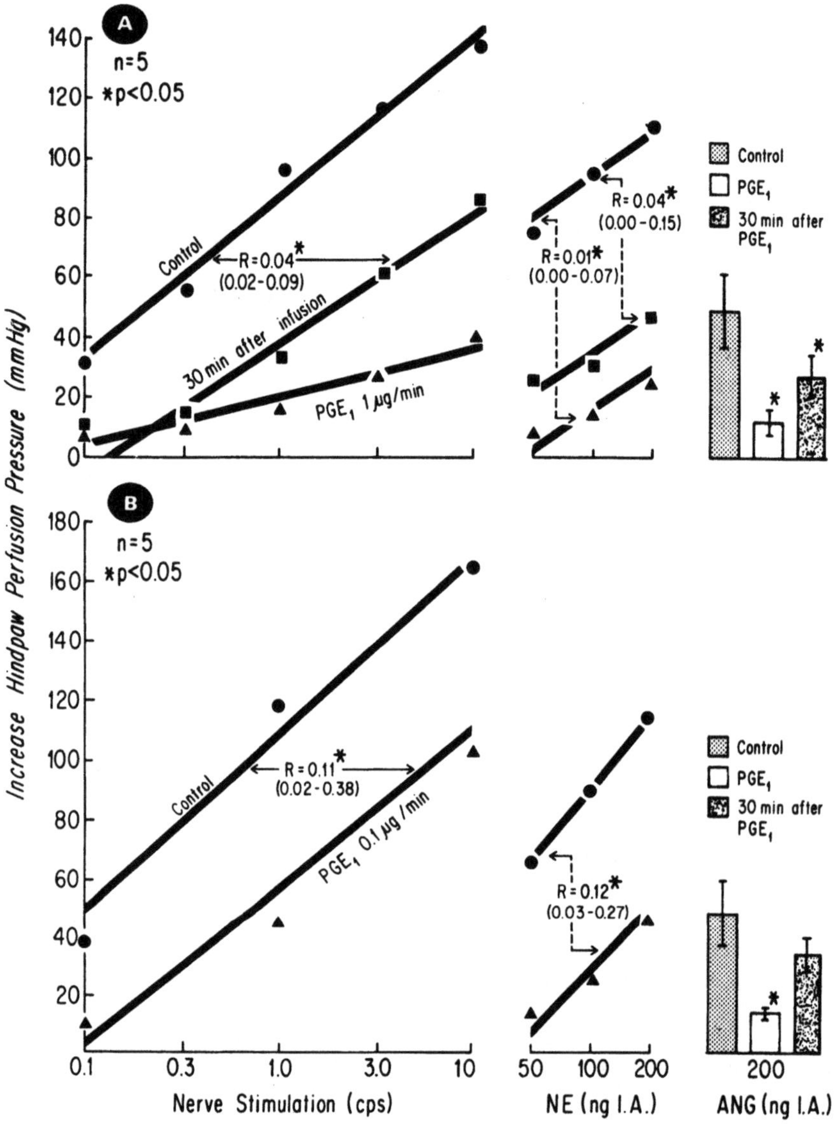

Figure 1. Effect of PGE(1) infusion, 1.0 microgram/min (panel A) and 0.1 microgram/min (panel B), on vasoconstrictor responses to sympathetic nerve stimulation, intra-arterial norepinephrine and angiotensin in the perfused hindpaw. Dose-response curves for norepinephrine, frequency response curves for nerve stimulation, and responses to angiotensin were obtained before and during prostaglandin infusion and again 30 min after termination of infusion, and these curves were analyzed by the parallel line bioassay method of Finney (1952). R shows the potency ratio with the 95% confidence limits in parentheses. A potency ratio is not shown for

perfusion pressures were measured with Statham transducers and recorded on a Grass Polygraph.

Intra-arterial injections of l-norepinephrine (Levophed bitartrate - Winthrop, dose in terms of base), angiotensin II amide (Hypertensin - Ciba), d,l-isoproterenol (Isuprel HCl - Winthrop), and tyramine monohydrochloride (Nutritional Biochemicals Corp) (doses all in terms of salt) were made in small volumes (0.03-0.10 ml) into the perfusion circuit. Prostaglandins E(1), E(2), A(1) and F(2-alpha) (Upjohn) were dissolved in 95% ethyl alcohol, 1 mg/ml, and stored in the freezer. On the day of use an aliquot of the stock solution was diluted to a volume of 10 ml with saline and infused with a Harvard infusion pump.

Data were analyzed using the Student t test for paired and group comparison (Snedecor, 1956) and parallel line bioassay method of Finney (1952). Relative potency ratios were calculated from parallel regression lines. The relative potency ratios were calculated from parallel regression lines. The relative potency ratio is considered to be significant if 1.00 does not lie in the 95% confidence limits and potency ratios for different curves are considered to differ significantly if both potency ratios do not fall within the other's 95% confidence limits. The 5% probability level was the criterion for significance in all experiments.

RESULTS

Effect of PGE(1) on Responses to Adrenergic Stimuli and Angiotensin in the Hindpaw

The effect of intra-arterial infusion of a wide range of concentrations of PGE(1) on vascular resistance and vasoconstrictor responses to adrenergic stimuli and angiotensin was examined in the cutaneous vascular bed of

---

the frequency response curve obtained during infusion of 1.0 microgram/min PGE(1) because the shift of the curve was not parallel, and curves for norepinephrine and nerve stimulation are not shown after termination of the 0.1 microgram/min infusion since they were not significantly different from control. N indicates number of animals. The shift of the curve to the right is considered to be significant if the upper limit does not exceed 1.00. (Redrawn from Kadowitz et al, in press).

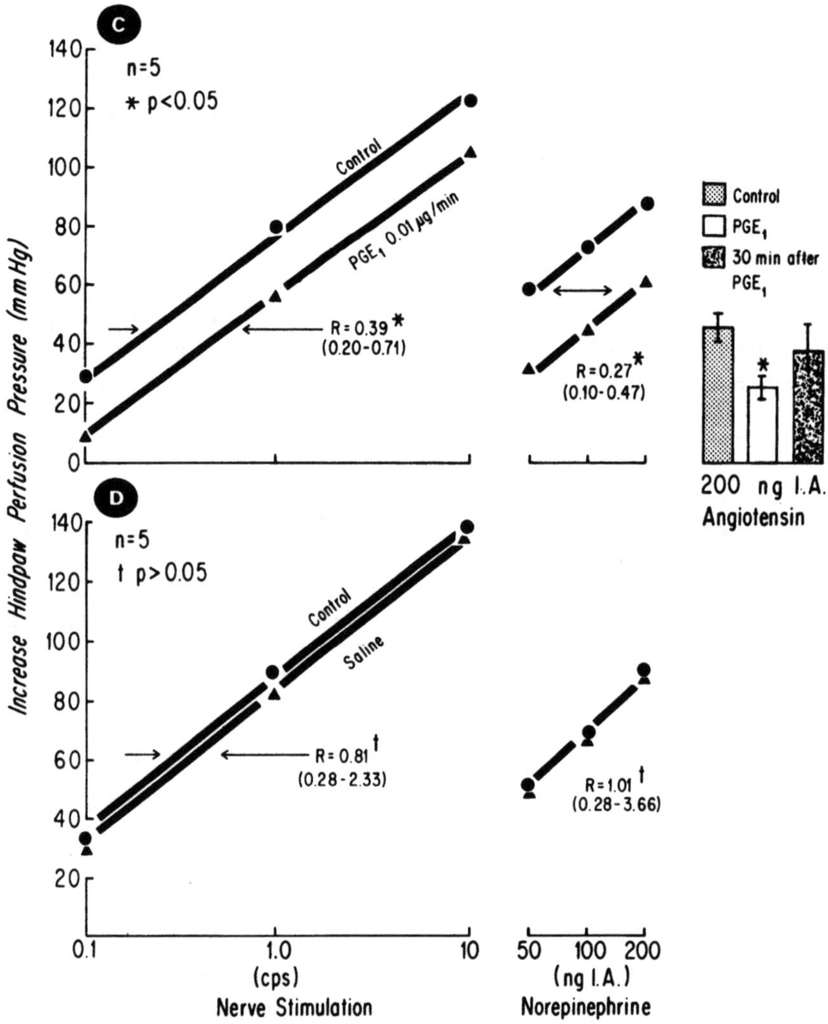

Figure 2. Effect of PGE(1) infusion, 0.01 microgram/min (panel C), on responses to nerve stimulation, norepinephrine and angiotensin; and the effect of physiological saline infusion (panel D) on responses to norepinephrine and nerve stimulation in the perfused hindpaw. Dose and frequency response curves and responses to angiotensin were obtained before and during infusion and again 30 min after termination of infusion and were compared by the parallel line bioassay method of Finney (1952). Curves for norepinephrine and nerve stimulation were not significantly different from control after termination of the infusions and are not shown. N indicates number of animals. The shift of the curve to the right is considered to be significant if the upper

the dog. Direct intra-arterial infusions of PGE(1) (0.01, 0.1 and 1.0 microgram/min) in three groups of dogs reduced perfusion pressure to 28 ± 5, 32 ± 6, and 34 ± 5 mm Hg but no significant change in systemic arterial pressure occurred (Table I). The reduction in hindpaw perfusion pressure was significant at each concentration evaluated ($p < 0.05$, paired comparison); however the decreases in hindpaw pressures were not significantly different from each other at the three dosages of PGE(1) ($p > 0.3$, group comparison).

The effect of PGE(1) on responses to adrenergic stimuli was evaluated in these three groups of animals by determining dose and frequency response curves for intra-arterial norepinephrine and sympathetic nerve stimulation before, during, and 30 min after constant intra-arterial infusion of this prostaglandin, each group of animals receiving only one concentration of PGE(1). Results of these experiments are shown in Fig. 1 and 2. Dose response curves for norepinephrine were shifted to the right in a parallel manner during infusion of the three concentrations of PGE(1). Analysis of potency ratios revealed that the shift of each curve was significant and that the magnitude of shift was related to the concentration infused (comparisons made using the parallel line bioassay method of Finney, 1952). Responses to norepinephrine returned toward control value when infusions were terminated and the curves for norepinephrine were not different from control 30 min after infusion of the two lower concentrations, whereas responses to norepinephrine were still depressed 30 min after the 1.0 microgram/min infusion (Fig. 1A). The curves for nerve stimulation were also shifted to the right by all concentrations of PGE(1); however the frequency response curve deviated from parallelism during infusion of the highest concentration (Fig. 1A). The shift of the curves at the 0.1 and 0.01 microgram/min infusion rates was significant and analysis of the potency ratios shows that magnitude of shift was related to the concentration of PGE(1) infused (Fig. 2B and 2C). Responses to nerve stimulation returned to control level 30 min after termination of the two lower infusion rates. There was only a small tendency for responses to nerve stimulation to recover toward control value after the 1.0 microgram/min infusion (Fig. 1A).

---

limit does not exceed 1.00. (Redrawn from Kadowitz et al, in press).

The influence of time alone on vascular resistance and responses to norepinephrine and nerve stimulation was evaluated in a fourth group of animals. Infusion of physiological saline, 0.1 ml/min, rather than prostaglandin, produced no significant change in hindpaw pressure or on vasoconstrictor responses to norepinephrine and nerve stimulation, and these responses were not different from control 30 min after saline infusion (Fig. 2D).

Hindpaw vasoconstrictor responses to intra-arterial angiotensin were depressed ($p < 0.05$, paired comparison) by all doses of PGE(1). Responses returned to control value 30 min after infusion of the two lower concentrations. The response to angiotensin was still smaller than control ($p < 0.05$) 30 min after the 1 microgram/min infusion (Fig. 1A, 1B, and 2C).

Effect of Phenoxybenzamine on Responses to PGE(1)

The influence of alpha-adrenergic blockade on the vasodilator and angiotensin blocking action of PGE(1) was examined in another series of experiments. Intravenous infusion of phenoxybenzamine (2.5 mg/kg IV) over a 5 min period resulted in a significant decrease in mean arterial pressure (123 ± 5 to 97 ± 5 mm Hg) but little change in hindpaw pressure (128 ± 3 to 123 ± 5 mm Hg). This amount of phenoxybenzamine decreased responses to all doses of norepinephrine studied ($p < 0.05$, paired comparison) but was without significant effect on responses to angiotensin in the hindpaw (Fig. 3). Subsequent infusion of PGE(1) (1 microgram/min) into the hindpaw of these same animals resulted in marked reduction in hindpaw perfusion pressure (123 ± 5 to 80 ± 6 mm Hg) ($p < 0.05$, paired comparison) but no significant change in systemic arterial pressure. This amount of PGE(1) greatly decreased constrictor responses to norepinephrine and angiotensin in the animals receiving phenoxybenzamine previously ($P < 0.05$, comparisons made by the method of Finney, 1952), (Fig. 3).

Effect of PGE(1) on Adrenergic Responses in Skeletal Muscle

The effect of PGE(1) on vascular resistance and responses to adrenergic stimuli was also evaluated in skeletal muscle. Close intra-arterial infusion of PGE(1) (1 microgram/min) into the gracilis muscle in 12 animals resulted in a 58 ± 5 mm Hg reduction in gracilis muscle

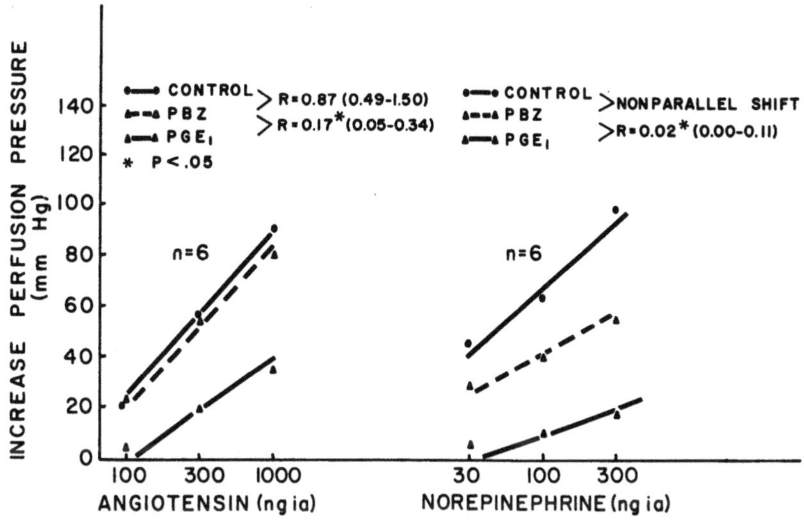

Figure 3. Effect of phenoxybenzamine (PBZ), 2.5 mg/kg IV, and prostaglandin E(1), 1 microgram/min IA, on dose response curves for IA antiotensin and norepinephrine in the perfused hindpaw. Responses to norepinephrine were decreased significantly by phenoxybenzamine ($p < 0.05$, paired comparison, all doses NE) and the curve for norepinephrine deviated significantly from parallelism. Phenoxybenzamine was without significant effect on responses to angiotensin. R shows the potency ratio with the 95% confidence limits in parentheses. The shifts of the curves for angiotensin and norepinephrine were parallel in the presence of PGE(1). Comparisons were made by the parallel line bioassay method of Finney (1952). The shift of the curve to the right is considered to be significant if the upper limit does not exceed 1.00. N indicates number of animals.

perfusion pressure (138 ± 9 to 81 ± 6 mm Hg) but no significant change in systemic arterial pressure. Vasoconstrictor responses to norepinephrine and nerve stimulation and vasodilator responses to isoproterenol were depressed by PGE(1) in the gracilis muscle. The shifts to the right of the curves for norepinephrine, isoproterenol and nerve stimulation were parallel, and analysis of the potency ratios reveals that PGE(1) is more effective in antagonizing responses to intra-arterial norepinephrine, and isoproterenol than nerve stimulation (Fig. 4).

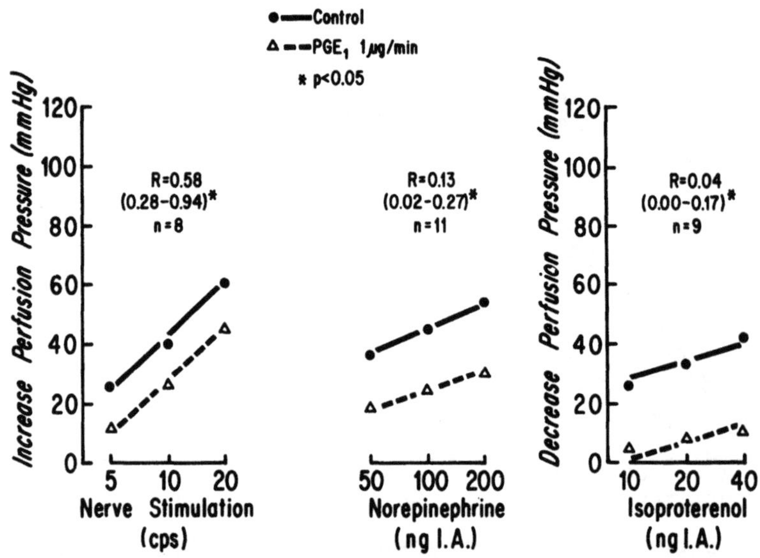

Figure 4. Effect of infusion of PGE(1), 1 microgram/min IA, on the frequency response curve for nerve stimulation, and on the dose response curve for norepinephrine and isoproterenol in the perfused gracilis muscle. Curves were obtained before and during infusion of PGE(1) and were analyzed by the parallel line bioassay method of Finney (1952). R shows the potency ratio with the 95% confidence limits in parentheses. The shift of the curve to the right is considered to be significant if the upper limit does not exceed 1.00. N indicates number of animals.

Effect of PGE(2) on Adrenergic Responses in the Hindpaw

Infusion of PGE(2) (1 microgram/min) into the hindpaw in eight animals resulted in a reduction in perfusion pressure of 20 ± 6 mm Hg ($p < 0.05$) but no significant change in mean arterial pressure. Responses to nerve stimulation and norepinephrine were not significantly different from control during infusion of this amount of PGE(2). However in four experiments when the infusion rate was increased to 2 microgram/min the responses to nerve stimulation were greatly enhanced, whereas the responses to norepinephrine remained unchanged (Fig. 5). The shift to the left of the frequency response curve was parallel in the presence of this prostaglandin.

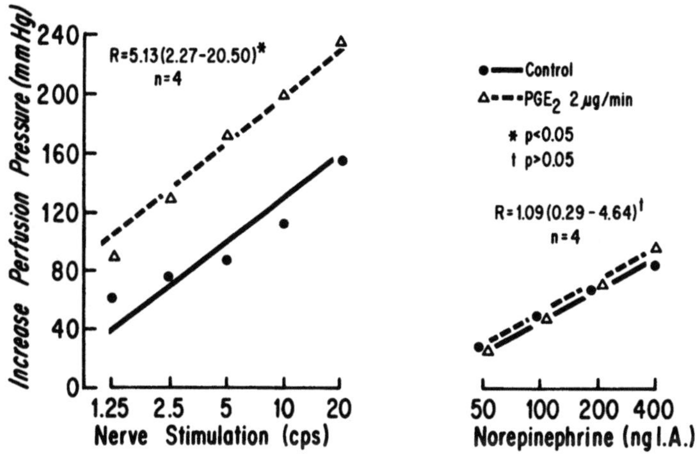

Figure 5. Effect of PGE(2), 2 microgram/min IA, on the frequency response curve for nerve stimulation and dose response curve for IA norepinephrine in perfused hindpaw. Responses were obtained before and during constant intra-arterial infusion of prostaglandin E(2) and the curves were analyzed by the parallel line bioassay method of Finney (1952). R shows the potency ratio with the 95% confidence limits in parentheses. The shift of the curve to the left is considered to be significant if the lower confidence limit exceeds 1.00. N indicates number of animals.

Effect of PGA(1) on Adrenergic Responses in the Hindpaw

Direct intra-arterial infusion of PGA(1) (0.1, 1.0, and 2.0 microgram/min) in three groups of animals resulted in reduction in hindpaw perfusion pressure of 31 ± 4, 35 ± 3, and 35 ± 5 (Table I). The reduction in hindpaw pressure was significant ($p < 0.05$, paired comparison); however these dilator responses were not significantly different from each other ($p > 0.2$, group comparison). Systemic arterial pressure was decreased by 27 ± 5 and 55 ± 7 mm Hg during infusion of PGA(1), 1.0 and 2.0 microgram/min (Table I).

The effect of PGA(1) on responses to norepinephrine and nerve stimulation was evaluated before, during and after termination of the PGA(1) infusion. Results of these experiments are shown in Fig. 6. Responses to norepinephrine and nerve stimulation were depressed during infusion of the two larger doses of PGA(1). The shift to the right of curves for norepinephrine and nerve

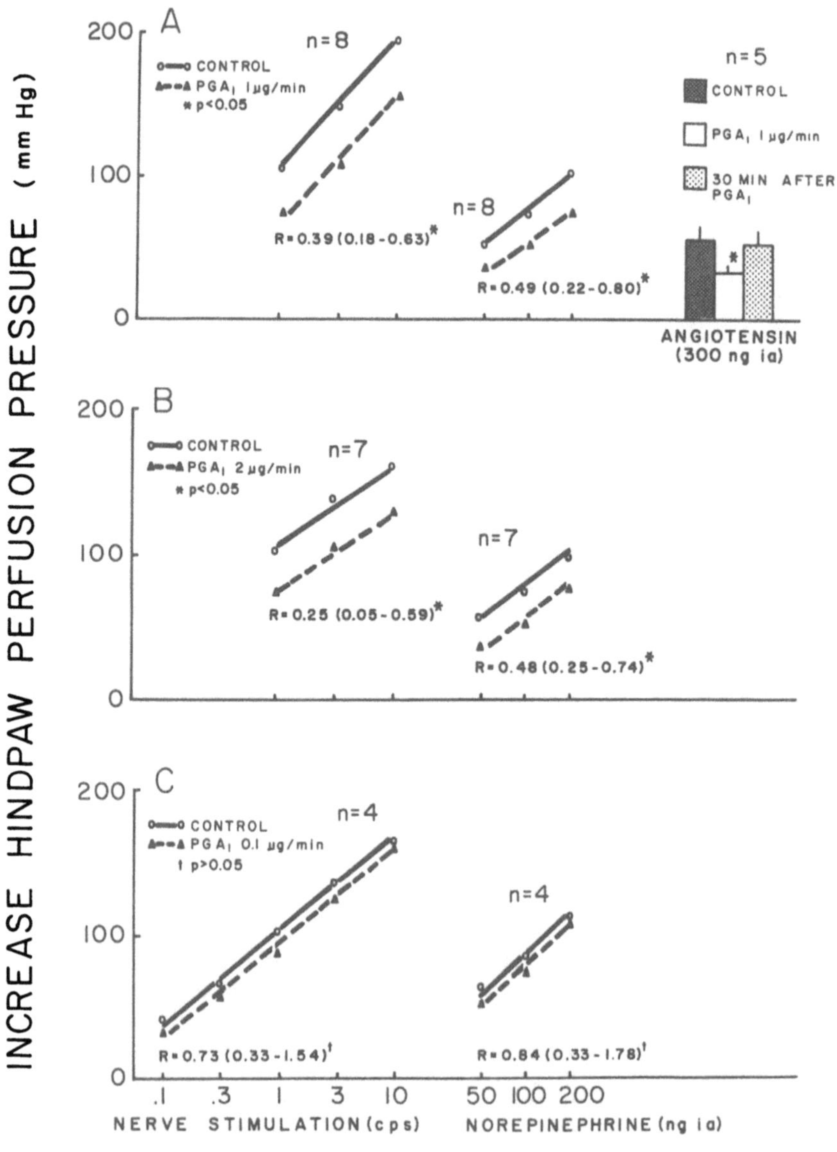

Figure 6. Effect of PGA(1) infusion, 1 microgram/min, panel A; 2 microgram/min, panel B; and 0.1 microgram/min, panel C, on responses to norepinephrine and nerve stimulation in the perfused hindpaw. Dose-response curves for norepinephrine and frequency response curves for nerve stimulation were obtained before and during prostaglandin infusion and again 30 min after termination of infusion. The curves were analyzed by the parallel line bioassay

stimulation was parallel. Analysis of the potency ratios reveals that the magnitude of shift is similar for norepinephrine and nerve stimulation at both the 1 and 2 microgram/min infusion rates and that the shift to the right is the same for all stimuli at both concentrations (Fig. 6A and 6B). The response to intra-arterial angiotensin was decreased ($p < 0.05$) by PGA(1), 1 microgram/min, in the five animals in which it was studied, and this response was not different from control 30 min after PGA(1) infusion (Fig. 6A). Responses to norepinephrine and nerve stimulation returned toward control level after the termination of the PGA(1) infusion and were not significantly different from control 30 min after termination of infusion. PGA(1) was without significant effect on responses to norepinephrine and nerve stimulation at the 0.1 microgram/min infusion rate (Fig. 6C).

Effect of PGF(2-alpha) on Adrenergic Responses in the Hindpaw and Saphenous Vein

Infusion of PGF(2-alpha) into the hindpaw, 0.01, 0.1, and 1.0 microgram/min (each animal receiving only one concentration), or 1.0 microgram/min into the saphenous vein produced no significant change in either hindpaw or saphenous vein perfusion pressure or in systemic arterial pressure (Table I). Dose response curves for tyramine and norepinephrine determined during and after infusion of PGF(2-alpha), 0.01, 0.1, and 1.0 microgram/min, into the hindpaw were not significantly different from control curves. Dose response curves for norepinephrine before and during infusion of PGF(2-alpha) are shown only for the 1.0 microgram/min infusion (Fig. 7). Responses to nerve stimulation were increased significantly at all stimulus frequencies evaluated (Fig. 7). The shift to the left of the frequency response curve was nonparallel at the 1.0 microgram/min concentration, and PGF(2-alpha) greatly increased the slope of the frequency response curve.

---

method of Finney (1952). R shows the potency ratio with the 95% confidence limits in parentheses. The shift of the curve to the right is considered to be significant if the upper limit does not exceed 1.00. The response to angiotensin was significantly smaller ($p < 0.05$) during infusion of PGA(1), 1 microgram/min. Curves for norepinephrine and nerve stimulation are not shown after termination of PGA(1) infusions since they were not significantly different from control. N indicates number of animals. (Redrawn from Kadowitz et al, in press).

There was only a small tendency for responses to return toward control values when the infusion was terminated and responses to nerve stimulation were still elevated ($p < 0.05$) 30 min after termination of the prostaglandin infusion. In another group of animals the curve for nerve stimulation was shifted to the left in a nonparallel manner during infusion of PGF(2-alpha), 0.1 microgram/min (Fig. 8). The slope was not as steep as observed with the 1.0 microgram/min concentration (Fig. 8). When the infusion rate was decreased to 0.01 microgram/min, the shift to the left of the curve for nerve stimulation was parallel and had a magnitude of 1/2 log unit (Fig. 8).

The effect of PGF(2-alpha) on responses to adrenergic stimuli in the venous bed was studied in the saphenous vein. Infusion of PGF(2-alpha) directly into the saphenous vein markedly enhanced venoconstrictor responses to intravenous norepinephrine, tyramine and sympathetic nerve stimulation. The shift to the left of the curves for norepinephrine, tyramine and nerve stimulation was parallel and of similar magnitude (Fig. 9).

Effect of PGA(2) and PGE(1) on Responses to Norepinephrine and Nerve Stimulation in the Perfused Hindlimb

Direct intra-arterial infusion of PGA(2), 1 to 4 microgram/min, resulted in a small nonsignificant reduction in perfusion pressure in five animals (Control 120 ± 13, PGA(2) 111 ± 16 mm Hg, $P > 0.2$). This amount of PGA(2) did, however, produce a significant reduction in systemic arterial pressure (Control 125 ± 4, PGA(2) 110 ± mm Hg, $P < 0.05$). Hindlimb vasoconstrictor responses to intra-arterial norepinephrine and nerve stimulation were decreased significantly at all doses of norepinephrine or the stimulus frequencies evaluated; all responses were not significantly different from control 30 min after the prostaglandin infusion (Table II).

Direct intra-arterial infusion of PGE(1) (1 to 4 microgram/min) in four dogs resulted in a 38 ± 9 mm decrease in hindlimb perfusion pressure (Control 111 ± 13, PGE(1) 74 ± 4 mm Hg, $P < 0.05$) but no change in systemic arterial pressure (Control 128 ± 3, PGE(1) 127 ± 5). Vasoconstrictor responses to intra-arterial norepinephrine and sympathetic nerve stimulation were reduced markedly during PGE(1) infusion (Table III).

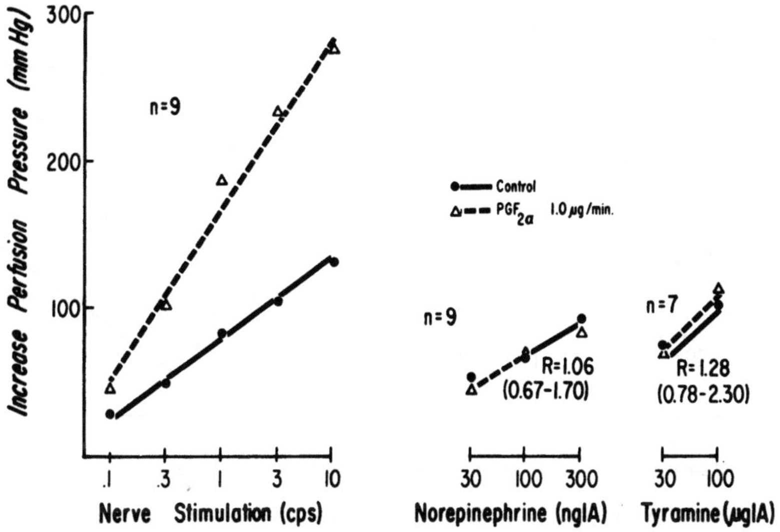

Figure 7. Effect of infusion of PGF(2-alpha), 1.0 microgram/min IA, on vasoconstrictor responses to sympathetic nerve stimulation, intra-arterial norepinephrine and tyramine in the perfused hindpaw. Dose and frequency response curves were obtained before and during prostaglandin infusion. Responses to all stimulus frequencies were significantly greater during PGF(2-alpha) infusion ($p < 0.05$, paired comparison); however the frequency response curve deviated significantly from parallelism. Dose response curves for norepinephrine and tyramine were not significantly different from control during PGF(2-alpha) infusion. R shows the potency ratio with the 95% confidence limits in parentheses. N indicates number of animals.

Effect of Reduced Renal Pressure on Adrenergic Responses in the Hindpaw

The effect of reduced renal pressure on responses to norepinephrine and nerve stimulation was examined in 15 hindpaw preparations perfused directly with renal venous blood. In order to exclude the influence of angiotensin released during reduced renal pressure, adrenergic responses were evaluated in preparations in which vasoconstrictor responses were facilitated maximally by angiotensin. Since this was the greatest effect obtainable with angiotensin the assumption was made that any further increase in response to sympathetic nerve stimulation observed during renal ischemia must have been due to the elaboration of another humoral factor.

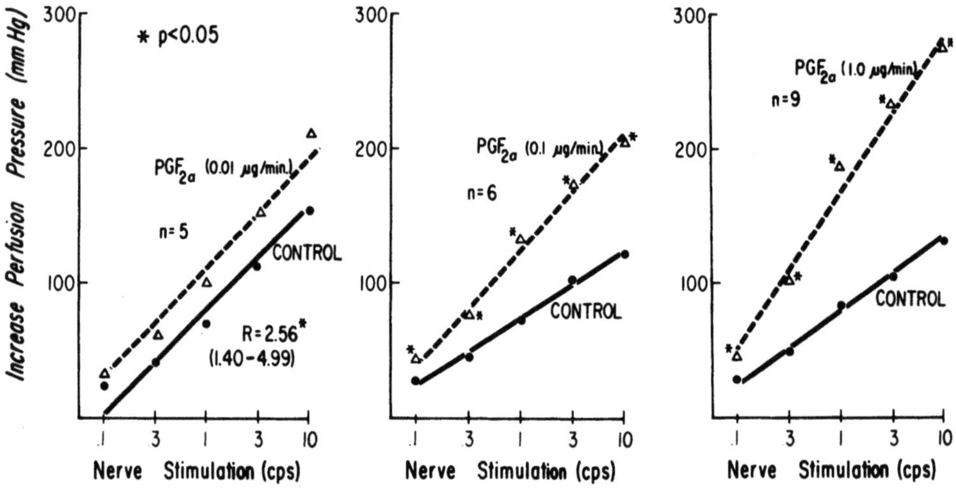

Figure 8. Comparison of the effect of a wide range of concentrations of PGF(2-alpha) on frequency response curves for nerve stimulation in three groups of animals. At the 0.01 microgram/min infusion rate the curve for nerve stimulation was shifted to the left in a parallel manner (left panel). At infusion rates of 0.1 (middle panel) and 1.0 (right panel), responses to nerve stimulation were increased significantly at each frequency studied ($p < 0.05$, paired comparison); however the frequency response curve deviated from parallelism during infusion of the 0.1 and 1.0 microgram/min concentrations. N indicates number of animals.

Results of these experiments are summarized in Table IV. Vasoconstrictor responses to sympathetic nerve stimulation but not norepinephrine were enhanced during angiotensin infusion. Subsequent reduction of renal pressure to 50% of control was accomplished by tightening a Blalock clamp placed on the aorta above both renal arteries. Responses to nerve stimulation were significantly greater during renal ischemia than during infusion of angiotensin alone. Responses to the two higher doses of norepinephrine were also increased significantly during the period of reduced renal pressure. In other experiments no enhancement of vasoconstrictor responses beyond that produced by angiotensin was observed when the angiotensin infusion was continued without reduction of renal pressure. Thus passage of time alone was not responsible for augmented responses seen during the time of reduced renal pressure.

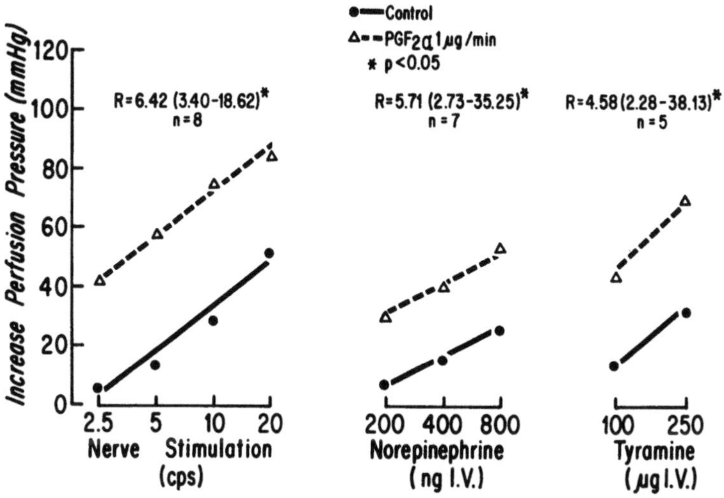

Figure 9. Effect of PGF(2-alpha), 1 microgram/min IV, on the frequency response curve for nerve stimulation and dose response curves for IV norepinephrine and tyramine in the perfused saphenous vein. Responses were obtained before and during constant IV infusion of PGF(2-alpha), and curves were analyzed by the parallel line bioassay method of Finney (1952). R shows the potency ratio with the 95% confidence limits in parentheses. The shift of the curve to the left is considered to significant if the lower limit exceeds 1.00. N indicates number of animals.

Interaction between PGE(1) and PGF(2-alpha) in the Hindpaw

The interaction between PGE(1) and PGF(2-alpha) was studied in seven hindpaw preparations. In these experiments vasoconstrictor responses to adrenergic stimuli were decreased by the infusion of PGE(1) (1 microgram/min) (Fig. 10). Subsequent infusion of PGF(2-alpha) returned responses to nerve stimulation but not norepinephrine to control value (Fig. 10). Responses to nerve stimulation were not different from control, whereas responses to norepinephrine were still depressed 30 min after the PGF(2-alpha) infusion (Fig. 10).

DISCUSSION

In the present study the effect of four prostaglandins on sympathetic transmission to vascular smooth muscle was evaluated in the perfused hindpaw. PGE(1) in all concentrations used decreased

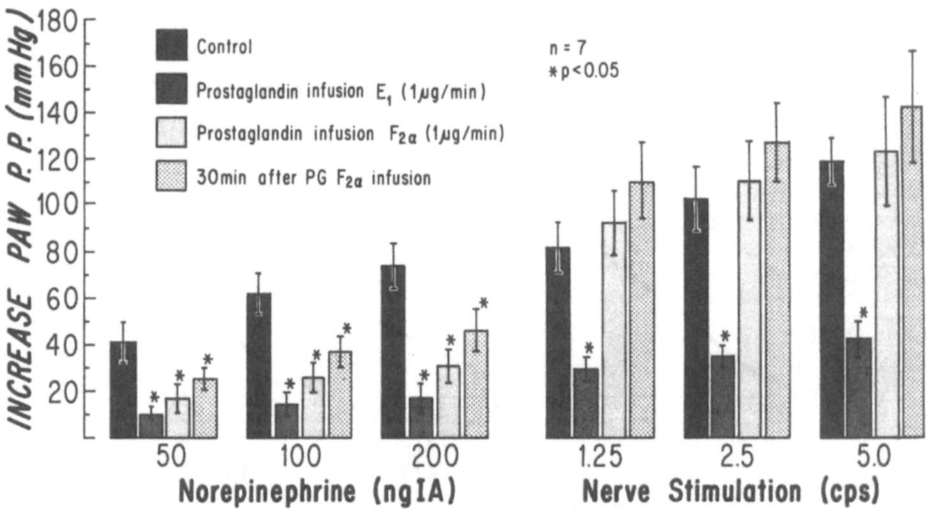

Figure 10. Interaction of prostaglandins E(1) and F(2-alpha) on adrenergic responses in the perfused hindpaw. Responses to norepinephrine and nerve stimulation were obtained 5 min after onset of infusion of PGE(1), again 5 min after onset of infusion of PGF(2-alpha) and 30 min after termination of the PGF(2-alpha) infusion. N indicates number of animals. (Redrawn from Kadowitz et al, 1971a).

vasoconstrictor responses to nerve stimulation, norepinephrine and angiotensin in the hindpaw. The ability of PGE(1) to antagonize responses to norepinephrine and sympathetic nerve stimulation was related to the concentration infused, and all dose and frequency response curves were shifted to the right in a parallel manner. However the curve for nerve stimulation deviated from parallelism during infusion of the 1.0 microgram/min concentration. The nonparallel shift of the curve for nerve stimulation in the presence of a parallel shift of the curve for norepinephrine may suggest that at high concentrations PGE(1) not only depresses smooth muscle sensitivity directly but that it may also interfere with release of norepinephrine from sympathetic nerve endings during nerve stimulation. It was recently reported that PGE(1) antagonized responses to norepinephrine and nerve stimulation and decreased the output of transmitter in the isolated cat spleen (Hedqvist and Brundin, 1969). Results of the present study with high doses of PGE(1) are consistent with these findings and support the conclusion of Hedqvist and Brundin (1969) that PGE(1) may act both on the nerve terminal and on postjunctional structures. The ability of PGE(1) to block responses to exogenous norepinephrine and nerve stimulation was similar at the two lower concentrations, and the antagonism was completely reversible. The similarity in effect of PGE(1) on responses to norepinephrine and nerve stimulation at lower doses indicates that at these concentrations the effect of this substance is due predominantly to direct depression of smooth muscle.

Antagonism between PGE(1) and angiotensin was related to the concentration of prostaglandin used. The vasodilatation in response to PGE(1) was not dose-related in the range of concentrations studied. The effect of PGE(1) and phenoxybenzamine, an alpha-adrenergic blocker, on the response to angiotensin was different. A dose of phenoxybenzamine that greatly attenuated the response to norepinephrine was without effect on the response to angiotensin, whereas PGE(1) depressed responses to both norepinephrine and angiotensin. Thus antagonism between PGE(1) and catecholamines is probably not due to specific alpha-adrenergic blockade but to a nonspecific action on the smooth muscle in hindpaw vessels.

The decreases in vascular resistance in response to the three concentrations of PGE(1) studied were not different and since these responses were similar to the maximum response obtained with glyceryl trinitrate, it appears that each dose of PGE(1) was capable of eliciting maximal vasodilation in the hindpaw.

PGE(1) antagonized responses to norepinephrine and nerve stimulation in muscle and skin. This prostaglandin was more effective against exogenous norepinephrine in muscle, whereas in skin its effect on nerve released and exogenous transmitter was equal. The differential effect of PGE(1) in skeletal muscle but not in skin may suggest a difference in special arrangement of the neuroeffector system in skin and skeletal muscle vessels. PGE(1) is less potent in its ability to antagonize catecholamine responses in skeletal muscle where 10 to 100 times more prostaglandin was required to produce the same approximate amount of depression of adrenergic responses as in skin. The decrease in response to isoproterenol is probably due to the decrease in resistance caused by PGE(1), since it is well known that the ability to vasodilate is related to the amount of tone present in the vessels. The decrease in vascular resistance in response to PGE(1) was greater in skeletal muscle than in skin and probably reflects a greater amount of tone in these vessels.

The ability of PGE(1) to depress responses to adrenergic stimuli in the hindlimb lies somewhere between its ability to decrease these responses in skin and skeletal muscle vessels alone. This finding is not unexpected in that the hindlimb is approximately 70% skeletal muscle. The same relationship was evident for the vasodilator effect of this prostaglandin in these tissues. PGA(2) also possessed the ability to depress responses to norepinephrine and nerve stimulation in the hindlimb but was less potent in this respect than PGE(1). The potencies of PGA(1) in the hindpaw and PGA(2) in the hindlimb were similar in regard to their ability to attenuate adrenergic responses. PGA(2) and PGA(1) both decreased the systemic arterial pressure when infused locally; however PGA(2) was much less effective as a vasodilator in the hindlimb. The reduction in arterial pressure observed in response to infusion of PGA(1) and PGA(2) but not in response to PGE(1) or PGE(2) is indicative of the resistance of the PGA compounds to clearance by the lung (McGiff et al, 1969; Ferreira and Vane, 1967).

Prostaglandin A(1) decreased vascular resistance and antagonized responses to norepinephrine, nerve stimulation, and angiotensin in the hindpaw but was about 100 times less potent in its ability to depress these responses than PGE(1). The ability of PGA(1) to inhibit adrenergic responses was equal at the highest concentrations used and the effect on responses to norepinephrine and nerve stimulation was similar, suggesting that its primary site of action is postjunctional.

Prostaglandin F(2-alpha), in contrast to PGE(1) and PGA(1), was found to enhance responses to sympathetic nerve stimulation in the absence of any apparent effect on hindpaw vascular resistance. The marked increase in response to nerve stimulation in the presence of PGF(2-alpha) appears to be the result of facilitation of transmitter release rather than an effect on the adrenergic neuroeffector or on the smooth muscle itself, since the dose-response curve for intra-arterial norepinephrine was not modified. The absence of an effect of PGF(2-alpha) on the response to tyramine, an agent that is believed to displace norepinephrine from the adrenergic nerve terminal (Burn and Rand, 1958), suggests that the facilitating action of PGF(2-alpha) is specific for the pool of transmitter releaseable by electrical excitation of the sympathetic nerves in the cutaneous vessels.

The curve for nerve stimulation was shifted to the left in a parallel fashion in the presence of the lowest concentration of PGF(2-alpha) used; however at higher concentrations, the curve for nerve stimulation deviated from parallelism and the slope appeared to increase with increasing concentrations of this prostaglandin. The nonparallel shift at higher concentrations in the absence of any apparent effect on responses to exogenous norepinephrine suggests that PGF(2-alpha) is more effective in facilitating release of transmitter at higher stimulus frequencies.

The observation that PGF(2-alpha) is capable of reversing the inhibitory effect of PGE(1) on vasoconstrictor responses to nerve stimulation but not norepinephrine in cutaneous vessels is consistent with the hypothesis that PGE(1) acts predominately on the smooth muscle to depress its sensitivity to transmitter, whereas PGF(2-alpha) acts mainly on the nerve terminal to facilitate release of transmitter.

The influence of PGF(2-alpha) on responses to adrenergic stimuli is different in the hindpaw and saphenous vein (Kadowitz et al, 1971a, 1971b). Whereas only the response to nerve stimulation was enhanced in the hindpaw, responses to norepinephrine, nerve stimulation and tyramine were increased equally in the saphenous vein, suggesting an effect on the metabolism of transmitter, the neuroeffector junction, or on the venous smooth muscle itself.

Prostaglandin E(2), in contrast to PGE(1), was found to enhance the response to nerve stimulation in the hindpaw in the absence of any apparent effect on the response to norepinephrine. PGE(2) appears to be similar to PGF(2-alpha) in that it probably increases the response to nerve stimulation by facilitating the release of transmitter from sympathetic nerve terminals in hindpaw vessels. It differs from PGF(2-alpha) in that much higher concentrations are required - 0.25 nanogram/ml for PGF(2-alpha) and 50 nanogram/ml for PGE(2) - and that it decreased vascular resistance in the hindpaw, whereas PGF(2-alpha) did not alter hindpaw pressure. The effect of PGE(2) is different in the isolated cat spleen and dog hindpaw. PGE(2) enhanced the response to nerve stimulation in the hindpaw but has been reported to reduce the response to nerve stimulation and decrease the efflux of transmitter in the isolated cat spleen (Hedqvist, 1970). The difference in results in these two studies may suggest that the effect of this prostaglandin is specific for certain organs or that it may be species-dependent.

The physiological role of prostaglandins in the cardiovascular system is entirely speculative at the present time (Bergstrom et al, 1968; Horton, 1969). The observation that they are released from the kidney and spleen during electrical excitation of the sympathetic nerves, infusion of vasoconstrictors or renal ischemia, and that they are capable of modifying adrenergic responses, has led to the hypothesis that these substances may be modulators of sympathetic neurotransmission to blood vessels (Davies et al, 1968; Sweet et al, 1971; Kadowitz, 1971a, 1971b; Hedqvist, 1970; Davies and Withrington, 1967). The possibility that prostaglandins of the E and F series are circulating hormones is remote, since the lung is so efficient in clearing these substances (McGiff et al, 1969; Ferreira and Vane, 1967). An alternate hypothesis is that prostaglandins may act locally in the organs from which they are released. The possibility that they may act on the venous side of the circulation cannot be excluded in view of the finding that PGF(2-alpha) has a marked effect on adrenergic venoconstrictor responses (Kadowitz et al, 1971b) and on venous capacitance (DuCharme et al, 1968).

Prostaglandins of the A series could theoretically act as circulating hormones since they are more resistant to clearance by the lungs (McGiff et al, 1969; Ferreira and Vane, 1967). PGA(1) is of special interest because of its potential use in the treatment of hypertension (Westura et al, 1970).

In conclusion, the data presented here show that prostaglandins have marked effect on sympathetic neurotransmission to blood vessels and suggest that the modulatory action may be at two different sites of the adrenergic neuroeffector system. In the first instance they may act to regulate the release of transmitter from sympathetic nerve terminals, and secondly they may act directly on smooth muscle to regulate its sensitivity to catecholamines and angiotensin. Results which show that some of the prostaglandins have marked effect on sympathetic neurotransmission at concentrations of less than 0.2 nanogram/ml arterial blood are consistent with and strongly support the postulate that these substances may have a physiological role in the regulation of blood pressure.

TABLE I

INFLUENCE OF INTRA-ARTERIAL INFUSION OF PROSTAGLANDINS E(1), A(1), E(2), AND F(2-ALPHA) ON MEAN ARTERIAL AND PERFUSION PRESSURE IN THE HINDPAW, SAPHENOUS VEIN AND GRACILIS MUSCLE

		Systemic Arterial Pressure (mm Hg)			Perfusion Pressure (mm Hg)		
	N	Control	5 min*	30 min after infusion†	Control	5 min*	30 min after infusion†
				Hindpaw			
Prostaglandin E(1)							
0.01 microgram/min	5	114±8	116±8	118±9	101±7	73±4□	86±9□
0.1 microgram/min	5	106±4	105±3	105±4	112±8	80±6□	86±7□
1.0 microgram/min	5	111±3	109±3	110±3	108±10	74±5□	75±6□
Saline							
0.1 ml/min	5	103±7	104±8	105±8	96±10	97±10	100±9
Prostaglandin A(1)							
0.1 microgram/min	4	129±8	123±10	134±9	128±11	98±9□	122±11
1.0 microgram/min	8	110±5	83±6□	110±3	108±4	74±4□	93±4□
2.0 microgram/min	7	133±3	76±6□	131±4	106±6	71±3□	94±4□

(Continued)

TABLE I (CONTINUED)

				Hindpaw			
Prostaglandin F(2-alpha)							
0.01 microgram/min	6	111±5	112±5	111±7	106±9	105±7	102±7
0.1 microgram/min	6	123±6	123±5	122±4	121±9	121±8	116±9
1.0 microgram/min	9	114±8	115±9	115±9	110±10	116±10	115±9
Prostaglandin E(2)							
2.0 microgram/min	4	120±9	118±10	124±11	124±13	105±10¤	118±14
				Gracilis Muscle			
Prostaglandin E(1)							
1.0 microgram/min	12	109±6	105±5	-	138±9	81±6¤	-
				Saphenous Vein			
Prostaglandin F(2-alpha)							
1.0 microgram/min	12	125±3	122±5	-	17±1	19±2	-

* 5 min after onset of constant infusion.
+ 30 min after termination of the infusion.
¤ Significantly different from control ($p < 0.05$) paired comparison.

TABLE II

EFFECT OF INFUSION OF PGA(2) ON RESPONSES TO NOREPINEPHRINE AND NERVE STIMULATION IN THE PERFUSED HINDLIMB

Increase Perfusion Pressure (mm Hg ± S. E.)

	Norepinephrine (micrograms IA)				Nerve Stimulation (CPS)				
	0.1	0.3	1	3	0.3	1	3	10	30
Control	22±2	42±1	62±4	88±2	28±5	49±5	64±6	83±6	102±9
PGA(2) (1-4 micro-gram/min	13±2*	26±5*	45±5*	66±9*	18±5*	32±6*	42±8*	66±8*	87±14*
30 min after PGA(2) infusion	23±4	45±7	61±6	86±6	22±6	44±9	60±10	74±12	103±16

* Significantly different from control ($P < 0.05$, $n = 5$)

TABLE III

EFFECT OF INFUSION OF PGE(1) ON RESPONSES TO NOREPINEPHRINE AND NERVE STIMULATION IN THE PERFUSED HINDLIMB

Increase Perfusion Pressure (mm Hg ± S. E.)

	Norepinephrine (microgram IA)				Nerve Stimulation (CPS)				
	0.1	0.3	1	3	0.3	1	3	10	30
Control	21±2	43±1	65±3	88±2	30±6	53±4	70±2	88±6	110±6
PGE(1) (104 micro-gram/min	6±2*	11±1*	20±2*	36±4*	9±2*	15±2*	26±3*	50±5*	77±6*

* Significantly different from control ($P < 0.05$, $n = 4$)

TABLE IV

EFFECT OF REDUCED RENAL PRESSURE ON RESPONSES TO NOREPINEPHRINE AND NERVE STIMULATION DURING INFUSION OF ANGIOTENSIN IN THE HINDPAW

	Increase Perfusion Pressure (mm Hg ± S.E.)					
	Norepinephrine (nanograms IA)			Nerve Stimulation (CPS)		
	50	100	200	5	10	20
Control	40	52	90	55	90	140
Angiotensin	47	63	101	72*	114*	170*
Reduced renal pressure plus angiotensin	49	75+	116+	88+	136+	215+

* Significantly different from control ($P < 0.05$, $n = 15$)

+ Significantly different from antiotensin ($P < 0.05$, $n = 15$)

ACKNOWLEDGEMENTS

The authors wish to thank Dr. John E. Pike and the Upjohn Company for the generous samples of prostaglandins, and Mr. Kao-Shing Huang and Dr. Paul Leaverton for help with the statistical analysis. We wish to thank the Williams and Wilkins Company for permission to reproduce Fig. 1, 2, 6, and 10.

This investigation was supported in part by USPHS grants HE 12964 and 5 T01 GM 00141 and by National Heart and Lung Institute Grant 5 T01 HE 5577 10.

REFERENCES

Bergstrom, S., Carlson, L. A., and Weeks, J. R., 1968, The prostaglandins: a family of biologically active lipids, Pharmacol. Rev. 20: 1.

Burn, J. H., and Rand, M. J., 1958, The action of sympathomimetic amines in animals treated with reserpine, J. Physiol. (London) 144: 314.

Davies, B. N., Horton, E. W., Withrington, P. G., 1968, The occurrence of prostaglandin E(2) in splenic venous blood of the dog following nerve stimulation, Brit. J. Pharmacol. 32: 127.

Davies, B. N., and Withrington, P. G., 1967, The effects of prostaglandins E(1) and E(2) on smooth muscle of the dog spleen and its responses to catecholamines, angiotensin and nerve stimulation, Brit. J. Pharmacol. 32: 136.

Dougherty, R. M., 1971, Effects of IV and IA prostaglandin E(1) on dog forelimb skin and muscle blood flow, Amer. J. Physiol. 220:392.

DuCharme, D. W., Weeks, J. R., and Montgomery, R. G., 1968, Studies on the mechanism of the hypertensive effect of prostaglandin F(2-alpha), J. Pharmacol. Exp. Ther. 160:1.

Dunham, E. W., and Zimmerman, B. G., 1970, Release of prostaglandin-like material from dog kidney during nerve stimulation, Amer. J. Physiol. 219: 1279.

Ferreira, S. H., and Vane, J. R., 1967, Prostaglandins: their disappearance from and release into the circulation, Nature (London) 216:868.

Finney, D. J., 1952, Statistical Methods in Biological Assay, Charles Griffin and Co. Ltd., London.

Gilmore, N., Vane, J. R., and Wyllie, J. H., 1968, Prostaglandins released by the spleen, Nature (London) 218:1135.

Hedqvist, P., 1970, Control by prostaglandin E(2) of sympathetic neurotransmission in the spleen, Life Sci. 9:269.

Hedqvist, P., 1970, Studies on the effect of prostaglandins E(1) and E(2) on the sympathetic neuromuscular transmission in some animal tissues, Acta Physiol. Scand. Suppl. 345:1.

Hedqvist, P., and Brundin, J., 1969, Inhibition by prostaglandin E(1) of noradrenaline release and of effector response to nerve stimulation in the cat spleen, Life Sci. 8:389.

Hedwall, P. R., Abdel-Sayed, W. A., Schmid, P. G., Mark, A. L., and Abboud, F. M., 1971, Vascular responses to prostaglandin E(1) in gracilis muscle and hindpaw of the dog, Amer. J. Physiol. 221:42.

Holmes, S. W., Horton, E. W., and Main, I. M. H., 1963, The effect of prostaglandin E(1) on responses of smooth muscle to catecholamines, angiotensin and vasopressin, Brit. J. Pharmacol. 21:538.

Horton, E. W., 1969, Hypotheses on physiological roles of prostaglandins, Physiological Reviews, 49:122.

Kadowitz, P. J., Sweet, C. S., and Brody, M. J., 1971a, Differential effects of prostaglandins E(1), E(2), F(1-alpha) and F(2-alpha) on adrenergic vasoconstriction in the dog hindpaw, J. Pharmacol. Exp. Ther. 177:641.

Kadowitz, P. J., Sweet, C. S., and Brody, M. J., 1971b, Potentiation of adrenergic venomotor responses by angiotensin, prostaglandin F(2-alpha) and cocaine, J. Pharmacol. Exp. Ther. 176:167.

Kadowitz, P. J., Sweet, C. S., and Brody, M. J., (in press), Blockade of adrenergic vasoconstrictor responses in the dog by prostaglandins E(1) and A(1), J. Pharmacol. Exp. Ther.

McGiff, J. C., Terragno, N. A., Strand, J. C., Lee, J. B., Lonigro, A. J., and Ng, K. K. F., 1969, Selective passage of prostaglandins across the lung, Nature (London) 223:742.

McGiff, J. C., Crowshaw, K., Terragno, N. A., Lonigro, A. J., Strand, J. C., Williamson, M. A., Lee, J. B., and Ng, K. K. F., 1970, Prostaglandin-like substances appearing in canine renal venous blood during renal ischemia, Circ. Res. 27:765.

Nakano, J., and McCurdy, J., 1967, Cardiovascular effects of prostaglandin E(1), J. Pharmacol. Exp. Ther. 156:538.

Nakano, J., 1968, Effects of prostaglandins E(1), A(1) and F(2-alpha) on the coronary and peripheral circulations, Proc. Soc. Exp. Biol. Med. 127:1160.

Snedecor, G. W., 1956, Statistical Methods, Iowa State College Press, Ames, Iowa.

Strong, C. G., and Bohr, D. F., 1967, Effects of prostaglandins E(1), E(2), A(1) and F(1-alpha) on isolated vascular smooth muscle, Amer. J. Physiol. 213:725.

Sweet, C. S., Kadowitz, P. J., and Brody, M. J., 1971, Another humoral substance that enhances adrenergic responsiveness during acute renal ischemia, Nature (London) 231:263.

Viguera, M. G., and Sunahara, F. A., 1969, Microcirculatory effects of prostaglandins, Can. J. Physiol. Pharmacol. 47:627.

Webb-Peploe, M. M., and Shepherd, J. T., 1968, Response of large hindlimb veins of the dog to sympathetic nerve stimulation, Amer. J. Physiol. 215:299.

Weiner, R., and Kaley, G., 1969, Influence of prostaglandin E(1) on the terminal vascular bed, Amer. J. Physiol. 217:563.

Westura, E. E., Kannegiesser, H., O'Toole, J. D., and Lee, J. B., 1970, Antihypertensive effect of prostaglandin A(1) in essential hypertension, Circ. Res. 18:(Suppl. 1):1-131.

Zimmerman, B. G., and Gomez, J., 1965, Increased response to sympathetic stimulation in the cutaneous vasculature in the presence of angiotensin, Int. J. Neuropharm. 4:185.

Zimmerman, B. G., and Whitmore, L., 1967, Effect of angiotensin and phenoxylbenzamine on release of norepinephrine in vessels during sympathetic nerve stimulation, Int. J. Neuropharm. 6:27.

DISCUSSION

<u>KALEY</u>: What is the effect of PGE(1) on isoproterenol?

<u>KADOWITZ</u>: The hindpaw will give a nice dilator response to isoproterenol, acetylcholine, and papaverine. However, if we remove the tone in the vessels by any means, the response to all of these agents will be lost.

<u>SUNAHARA</u>: If one looks at the heart instead of the hind limb one gets a completly different picture. With the infusion of PGE(1) one gets an initial vasodilation. However, after the coronary flow is allowed to return to normal by reducing PGE(1) infusion rate and then noradrenaline is administered, the metabolically-induced vasodilator response will be either inhibited or blocked by the PGE(1). So far we have only studied PGE(1).

<u>KADOWITZ</u>: Were these hearts driven or running free?

<u>SUNAHARA</u>: These hearts (Langendorff preparations) were beating at their own intrinsic rate. However, we found that the inhibitory response of PGE(1) was the same whether the hearts were paced or not. There was no inotropic effect or chronotropic effect of PGE(1).

<u>KADOWITZ</u>: Did the prostaglandins modify the cardiostimulant action of catecholamines as Hedquist has shown?

<u>SUNAHARA</u>: No. Inotropic and chronotropic effects of either exogenously applied or endogenously released catecholamines were not altered by the previous infusion of prostaglandins.

<u>BUCKLES</u>: In the presentations this afternoon it seems that you have ignored the problem of closing the loops. The prostaglandins can be made in all tissues and destroyed in all tissues, but I have not heard anyone discuss what we engineers call "mass balance". If it is possible for synthesis to occur, it is possible for change to occur, and just because you infuse PGE(1) and get an effect, it is attributed to PGE(1). Yet it is not at all clear that it is PGE(1) which is causing the action. It may be a difficult experiment to conceive of, whereby you take a particular organ and attempt to do a mass balance on all of the prostaglandins and their precursors, but it seems to me that this is what is necessary. Questions have been asked as to whether prostaglandins act within

the cells, and if so, whether they act on adenylate cyclase or an adrenergic receptor, but these are just fragments of models. It is particularly interesting that people are trying to separate mechanisms that are either cellular in that they deal with the tissue, or that they deal with circulating elements. I would hope that people are looking at the vascular and membrane permeability to prostaglandins as well as partition coefficients for the prostaglandins between plasma and tissues. Work should also be done on transport solubility characteristics in the cell.

INDEX

MIX
Papier aus verantwortungsvollen Quellen
Paper from responsible sources
FSC® C105338

If you have any concerns about our products,
you can contact us on
ProductSafety@springernature.com

In case Publisher is established outside the EU,
the EU authorized representative is:
Springer Nature Customer Service Center GmbH
Europaplatz 3, 69115 Heidelberg, Germany

Printed by Libri Plureos GmbH
in Hamburg, Germany